T0313325

Photonic Interconnects for Computing Systems
Understanding and Pushing
Design Challenges

RIVER PUBLISHERS SERIES IN OPTICS AND PHOTONICS

Series Editors

Manijeh Razeghi
Northwestern University
USA

Kevin Williams
Eindhoven University of Technology
The Netherlands

Indexing: All books published in this series are submitted to Thomson Reuters Book Citation Index (BkCI), CrossRef and to Google Scholar.

The "River Publishers Series in Optics and Photonics" is a series of comprehensive academic and professional books which focus on the theory and applications of optics, photonics and laser technology.

Books published in the series include research monographs, edited volumes, handbooks and textbooks. The books provide professionals, researchers, educators, and advanced students in the field with an invaluable insight into the latest research and developments.

Topics covered in the series include, but are by no means restricted to the following:

- Integrated optics and optoelectronics
- Applied laser technology
- Lasers optics
- Optical Sensors
- Optical spectroscopy
- Optoelectronics
- Biophotonics photonics
- Nano-photonics
- Microwave photonics
- Photonics materials

For a list of other books in this series, visit www.riverpublishers.com

Photonic Interconnects for Computing Systems
Understanding and Pushing Design Challenges

Editors

Mahdi Nikdast
Polytechnique Montréal
Canada

Gabriela Nicolescu
Polytechnique Montréal
Canada

Sébastien Le Beux
Ecole Centrale de Lyon
France

Jiang Xu
Hong Kong University of Science and Technology
China

River Publishers

Routledge
Taylor & Francis Group

LONDON AND NEW YORK

Published 2017 by River Publishers
River Publishers
Alsbjergvej 10, 9260 Gistrup, Denmark
www.riverpublishers.com

Distributed exclusively by Routledge
4 Park Square, Milton Park, Abingdon, Oxon OX14 4RN
605 Third Avenue, New York, NY 10017, USA

Photonic Interconnects for Computing Systems: Understanding and Pushing Design Challenges / by Mahdi Nikdast, Gabriela Nicolescu, Sébastien Le Beux, Jiang Xu.

Routledge is an imprint of the Taylor & Francis Group, an informa business

ISBN 978-87-93519-80-0 (print)

While every effort is made to provide dependable information, the publisher, authors, and editors cannot be held responsible for any errors or omissions.

Contents

Mahdi Nikdast, Gabriela Nicolescu, Sebastien Le Beux and Jiang Xu

PART I: Design, Application and Implementation

Peng Yang, Xiaowen Wu, Yaoyao Ye and Jiang Xu

2   Design and Optimization of Vertical Interconnections for Multilayer Optical Networks-on-Chip   41

Alberto Parini and Gaetano Bellanca

3   Optical Interconnection Networks: The Need for Low-Latency Controllers   73

Felipe Gohring de Magalhães, Mahdi Nikdast, Fabiano Hessel, Odile Liboiron-Ladouceur and Gabriela Nicolescu

**PART II: Developing Design Automation Solutions
and Enabling Design Exploration**

PART III: Challenges in Performance Analysis
and Design Solutions

8   Thermal Management of Silicon Photonic NoCs
in Many-core Systems   **227**

Tiansheng Zhang, Jonathan Klamkin, Ajay Joshi and Ayse K. Coskun

9   Thermal-Aware Design Method for On-Chip Laser-based
Optical Interconnect   **249**

Hui Li, Alain Fourmigue, Sébastien Le Beux, Xavier Letartre,
Ian O'Connor and Gabriela Nicolescu

PART IV: On the Impact of Fabrication Non-Uniformity

**12 Impact of Fabrication Non-Uniformity on Silicon Photonic
Networks-on-Chip   355**

Mahdi Nikdast, Gabriela Nicolescu, Jelena Trajkovic
and Odile Liboiron-Ladouceur

List of Contributors

Abderazek Ben Abdallah, *Department of Computer Science and Enginee-ring, University of Aizu, Aizu-Wakamatsu, Fukushima, 965-8580 Japan*

Ajay Joshi, *Dept. of ECE, Boston University, Boston, MA, U.S.A.*

Alain Fourmigue, *École Polytechnique de Montréal, Montréal, Canada*

Alan Mickelson, *Electrical, Computer and Energy Engineering, University of Colorado at Boulder, Boulder 80309-0425 CO, USA*

Alberto Parini, *FOTON Laboratory, CNRS UMR 6082, University of Rennes 1, ENSSAT, F-22305 Lannion, France*

Alessandro Cilardo, *Department of Electrical Engineering and Information Technologies, University of Naples Federico II, Naples, Italy*

Andrea Peano, *Dipartimento di Ingegneria, University of Ferrara, Italy*

Ayse K. Coskun, *Dept. of ECE, Boston University, Boston, MA, U.S.A.*

Davide Bertozzi, *Dipartimento di Ingegneria, University of Ferrara, Italy*

Edoardo Fusella, *Department of Electrical Engineering and Information Technologies, University of Naples Federico II, Naples, Italy*

Fabiano Hessel, *Pontifical Catholic University of Rio Grande do Sul – GSE, Porto Alegre, Brazil*

Felipe Gohring de Magalhães, *1) Pontifical Catholic University of Rio Grande do Sul – GSE, Porto Alegre, Brazil
2) École Polytechnique de Montréal, Montréal, Canada*

Gabriela Nicolescu, *Polytechnique Montréal, Montréal, Canada*

Gaetano Bellanca, *Department of Engineering, University of Ferrara, I-44122 Ferrara, Italy*

Hui Li, *Lyon Institute of Nanotechnology, INL-UMR5270, Ecole Centrale de Lyon, Ecully, F-69134, France*

Ian O'Connor, *Lyon Institute of Nanotechnology, INL-UMR5270, Ecole Centrale de Lyon, Ecully, F-69134, France*

Ishan G. Thakkar, *Colorado State University, Forty Collins, Colorado, United States*

Jelena Trajkovic, *Concordia University, Montréal, Canada*

Jiang Xu, *Hong Kong University of Science and Technology, China*

Jonathan Klamkin, *Dept. of ECE, University of California Santa Barbara, Santa Barbara, CA, U.S.A.*

José Flich, *Department of Computer Engineering (DISCA), Universitat Politècnica de València, Valencia, Spain*

Keyon Janani, *Electrical, Computer and Energy Engineering, University of Colorado at Boulder, Boulder 80309-0425 CO, USA*

Luca Ramini, *Dipartimento di Ingegneria, University of Ferrara, Italy*

Maddalena Nonato, *Dipartimento di Ingegneria, University of Ferrara, Italy*

Mahdi Nikdast, *Polytechnique Montréal, Montréal, Canada*

Mahdi Tala, *Dipartimento di Ingegneria, University of Ferrara, Italy*

Marco Balboni, *Dipartimento di Ingegneria, University of Ferrara, Italy*

Marta Ortín Obón, *Departamento de Informática e Ingeniería de Sistemas, University of Zaragoza, Spain*

Michael Meyer, *Department of Computer Science and Engineering, University of Aizu, Aizu-Wakamatsu, Fukushima, 965-8580 Japan*

Moustafa Mohamed, *Electrical, Computer and Energy Engineering, University of Colorado at Boulder, Boulder 80309-0425 CO, USA*

Nikos Hardavellas, *Northwestern University, USA*

Odile Liboiron-Ladouceur, *McGill University, Montréal, Canada*

Paolo Grani, *Dipartimento di Ingegneria dell'Informazione e Scienze Matematiche, Università degli Studi di Siena, Italy*

Peng Yang, *Hong Kong University of Science and Technology, China*

Sébastien Le Beux, *Lyon Institute of Nanotechnology, INL-UMR5270, Ecole Centrale de Lyon, Ecully, F-69134, France*

Sai Vineel Reddy Chittamuru, *Colorado State University, Forty Collins, Colorado, United States*

Sakshi Singh, *Electrical, Computer and Energy Engineering, University of Colorado at Boulder, Boulder 80309-0425 CO, USA*

Sandro Bartolini, *Dipartimento di Ingegneria dell'Informazione e Scienze Matematiche, Università degli Studi di Siena, Italy*

Sudeep Pasricha, *Colorado State University, Forty Collins, Colorado, United States*

Tiansheng Zhang, *Dept. of ECE, Boston University, Boston, MA, U.S.A.*

Victor Viñals-Yufera, *Departamento de Informática e Ingeniería de Sistemas, University of Zaragoza, Spain*

Xavier Letartre, *Lyon Institute of Nanotechnology, INL-UMR5270, Ecole Centrale de Lyon, Ecully, F-69134, France*

Xiaowen Wu, *Huawei Co ltd, China*

Yaoyao Ye, *Shanghai Jiao Tong University, China*

Yigit Demir, *Intel, USA*

List of Figures

List of Tables

List of Abbreviations

2D	Two-dimensional
3D-FDTD	Three dimensional finite-difference in time-domain
ADF	Add-drop filter
AFT	Adaptive frequency tuning
ASP	Answer set programming
BEOL	Back-end-of-line
BER	Bit-error-rate
BGA	Ball grid array
CAD	Computer-aided design
CD-SEM	Critical dimension scanning electron microscopy
CG	Communication graph
CMOS	Complementary metal oxide semiconductor
CMP	Chip multi processor
COP	Constrained optimization problem
CPSE	Crossing photonic switch element
CPU	Central processing unit
CWDM	Coarse wavelength division multiplexing
D2D	Die-to-die
DBCTM	Double bit crosstalk mitigation mechanism
DC	Directional coupler
DFB	Distributed feedback
DOR	Dimension order routing
DVFS	Dynamic voltage and frequency scaling
DVOPD	Dual video object plane decoder
DWDM	Dense wavelength division multiplexing
E/O	Electrical-to-optical
EBeam	Electron beam
EDA	Electronic design automation
EDP	Energy-delay product
EM	Electromagnetic model
Emesh	Electrical mesh

ENoC	Electrical networks-on-chip
EO	Electrical-optical
E-O	Electro-optical
EO/OE	Electrical-optical/optical-electrical
EO/OE	Electro-optical/opto-electrical
EPI	Energy per instruction
FCA	Free carrier absorption effect
FDTD	Finite-difference time-domain
FFT	Fast Fourier transform
FIFO	First-in-first-out
FP	Fabry-Perot
FPGA	Field programmable gate array
FSR	Free spectral range
FWHM	Full width at half maximum
GA	Genetic algorithm
Gbps	Gigabit-per-second
GTECH	Generic technology library
GWOR	Generic wavelength-routed optical router
I/O	Input/output
I/Os	Inputs/outputs
ILP	Integer linear programming
IP core	Intellectual property core
ISA	Instruction set architecture
ITRS	International technology roadmap for semiconductors
L2	Level two
LA	Limiting amplifier
LAN	Local area network
L-CBF	Low-power counting bloom filter
LDPC	Low-density parity-check
LLC	Last level caches
MESI	Modified exclusive shared invalid
MOESI	Modified owned exclusive shared invalid
MLf	Machine learning for financial market
MMI	Multi-mode interference
MP3	Moving picture expert group-1/2 audio layer 3
MPSoCs	Multiprocessor systems-on-chip
MR	Microring resonator
MRR	Micro-ring resonator
MRs	Microring resonators

MSHR	Miss-handling register
MWD	Multi-window display
MWMR	Multiple writers multiple readers
MWSR	Multiple writers single reader
MZI	Mach-zehnder interferometer
MZIs	Mach-zehnder interferometers
NI	Network interface
NIC	Network interface card
NoC	Network-on-chip
NUCA	Non-uniform cache architecture
O/E	Optical-to-electrical
OE	Optical-electrical
O-E	Opto-electrical
OINs	Optical interconnection networks
ONI	Optical network interface
ONoC	Optical network-on-chip
OoO	Out-of-order
OS	Operating system
OSNR	Optical signal to noise ratio
P&R	Place and route
PCB	Printed circuit board
PCI	A cable transmission standard
PD	Photodetector
PIC	Photonic integrated circuit
PIP	Picture-in-picture
PNoC	Photonic network-on-chip
PPSE	Parallel photonic switch element
PSE	Photonic switching element
PSM4	A standard for a transmitter with four lasers, four modulators and four fibers
PV	Process variations
Rconv	Convective thermal resistance
ROB	Reorder buffer
ROM	Read-only memory
RR	Round robin
RS	Random search
RSD	Reed-Solomon decoder
R-SWMR	Reservation-assisted single-writer multiple-reader
RTL	Register transfer level

SCC	Single-chip cloud computer
SD	Shortest distance
SE	Spectroscopic ellipsometry
Si	Silicon
SiEPIC	Silicon electronic-photonic integrated circuits program
SiO$_2$	Silicon dioxide
SiP	Silicon photonics chip
SiPh	Silicon photonic
SLED	Super-luminescent light emitting diode
SNR	Signal to noise ratio
SoC	System-on-chip
SOI	Silicon on insulator
SPF	Shortest path first
SWMR	Single writer multiple readers
SWSR	Single writer single reader
TDM	Time division multiplexing
TE	Transverse electric
TF	Temperature first
TG	Topology graph
TIA	Transimpedance amplifier
TIA	Trans-impedance amplifier
TM	Transverse magnetic
TOC	Thermo-optic coefficient
TSV	Through Silicon Via
VCSEL	Vertical cavity surface emitting laser
VOPD	Video object plane decoder
WAN	Wide area network
WBA	Weight based arbiter
WCPL	Worst-case power loss
WDM	Wavelength division multiplexing
WG	Waveguide
WID	Within-die
WRG	Wavelength resolution graph
WRONoC	Wavelength-routed optical network-on-chip

Introduction

Mahdi Nikdast[1], Gabriela Nicolescu[1], Sébastien Le Beux[2] and Jiang Xu[3]

[1]Polytechnique Montréal, Canada
[2]Ecole Centrale de Lyon, France
[3]Hong Kong University of Science and Technology, China

The purpose of this book on Photonic Interconnects for Computing Systems is to provide a comprehensive overview of the current state-of-the-art and the latest advances in the field of silicon photonics for interconnection networks and high-performance computing. Achieving this objective, the book presents a compilation of 13 outstanding contributions from several leading research groups in the field. The contributions are grouped into four main sections. The first section comprises the contributions that focus on the design and development of silicon photonic interconnects for computing systems, as well as comparisons with electrical interconnects. The second section is composed of contributions discussing design automation solutions and design exploration in silicon photonic interconnects. The third section contains contributions focus on challenges in silicon photonic interconnects, as well as presenting designing techniques to improve reliability, energy efficiency, and fault-tolerance in such systems. Finally, the last section includes contributions evaluating the impact of fabrication non-uniformity in silicon photonic interconnects and enhancing their reliability under such variations.

Computing systems play an increasingly important role in our daily lives, from automobiles and cellphones to industrial infrastructures and data centers. These systems are rapidly growing to keep the pace with the ever growing computation and communication requirements driven by the emergence of new applications, such as internet-of-things (IOT) and big data applications. Furthermore, rapid advances of the semiconductor technology roadmap into the deeper sub-micron domain have enabled the integration of a large number of processing cores in computing systems, realizing multi-processor systems-on-chip (MPSoCs) as a standard platform for multicore and many-core computing systems. As a result, the overall performance of computing systems consisting of a large number of integrated processor cores are not only determined by the computational efficiency of the cores, but most importantly by how efficiently these many cores are communicating with one another.

Networks-on-chip (NoCs) have been introduced to the communication infrastructure in MPSoCs to outperform the traditional interconnection networks in such systems [3, 4, 8]. NoCs are based on exchanging information through routing packets in the network, replacing the routing design-specific global interconnects. As a result, comparing to the traditional bus or ad-hoc architectures, NoCs indicate a better scalability and design utilization when applied to MPSoCs [9]. In deep sub-micron VLSI technologies, metallic interconnects are susceptible to parasitic resistance and capacitance [11]. Consequently, as MPSoCs scale, integrating a large number of processing cores, the metallic interconnects in NoCs can no longer satisfy the high-bandwidth and low-latency requirements while respecting the overall system's power budget.

On the other hand, silicon photonics presents a promising solution to outperform the low-bandwidth, high-latency, and high-power dissipating metallic interconnects in MPSoCs [1, 2]. Moreover, 3D integration technology facilitates the integration of photonic interconnects into MPSoCs. An interesting advantage of using photonic interconnects over metallic ones is that different optical wavelengths can simultaneously transmit through an optical waveguide (i.e., wavelength division multiplexing), further enhancing the bandwidth performance in photonic interconnects. Photonic interconnects are based on employing CMOS-compatible silicon photonic devices to modulate, transmit, switch, and detect optical signals. Generally, at the source processor core, digital data (0 or 1) are modulated on an optical signal through optical modulators, and then the modulated optical signal passes through a photonic network consisting of different switches and filters, and ultimately the photodetectors at the end of the communication line in the destination processor core detect the optical signal, where it is converted back to the digital data. Several MPSoCs integrating photonic interconnects have been proposed [5, 7, 10, 13, 14]. Moreover, a microprocessor system that uses on-chip photonic devices to directly communicate with other chips using light has been developed and implemented in [12]. In the context of interconnects in data centers, Intel has recently announced the volume production and shipping of 100 Gbps optical transceivers for switch-to-switch optical interconnects in data centers [6].

Comparing to conventional metallic interconnects, employing photonic interconnects for both chip-scale and data center communications requires the development of new interconnect architectures, communication protocols, routing techniques, etc., under which the photonic interconnect can be efficiently utilized. For instance, optical signals cannot be buffered (optical buffers are extremely costly), and optical-electrical (O-E) and electrical-optical (E-O) conversions impose a relatively high cost and are power hungry, necessitating the development of photonic interconnects along with communication protocols that require minimum number of O-E/E-O conversions. Furthermore, there are several challenges for employing photonic interconnects, such as power loss, thermal noise, fabrication non-uniformity, etc. As a result, analytical models and simulation platforms are required to help system designers evaluate the performance of the designed systems considering those issues, enabling them to develop novel design techniques to mitigate the impact of such issues.

This book presents a comprehensive overview of the design, advantages, challenges, and requirements of photonic interconnects for computing systems. The selected contributions present important discussions and approaches related to the design and development of novel photonic interconnect architectures for computing systems, as well as various design solutions to improve the performance of such systems under different challenges.

Organization

This book is organized in four sections: (i) design, application, and implementation of silicon photonic interconnects, (ii) developing design automation solutions and enabling design exploration in photonic interconnects, (iii) challenges, performance analysis, and design solutions for high-performance photonic interconnects, and (iv) impact of fabrication non-uniformity on silicon photonic interconnects. Here, we present an overview of the four sections along with a brief introduction of the contents of each of the individual chapters.

Design, Application, and Implementation

The first part of this book highlights the importance of developing efficient interconnects for many-core computing systems. The section starts by discussing the design and development of a photonic inter- and intra-chip interconnection network for many-core computing systems, followed by the second chapter that discusses the implementation of multilayer optical networks-on-chip. The importance of electrical controllers for photonic interconnects along with the design of a low-latency controller are discussed in the third chapter. Finally, the fourth chapter presents an overview of both metallic and photonic interconnects to be employed for different scale communications, from the chip-scale to the data center-scale.

The chapter *Unified Inter- and Intra-Chip Optical Interconnect Networks* proposes unified inter- and intra-chip optical networks for many-core process systems. The proposed photonic interconnection network replaces the electrical interconnects for both inter- and intra-chip communications. It is composed of intra-chip and inter-chip networks, which can coordinate closely to not only fulfill the intra-chip communication demands, but also relieve the off-chip bandwidth limitation. The evaluations in the chapter demonstrate that the proposed photonic interconnect networks can achieve great improvement on network performance and energy efficiency compared to their electrical counterparts.

The chapter *Design and Optimization of Vertical Interconnections for Multilayer Optical Networks-on-Chip* presents an overview of material platforms and vertical interconnection schemes enabling the implementation of multilayer optical networks-on-chip. Issues related to the design of staked photonic circuits, with particular regard to optical isolation, are evidenced and discussed. Among the various interconnection solutions outlined in the chapter, two significant ones are detailed: the former exploiting directional couplers with inverse adiabatic tapers; the latter relying on multimode interference devices. Results showing the feasibility and optimization of these approaches are presented and discussed.

The chapter *Optical Interconnection Networks: The Need for Low-Latency Controllers* discusses the importance of electrical controllers to realize high-performance photonic interconnects. It overviews some of the common control techniques employed in photonic interconnection networks. Moreover, a low-latency controller is presented, which relies on a centralized core and pre-calculated routes, reducing the overall system communication latency through granting different requests within one clock cycle.

The chapter *Interconnects and Data System Throughput* discusses and analyzes the role of optics and optical devices in data centers. Reasoning behind the present use of optics for interconnections of distances exceeding a few meters, of copper for micron to meter distances, and doped semiconductor for the shortest distances are elucidated in this chapter. Discussion of latency, insertion loss, pulse distortion and dissipation of optical in comparison with electrical components indicates why the performance of optical components cannot be evaluated by present data center modeling techniques. The potential role of optics in future disaggregated architectures is the last topic of the chapter.

Developing Design Automation Solutions and Enabling Design Space Exploration

The second part of this book discusses the importance of developing design automation solutions for photonic interconnects, as well as how design space exploration impacts the performance of such systems. It comprises three chapters that focus on providing a synthesis methodology for designing wavelength-routed optical networks-on-chip, and exploring the design space of such systems to improve their performance.

The chapter *Design Automation Beyond Its Electronic Roots: Toward a Synthesis Methodology for Wavelength-Routed Optical Networks-on-Chip* reviews the importance of tools and methodologies in order to bridge the gap between photonic interconnect system designers and technology developers. Particularly, it provides an early-phase synthesis methodology for wavelength-routed optical networks-on-chip, capturing all design points in a unified design framework, and refining them into an actual implementation.

The chapter *Application-Specific Mapping Optimizations for Photonic Networks-on-Chip* discusses the impact of application mapping in photonic networks-on-chip to fully exploit photonic interconnects. It presents a methodology which automatically maps the tasks onto a generic regular photonic network-on-chip architecture such that the worst-case insertion loss and crosstalk noise are minimized, allowing a higher network scalability.

The chapter *Integrated Photonics for Chip-Multiprocessor Architectures* explores the design space of photonic interconnection networks for their employment in chip-multiprocessors. This chapter discusses the vertical interaction between highly abstract issues (e.g., shared memory and parallel applications) and very detailed physical facets of the integrated photonic interconnections.

Challenges, Performance Analysis, and Design Solutions

The third part of the book includes chapters discussing different challenges in photonic interconnects, and presents design techniques to compensate for those challenges. It consists of four chapters that discusses the impact of thermal and process variations on the performance of photonic interconnects, while presenting designing techniques to mitigate the impact of such variations. Also, discussions related to improving fault-tolerance and energy efficiency in photonic interconnects are covered in this section.

The chapter *Thermal Management of Silicon Photonic NoCs in Many-core Systems* summarizes basic components in photonic networks-on-chip, as well as their operating mechanisms and sensitivity to thermal and process variations. It discusses existing design-time and runtime techniques for photonic networks-on-chip thermal management. Particularly, the chapter focuses on techniques with a design automation and/or software optimization component.

The chapter *Thermal-Aware Design Method for On-Chip Laser-based Optical Interconnect* proposes a methodology enabling thermal-aware design for on-chip photonic interconnects. Thermal analysis allows the design of photonic networks-on-chip interfaces with low gradient temperature, by describing the studied architecture from device level to system level. Analytical models allow the influence of thermal effects to be evaluated, in particular on the signal-to-noise ratio (SNR), based on the obtained thermal map.

The chapter *Fault-tolerant Photonic Network-on-Chip* describes a fault-tolerant photonic NoC architecture suffering from temperature variations and manufacturing errors. The system is based on a fault-tolerant path-configuration and routing algorithm, a microring fault-resilient photonic router, and uses minimal redundancy to assure accuracy of the packet transmission even after faulty microrings are detected.

The chapter *Techniques for Energy Proportionality in Optical Interconnects* reviews some of the recently proposed techniques to achieve energy proportionality in photonic interconnects. In particular, it presents a technique to turn off the laser source (i.e., power gate the laser) when the network is idle, and a method to minimize the energy consumed to keep the microring resonators within a narrow temperature zone.

On the Impact of Fabrication Non-Uniformity

The last section of this book is devoted to a critical issue in photonic interconnects, fabrication non-uniformity. The first chapter presents a comprehensive study on the impact of fabrication non-uniformity in photonic interconnects, followed by the second chapter that discusses a novel encoding mechanism to mitigate the impact of fabrication non-uniformity in photonic interconnects.

The chapter *On the Impact of Fabrication Non-Uniformity in Silicon Photonic Networks-on-Chip* presents a comprehensive study on the impact of fabrication non-uniformity on photonic networks-on-chip. It develops a computationally efficient and accurate analytical method to enable such study in large-scale photonic networks-on-chip. Furthermore, the chapter explores different characteristics of fabricated photonic devices under fabrication non-uniformity.

The chapter *Enhancing Process Variation Resilience in Photonic NoC Architectures* proposes a novel encoding mechanism that intelligently adapts to on-chip fabrication process variations, and improves worst-case optical SNR by reducing crosstalk noise in microresonators used within dense wavelength division multiplexing (DWDM)-based photonic networks-on-chip.

References

[1] Y. Arakawa, T. Nakamura, Y. Urino, and T. Fujita. Silicon photonics for next generation system integration platform. *IEEE Communications Magazine*, 51(3):72–77, March 2013.

[2] S. Rumley, M. Bahadori, R. Polster, S. D. Hammond, D. M. Calhoun, K. Wen, A. Rodrigues, and K. Bergman. Optical interconnects for extreme scale computing systems. *Parallel Computing*, pages 65–80, 2017. doi: 10.1016/j.parco.2017.02.001

[3] L. Benini and G. De Micheli. Powering networks on chips. In *International Symposium on System Synthesis*, pages 33–38, September 2001.

[4] L. Benini and G. De Micheli. Networks on chip: a new paradigm for systems on chip design. In *Proceedings Design, Automation and Test in Europe Conference and Exhibition*, pages 418–419, 2002.

[5] E. Fusella and A. Cilardo. H^2ONoC: A hybrid optical-electronic NoC based on hybrid topology. *IEEE Transactions on Very Large Scale Integration (VLSI) Systems*, 25(1):330–343, January 2017.

[6] Intel. http://www.intel.com

[7] S. Koohi and S. Hessabi. All-optical wavelength-routed architecture for a power-efficient network on chip. *IEEE Transactions on Computers*, 63(3):777–792, March 2014.

[8] S. Kumar, A. Jantsch, J. P. Soininen, M. Forsell, M. Millberg, J. Oberg, K. Tiensyrja, and A. Hemani. A network on chip architecture and design methodology. In *Proceedings IEEE Computer Society Annual*

Symposium on VLSI. New Paradigms for VLSI Systems Design. ISVLSI 2002, pages 105–112, 2002.

[9] H. G. Lee, N. Chang, U. Y. Ogras, and R. Marculescu. On-chip communication architecture exploration: A quantitative evaluation of point-to-point, bus, and network-on-chip approaches. *ACM Trans. Des. Autom. Electron. Syst.*, 12(3):23:1–23:20, May 2008.

[10] Y. Pan, J. Kim, and G. Memik. Flexishare: Channel sharing for an energy-efficient nanophotonic crossbar. In *The Sixteenth International Symposium on High-Performance Computer Architecture*, pages 1–12, January 2010.

[11] S. Pasricha and N. D. Dutt. Trends in emerging on-chip interconnect technologies. *IPSJ Trans. System LSI Design Methodology*, 1:2–17, 2008.

[12] C. Sun, M. T. Wade, Y. Lee, et al. Single-chip microprocessor that communicates directly using light. *Nature*, 528(7583):534–538, 2015.

[13] D. Vantrease, R. Schreiber, M. Monchiero, M. McLaren, N. P. Jouppi, M. Fiorentino, A. Davis, N. Binkert, R. G. Beausoleil, and J. H. Ahn. Corona: System implications of emerging nanophotonic technology. In *International Symposium on Computer Architecture*, pages 153–164, June 2008.

[14] X. Wu, J. Xu, Y. Ye, Z. Wang, M. Nikdast, and X. Wang. Suor: Sectioned undirectional optical ring for chip multiprocessor. *Journal of Emerging Technologies in Computing Systems*, 10(4):29:1–29:25, June 2014.

PART I
Design, Application and Implementation

1

Unified Inter- and Intra-chip Optical Interconnect Networks

Peng Yang[1], Xiaowen Wu[2], Yaoyao Ye[3] and Jiang Xu[1]

[1]Hong Kong University of Science and Technology, China
[2]Huawei Co ltd, China
[3]Shanghai Jiao Tong University, China

Abstract

The demand for high performance and energy efficiency makes many-core processor system a promising platform for modern computing system. The communication of inter/intra-chip is becoming as important as the computation speeds of single processing cores. The huge communication demands also aggravate the power dissipation issue on the increasingly high integration density of individual chips. Spreading the processing cores to different chips can relieve the on chip communication and power problem to some extent. But the overall communication efficiency would suffer from the limited off-chip bandwidth. To this end, we propose unified inter- and intra-chip optical networks, called SUPERB to replace the electrical interconnects for both inter/intra-chip communications. SUPERB is composed of intra-chip networks and inter-chip networks, which can coordinate closely to not only fulfill the intra-chip communication demands, but also relieve the off-chip bandwidth limitation. The evaluations demonstrate that the optical interconnect networks can achieve great improvement on network performance and energy efficiency compared to their electrical counterparts.

1.1 Introduction

Modern computer systems have to be increasingly complex to meet the growing performance requirements by various applications, such as cloud computing, big data. More than billions of transistors have been integrated on single chip to fulfill the performance demand, which brings along serious power issue. Many-core processor is becoming an attractive platform providing high performance and reducing power density. It is projected that hundreds or even thousands of processing cores would be integrated on the chip. In a many-core computer system, cores need to cooperate with their peers. The cooperation efficiency is greatly decided by the complex communication system among cores and memory subsystem, which would involve a lot in data communication in the manycore processor system. If the communication system is inefficient, the cores may waste much time on waiting for the required information from other cores or the memory, consuming extra power and worsening the power density. [1] estimated that more than 50% cores on the chip at 8 nm will not be used in consequence of power constraint. The complexity and size of single chip will also limit the scalability and yield of future manycore processors. Distributing the heavy workload from a single complex chip to multiple smaller processors may alleviate the power dissipation and lift the yield. This strategy brings along the challenge of inter-chip communication considering the already limited off-chip bandwidth and pin constraint in electrical interconnect. 3D technology can be used to stack the chips and support low-latency inter-chip communication but the power density becomes higher [2].

Photonic technologies have been successfully used in WAN, LAN, and board level, showing strengths in multicomputer systems and Internet core routers. With the recent progress in silicon photonics, optical interconnects may be adopted to address the communication issues for both inter- and intra-chip effectively. It is CMOS compatible and can be easily integrated and fabricated by the existing foundry with slight modification. Thanks to the silicon photonics devices and optical signal transmission, optical interconnects promise ultra-high bandwidth, low latency and low energy consumption. They can address both the intra-chip and inter-chip communication requirements with limited power budget. Wavelength Division Multiplexing (WDM) can utilize a bundle of optical signals to transmit signals together, which can boost the bandwidth significantly. For example, a silicon waveguide on the chip can support a data rate of 10 Gbps for each light wavelength, and multiple wavelengths can be multiplexed into the single waveguide to achieve extremely high bandwidth. The waveguide can also be connected with off-chip waveguide passively to support ultra-high off-chip bandwidth. Traditionally, optical devices were thought to be too big to be used for chip level communication and the switching time was also too slow for high speed communication. The recent progress has reported that the size of microresonator (MR) can be as small as 3 μm in diameter and 30 ps switching time has been demonstrated. Vertical-cavity surface-emitting lasers (VCSELs) are bonded on the chip to supply optical power [3, 4]. All the optical devices can be connected through the processor die with through-silicon-via (TSV) [5].

Optical interconnects are promising to address the issues on communication and power density for its ultra-high bandwidth, low latency and low energy consumption. Optical Network-on-Chip (ONoC), constructed by optical interconnects, has been put up with to replace electrical NoC for intra-chip communications by many studies [6, 7, 8, 9, 10, 11]. In [12], it only deals with the inter-chip communication. Both intra-chip and inter-chip communications can be implemented with optical waveguides and connected by passive couplers. The bandwidth of waveguides are broad enough for real applications. The low latency of optical interconnect makes the delay difference for on-chip and off-chip links minor. For inter-chip communication, polymer waveguides are fabricated on a PCB as transmission medium. Integrated chips are directly surface mounted to a PCB by a conventional ball grid array (BGA) solder process similar to [13]. Mirrors and lens arrays are incorporated to couple the light between silicon waveguides on-chip and the polymer waveguides on board. Considering there existing works on intra-/inter-chip communication and optical interconnect properties, it is natural to design both inter-chip and intra-chip optical networks simultaneously to facilitate both communications and result in a more efficient all-optical network for the whole system. In this chapter, we will propose a unified intra-/inter-chip optical network design, called SUPERB. SUPERB is a Sectioned Undirectional Photonic-Electric Ring Bus architecture. It is an all-optical network trying to take advantage of both intra-chip and inter-chip optical interconnects, at the same time removing the performance gap of on-chip and off-chip communications. What's more, there is no Electrical-Optical/Optical-Electrical (EO/OE) conversion between the intra-chip and inter-chip optical networks, saving extra power and electrical buffer resources and reducing delay on the EO/OE conversion. The optical data channels are composed of closed-loop optical waveguides, connecting processing elements and memory together. To make better use of the network resources, we virtually divide the data channels into non-overlapping independent sections so that transactions on different sections can be supported simultaneously. Bidirectional transmission is also supported thanks for the specially designed optical transceiver. These techniques will apply to both intra-chip and inter-chip optical networks. The inter-chip and intra-chip network will work together to serve the data transmission from one chip to another chip. For the local communication within a chip, the intra-chip optical network will be responsible for it. The proposed all-optical intra and inter chip network shows advantages on both cost and performance compared to two alternative designs.

1.2 Related Work

With the development of silicon photonic technologies, optical interconnects are proposed for the multicore on-chip communication. Pan *et al.* [14] proposed Firefly – a scalable, hybrid, hierarchical network architecture. It used traditional, electrical signaling to connect nodes within clusters while nanophotonics for inter-cluster communication since it could exploit the benefits of electrical signaling for short distance communication and nanophotonics for long distance, global communication. Cianchetti *et al.* [8] proposed a packet-switched hybrid electrical/optical routing network, called Phastlane, to decrease the latency. FlexiShare – a nanophotonic crossbar architecture that minimizes static power consumption by fully sharing a reduced number of channels across the network is proposed in [10]. In addition to the crossbar architecture, the ring/loop based architecture were also studied for their better scalability and smaller switches. Pasricha *et al.* [15] proposed an optical ring bus (ORB) based on-chip communication architecture for next generation MPSoCs, which replaces global pipelined electrical interconnects with an optical ring waveguide while preserving the interface with today's bus protocol standards such as AMBA AXI. Kirman *et al.* [16] presented an opto-electrical hierarchical bus for future manycore processors with cache-coherence supported. An optical loop at the top is for global communication and the bottom electrical wires are used for local interconnects. Koohi *et al.* [17] proposed a scalable all-optical NoC, referred to as Two Dimensional Hierarchical Expansion of Ring Topology (2D-HERT), which offers passive routing of optical data streams based on their wavelengths. Olympic, an all-optical hierarchical NoC architecture, is proposed in [18]. It employs a hierarchical clustered network topology in which clusters are interconnected through a global photonic ring and tiles inside each cluster are connected through a local photonic ring. LumiNOC, the novel nanophotonic NoC architecture in [19], employs partitions of network, a purely photonic, in-band, distributed arbitration scheme and a channel sharing arrangement utilizing the same waveguides and wavelengths for arbitration to achieve high performance and power-efficiency. The widely studied electrical NoC architectures, mesh, torus, fattree attract much attention in the optical interconnect domain. Morris *et al.* [20] utilized emerging nanophotonic technology to design a scalable low-power 64-core NoC called PROPEL, which strikes a balance between cheaper electronics and more expensive optics by facilitating nanophotonic interconnects for long distance inter router communication and electrical switching for routing and flow control. A regular multiplexing schedule for all-to-all connectivity was proposed in [21]. Mo *et al.* [22] proposed a hybrid optical mesh NoC, which also utilizes optical waveguides as well as metallic interconnects in a hierarchical way. Ye *et al.* [23] proposed a torus-based hierarchical hybrid optical-electronic NoC, called THOE, which utilized both electronic and optical interconnects in a hierarchical manner through novel hybrid optical-electrical router designs.

The fat tree network is a universal network for provably efficient communication. The links in a fat-tree become "fatter" as one moves up the tree towards the root so that it can provide higher bandwidth for the bandwidth hungry part of the "tree". Wang *et al.* [24] proposed a method to optimize the fat-tree floorplan, which can effectively reduce the number of crossings and minimize the interconnect length. Two types of floorplans are proposed in [24], which could be applied to fat-tree based networks of arbitrary size. Li *et al.* [25] also proposed a hybrid electrical/optical cluster-based hierarchical on chip network architecture, which connects the lowest level cluster of IP cores with electrical interconnects while optical interconnects are used for inter-cluster communication to improve the efficiency of the network.

In most of these designs, only one chip is considered and the optical networks are dedicated for on-chip communication, whether in a hierarchical or all-optical way. Only a few works get to the off-chip communication but without comprehensive consideration on inter-chip optical networks. The Dragonfly topology was proposed for large-scale, off-chip networks to exploit the availability of economical, optical signaling technology and high-radix routers to create a cost-efficient topology in [26]. Corona [6], a 3D many-core architecture, uses nanophotonic communication for both inter-core and off-stack communication to memory or I/O devices. Pasricha *et al.* [27] proposed a hybrid photonic 3D NoC combining low cost photonic rings on multiple photonic layers with a 3D ring-mesh NoC in active layers to significantly reduce on-chip communication power dissipation and packet latency.

In SUPERB, we consider both intra-chip and inter-chip communication together so that the on-chip and off-chip optical network can cooperate together efficiently and make full use of the features of optical interconnects. The on-chip subnetwork in SUPERB is an optical network with ring topology. [6, 10, 14, 28, 29] have also proposed similar topologies. In crossbar design with large network resources, link sharing is important to reduce the resource requirements. Multiple writers and a single reader (MWSR) channel design is adopted by the crossbar in Vantrease *et al.* [6]. On the other hand, Pan *et al.* [14] proposed a design that a waveguide is shared by single writer and multiple readers (SWMR). In some other designs, the data channel is further shared by multiple writers and multiple readers (MWMR). Xu *et al.* [29] proposed a channel borrowing technology to improve the channel utilization and also reduce the power consumption. In all these designs, a waveguide is unidirectional, and at any time, there can be no more than one transaction with same wavelengths in a single waveguide. By using ring topology in SUPERB and the special design of optical transceiver, bidirectional transmission is supported and the longest transmission distance is reduced. The resource sharing is even more improved by virtually channel segmentation in SUPERB and allowing concurrent transactions with the same set of wavelengths on a single waveguide but different sections.

ORNoC, a contention-free architecture based on optical network on chip is presented in [28], which provides wavelength sharing technique to reduce the required waveguide and support multiple transactions to improve the performance. In their design, the wavelength is statically assigned based on the connectivity requirements. While in SUPERB, wavelength division multiplexing (WDM) is utilized and a bunch of wavelengths are used together to fulfill data transfer on a single waveguide but on different sections the transactions can happen simultaneously, which is dynamically decided by the arbitration. Morris [30] proposed an optical network with 3D stacking technology. A large crossbar is decomposed into multiple small crossbars on different layers to reduce the power. The idea of decomposing a long link into some shorter links is also adopted in SUPERB. Datta *et al.* [31] proposed segmented optical bus. Buses are segmented to reduce power consumption and they are interconnected by electrical routers. We need not physically divide the data channel but virtually segmentation. Only one optical layer is required. The channel segmentation in SUPERB is in more depth and the throughput and resource utilization is higher with efficient arbitration and more independent channel sections. No need of electrical switches and buffers will save much power. The shorter transmission distance also means the lower required optical laser power. In SUPERB, link sharing is explored to improve the resource utilization and performance, but it doesn't impact much on the arbitration. By setting accessing rules to data channel, we can reduce the arbitration overhead dramatically without sacrificing much network flexibility. The detailed illustration will be given during the following sections.

1.3 Architecture Overview

SUPERB is dedicated to optically connected multicore systems with multiple processor and memory chips. It should be noticed that cache/memory subsystem is an important part for manycore computer system. We take fair consideration on how to design it and to make it serve the computing elements efficiently. To this end, we have individual cache nodes on chip and separate memory chips to provide enough memory space for the whole system. On each chip, there are one optical layer and one electrical layer. They are stacked together with 3D stacking technology. The computation elements, including processing core, cache/memory, buffer and control unit are all located in the electrical layer. Optical layer contains the silicon photonic devices used for optical interconnects, lasers, optical routers, waveguides and photodetectors. VCSELs are chosen for on-chip laser source, bonded on the chip. Through-silicon-via (TSV) is used for communication between the electrical layer and optical layer. Further, chips are bonded on the board and connected by the inter-chip optical interconnects. It is also called the board-level optical interconnects. The intra-chip and inter-chip optical networks cooperate to facilitate both the on-chip communication among cores and caches on the same chip, and the board-level communication among elements on different chips.

To provide a clear concept of this architecture, we give the logical view of SUPERB as shown in Figure 1.1. It contains an inter-chip network and several intra-chip networks. The nodes on the same chip are connected by the corresponding intra-chip network. On the other hand, the inter-chip network thread the chips in the third dimension by connecting to the nodes on the intra-chip network. The chips are virtually stacked similar to a 3D chip. So all elements are fully connected together by the intra-chip and inter-chip network. Electrical links need repeater to ensure that the electrical signal can be transmit to a long distance. While optical signals can tolerate much longer distance thanks to relatively low power and latency overheads. That's why the chips are physically palced far away from each other. It can relieve the power

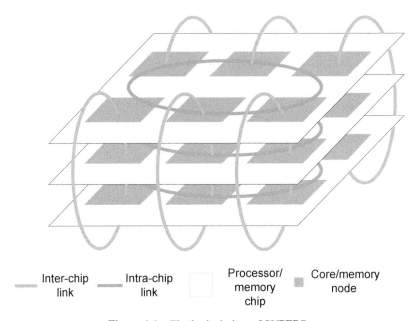

Figure 1.1   The logical view of SUPERB.

density issue, which has been one of the major issues for modern computer systems.

The intra-chip and inter-chip architecture targets for heterogenous many-core system. Even on each chip, the node is not necessary to be with the same organization, which can vary on the cache and memory configurations or even the processing cores. Each core will have private L1 data and instruction cache, and every four cores form a core cluster with a sharing L2 cache. The system will also have individual memory interface for the easy of memory access. As mentioned above, the node can mean either the core cluster node or the memory node. From the side of network, these nodes all obey the same rule to communication with other nodes, either in the same chip or a different chip. We denote the j-th node in i-th chip by $N(i,j)$ for clearness, and there are M chips in total for the whole system and there are W_i nodes for the i-th chip. The node can be either core cluster or memory node. But it won't matter much on our consideration of the whole network architecture design. All nodes on a chip are interconnected by the intra-chip network. The chips are interconnected by parallel circular inter-chip optical links, which are controlled by "arbiter chip". Logically, these links are perpendicular to the chip plane, and thus they address the vertical communication in the system. That is to say, The intra-chip and inter-chip networks intersect at the nodes. Each such a node is a buffering point for inter-chip communications. If one transaction involves both intra-chip and inter-chip network, we make the packets always take intra-chip path first to avoid deadlock.

1.4 Intra-chip Network Design

There are intra-chip networks for each chip to facilitate the communication within this chip and also play a indispensable role in inter-chip communication. The overview of intra-chip network is as shown in Figure 1.2. Processor and memory nodes are connected by the closed waveguides loop and the control module is located in the central of the chip, communicating with each node via dedicated electrical wires. Within the processor and memory nodes, different components communicate with each other by electrical links. The optical data channels composed of optical components are responsible for communication among nodes. As a result, we make use

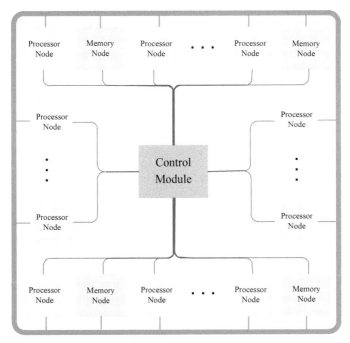

Figure 1.2 Overview of intra-chip network.

of hybrid of optical and electrical domain as communication media. The memory nodes are evenly distributed among the processor nodes, reducing the memory access latency and avoiding access to far away memories. To give a more clear view of the architecture organization, we give the floor-plan in Figure 1.3. 3D technology is used to stack an optical layer on the top of electrical layer. On the optical layer, the waveguides snake acorss the chip, connecting all cluster nodes and memory nodes through optical transceivers. Each node has a node agent, which locates in the center of the chip and communicate via electrical links, as mentioned above. In this way,

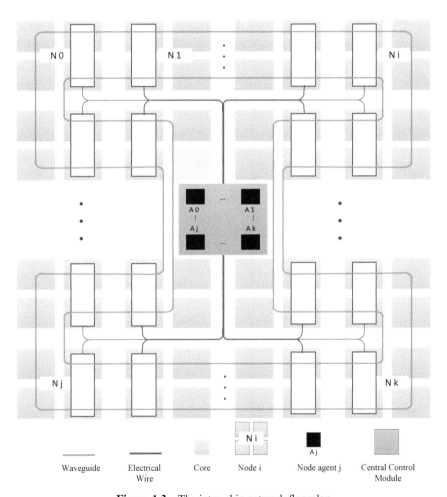

Figure 1.3 The intra-chip network floorplan.

the communication in optical domain and electrical domain will be isolated from each other.

All processor nodes are identical in the chip, that is to say, with the same organization. In each processor node, there are four identical cores and a private L1 I/D cache for each core. A L2 cache bank is located in each processor node. It is shared L2 cache, meaning that it can be accessed by both the L1 cache within and outside of local processor node. The optical network interface is to connect the local elements and the optical data channels, enabling the communication of core or cache in local node and that of other nodes. In Figure 1.2 and Figure 1.3, we just omit some nodes for simplicity. Furthermore, each core has a private L1 instruction and data cache and the four cores within a node have a shared L2 cache. Each node also includes a optical network interface to transmit and receive data and control signals from data channels and control system, respectively. The network interface is an important component in the system. In network interface, the signals can be transformed from electrical domain to optical domain, or in a reverse way. The optical transceiver will be introduced in next subsection.

1.4.1 Data Channel Design

Data channels are used to transmit optical signals among different nodes through the parallelly aligned waveguides. The waveguides are closed loops, which can facilitate bidirectional transmission together with the dedicated design of optical network interface and optical transceiver. Every node is attached to the data channels by the optical network interface and one data channel is accessible for multiple nodes. Concurrent transactions between different sender-receiver pairs are fully supported on a single data channel thanks to the virtual channel segmentation. The on-chip lasers, VCSELs are bonded on the chip in the network interfaces as light sources of data channels. Wavelength Division Multiplexing (WDM) is used to increase the data channel bandwidth, as a result a set of MRs are needed to deal with the optical signals in a single waveguide. Channel partition is also proposed to reduce the arbitration overhead of control system at the sacrifice of limited flexibility.

1.4.1.1 Optical network interface

Network interface, NI in short, is responsible for both the on-chip and off-chip communication. There is a NI for each node and it is used to interconnect the corresponding node with the data channels on the intra-chip and inter-chip networks. As a result, it is a bridge between the intra-chip and inter-chip network and is a buffering point for the inter-chip communications. The data transfer requests from the inter-chip network and the local node will be treated equally by the intra-chip control subsystem so that the way to deal with inter-chip communication can be simplified. Because of channel partition for the intra-chip optical network, which will be explained later, in some data channels, the network interface will function as an optical transceiver, but in other data channels, the network interface only needs to work as an optical receiver. As a result, we can decrease the complexity of network interface to some extent. As described above, the control system is in electrical domain. Two dedicated metal signal channels are implemented to send request to and receive grant information from the corresponding agent. The detailed design of control system will be introduced later. It should be noticed even though we have two types of nodes, processor node and memory node, the network interface and corresponding node agent are the same. This is mainly because the network interface and node agent are responsible for data transmission and request arbitration, there is nothing to do with the specific source and destination of request. Optical network interface contains multiple optical transceivers, which are fabrics to fulfill optical signal transmission.

1.4.1.2 Optical transceiver

Optical transceiver is composed of VCSELs, waveguides, photodetectors (PDs), and microresonators (MRs), as shown in Figure 1.4. It is able to send

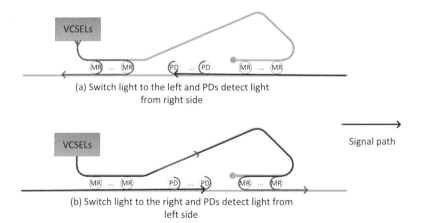

(a) Switch light to the left and PDs detect light from right side

(b) Switch light to the right and PDs detect light from left side

Figure 1.4   Optical transceiver.

data to and receive data from data channels by power on/off MRs under the control signals. The on-chip laser sources, VCSELs, can be powered off when there is no data transfer. This will dramatically reduce the power consumption compared with the off-chip laser. Another advantage is that the output power can be controlled base on the loss along the path, which will avoid redundant power when the path loss is low. The obvious disadvantage of on-chip laser is that it would worsen power dissipation for the chip. However, this overhead would be well compensated by saved power. The VCSELs can be directly modulated with a rate of 40 Gbps and the turn-on delay can be as low as 1 ns, which would be hidden by path-setup delay and won't impair much on the network performance. MRs are switching elements to bridge the transceiver and the data channels. By switching the optical signals from different directions, the transceiver can realize bidirectional transmission. MRs can also work as integrated filters in photodetectors, selecting optical signals on resonant frequency. The optical transceiver can be designed to support more than two data channels by duplicating the switching elements and sharing the same set of VCSELs and PDs, reducing power and area overhead.

1.4.1.3 Channel segmentation

Channel segmentation means the channel is segmented into multiple individual and independent sections. In this way, data transmission can be executed bidirectionally and concurrently in different sections of the same data channel, thus channel utility can be lifted. Here we introduce this idea but segment the data channel virtually. Here "virtually" means that in the logic view, the long link S[0,n] (S[i,j] denotes the data channel section between N_i and N_j), is replaced by several independent shorter sections. It makes better use of the available resources. The detailed realization relies on the optical transceiver and the special design of control scheme. It is natural to come to mind that channel segmentation will make the control and arbitration more complicated because the arbiter has to consider more factors when doing arbitration. This is true and may introduce much arbitration overhead if not designing the control system carefully. But as we can see in next subsection, this issued can be addressed by our control system. More specifically, the segmentation can be illustrated as Figure 1.5. For nodes N_i, N_j, N_m and N_n, they can use the channel sections between two nodes to transmit data without disturbed by other transactions. The segmentation for different data channels can be different. For example, on the outermost data channel, N_i and N_j can use the section between them, N_m and N_n using the other section

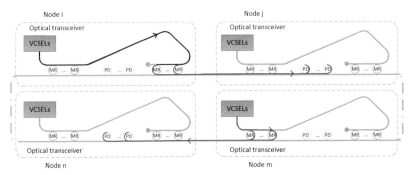

Figure 1.5   A detailed view on channel segmentation and bidirectional transmission.

between them, while on the innermost data channel, N_j and N_n can use the section between them, N_i and N_m using the other section between them.

1.4.2 Control Subsystem

A relatively distributed control scheme is proposed for the intra-chip optical network. Since the on-chip optical network is the same for different chips, the control subsystems are also the same and work independently from chip to chip. Each node is assigned with a dedicated node agent which is responsible for processing the requests of accessing the data channels. The arbitrations are made in a relative independent fashion, except of sharing some limited information including the memory states and the channel states. We moved all the agents to the center of chip so that they can share limited information with short latency by using the very-short local electrical wires. There are much more data channels and processor/memory nodes attached to each channel compared to inter-chip network. The channel segmentation also introduces more possible traffic patterns. Therefore, the intra-chip network resources are more abundant but the arbitration overhead will also be much higher. The limited area budget for on-chip network controller restricts the complexity of the arbitration scheme further. Under these circumstances, channel partition is proposed to effectively reduce the arbitration complexity. Credit-based flow control is used in SUPERB. The flow control scheme is achieved by different control units collaboratively. Each node agent has the initial number of tokens corresponding to the number of available buffer slots of the receiver. It counts down the tokens every time a packet is grant. On the other hand, the receiver node will send the new tokens back to the agent if the number of buffer slots is changed. If no token left, the requests will not be processed.

1.4.2.1 Channel partition

Channel partition is to divide all data channels into different groups, within each group only specific traffic patterns are allowed. All possible traffic patterns are spread among these groups. These accessing rules are set at the first place. The detailed accessing rules and channel partition are illustrated in Figure 1.6. In the i-th channel group, there are 2^i data channels and only transactions with distances $\in (2^{i-1}, 2^i]$ hops are allowed. The maximum number of groups is decided by the longest transaction distance. Different groups only need to deal with transactions with different distances so that collisions between groups are avoided and the corresponding arbitration is only limited to these allowed traffic patterns. For example, according to this classification the traffic distance with 12 hops will be assigned to group 4, and will not be interfered by the transactions in other groups. As shown in Figure 1.6, the data channels are further classified based on the allowed intervals within each group. In group i, all nodes should

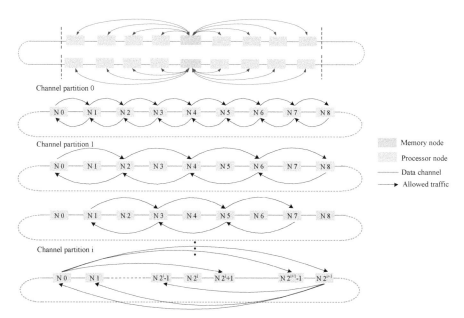

Figure 1.6   Channel partition. Channels that are classified into different groups would deal with the allowed traffic patterns.

be able to send data to the nodes with distances allowed for this group. A way to reduce the potential collisions on a data channel is to set the interval between senders to be 2^i so that the data channel is divided into $N/2^i$ sections with length 2^i hops and no cross-section collisions exist. To be specific, on the j-th channel in group i, the allowed senders are the nodes with labels $j + k \times 2^i \% N$ where k is any nonnegative integer and N is the total number of core nodes. We can easily get that there are $N/2^i$ senders on each data channel for group i. Since there are individual memory nodes, we give a special partition for memory nodes so that the request to memory can be dealt with more efficiently. In real applications, the demand for different traffic pattern varies, so there will be multiple identical groups for each channel group to improve the network's throughput.

1.4.2.2 Node agent design

Each processor or memory node is electrically connected to the corresponding node agent. The node would not only send requests, but also send buffer tokens to the dedicated agents via optical network interface since the buffer information will be used for arbitrating. One bit is enough to identify the two types of packets. There are also two types of information from the agent to the node: the grants answering the requests; the grants informing the receiver that the new packets are coming.

The node only needs to negotiate with neighboring agents for each transaction thanks to the accessing rules in channel partition. The complexity of arbitration algorithm is reduced to $O(1)$. The algorithm in node agent is shown in Algorithm 1. The packets from nodes would be decoded by the agent first to determine the packet type. If it is a packet with buffer tokens, the tokens would be sent to the related sender agents. If it is a request packet, the request would be pushed in the request pool. For each request in the request pool, it has to undergo three steps before granted. The first step is to check whether the destination buffer is full, which is implemented in flow controller unit. It is followed by the channel collision solver which checks whether the channel segment is available and try to reserve a link section for the current request. Lastly, the state of the destination agent is checked in the destination checker. It is to make sure that the destination agent is able to inform the destination node to open the detector on time. If the three steps are passed successfully, the request is granted and the grant information is sent back to the node. Otherwise, the request would be stored back into the request pool. Multiple requests are processed in parallel and the

---

Algorithm 1 Node agent arbitration algorithm

Require: the incoming packet P from node

1: **if** $IsRequest(P) == false$ **then**
2:      Buffer change packet
3:      $InformOtherAgents$
4: **else**
5:      Decide channel group for (P)
6:      $RequestBuffer.Add(P)$
7:      **if** $(I) <= Processing_parallel_level$ **then**
8:          $RequestBuffer.Select(I) = P$
9:      **end if**
10: **end if**
11: $TryReserveLink(I)$
12: $CheckDestAgent(I)$
13: **if** $(CheckDestinationBusy(I) == false)AND$ $(TryReserveLink(I) == success)AND$ $(CheckDestinationAgent(I) == success)$ **then**
14:     $Grant == True$
15: **else**
16:     $Grant == flase$
17:     $RequestBuffer.Add(P)$
18:     Clear Reserved Link
19: **end if**
20: **return** $Grant$

---

process of each request is well pipelined to improve the throughput of the controller.

1.5 Inter-chip Network Design

Conventional electrical wires are unable to promise the large bandwidth density, low latency and low power consumption requirements for off-chip communication. The inter-chip optical network is to address the communication issues among chips. The inter-chip network is composed of data channels and the corresponding control fabrics. There are N (the maximum number of nodes on a chip) data channels which are parallel to each other with the same design and connection. The control fabrics is composed of the control channels and an arbiter chip. Before accessing the data channel, the nodes are required to send requests to the arbiter chip through the control channels. The arbiter chip will make the arbitration and also configure the data channels by sending out control information to the data channels.

1.5.1 Inter-chip Data Channel

The inter-chip data channels are homogeneous and parallel to each other without waveguide crossings. We also pack multiple wavelengths into the waveguide for each transaction for inter-chip network. The design of data channel is similar to that of intra-chip network. The channel is composed of closed-loop waveguides threading different chips with optical transceivers attached to them. The on-chip optical transceivers interact with the silicon waveguides to get the light out of the data channel or inject the light into the channel. The difference is that each closed-loop waveguides in inter-chip network is built by bridging silicon waveguides on chip and the polymer waveguides on board, while the on-chip channel is a single closed-loop silicon waveguide. The similarity between intra-chip and inter-chip data channel implies that inter-chip channel also has the bidirectional transmission and channel segmentation properties. The designs of optical transceivers for intra-chip network and inter-chip network are the same. The benefits of the properties are obvious compared to MWSR [6] and MWMR [10] channel design, as shown in Figure 1.7. For MWSR, each channel is accessible to all writers and only one reader, but only one transaction can be supported. Each channel is accessible to all writers and readers for MWMR, the same as our design. But it can only support one transaction at a time. In contrast, the

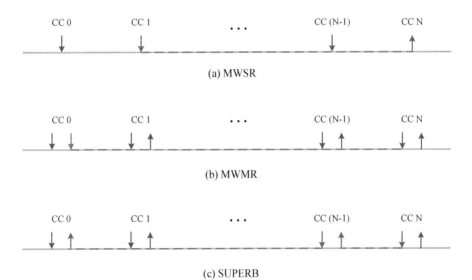

Figure 1.7   Illustration of different data channel accessing designs. The red lines with arrow stand for one possible transaction on the data channel.

SUPERB data channel can support multiple transactions on different channel sections simultaneously. In best case, the SUPERB data channel can support M concurrent transactions, improving the throughput by M times.

1.5.2 Control Subsystem

The data channel requires a conflict resolution scheme to avoid two trans-actions overlapping at the same channel section. The path set-up is required before the transmission for circuit switching in inter-chip optical network, which is similar to intra-chip network. The similarities of intra-chip and inter-chip data channel design and behavior make it possible to use the same arbitration scheme as the intra-chip network. The difference is that the inter-chip network resources are much less than intra-chip network so that the arbitration overhead is not as high as intra-chip. There is no need to do the channel partition as intra-chip network. The arbitration can be done in a single arbitration chip for all channels. Since each channel is independent with the others, we control each inter-chip channel in a separate way with individual control unit, reducing the arbitration complexity effectively. All the processor and memory chips are optically connected to the arbiter chip via on-board waveguides.

Although in most cases the inter-chip network needs to cooperate with intra-chip network to fulfill data transfer, they can work independently because the buffering elements in network interface can isolate the commu-nications in the two domains. That is to say, the same transaction request will be dealt with separately in intra-chip and inter-chip networks. The overall inter-chip communication can be formulated in this way. We denote a sender-receiver pair by following the previous definition to demonstrate the process. The source node is $N(i, j)$, the j-th node in i-th chip. The destination node is $N(m, n)$. Firstly, the data packets generated by the source node will be sent to $N(i, n)$ and buffered in $N(i, n)$ waiting for the grant information from the inter-chip arbitration unit. Secondly, once the n-th inter-chip data channel is grant to this request, the data will be transferred from $N(i, n)$ to $N(m, n)$, which is the destination. It is the end of the conceptual view of data transmission via intra-chip and inter-chip network.

As mentioned before, there is an individual control unit for each inter-chip data channel. The control unit can deal with all requests from the nodes of different chips on the corresponding data channel. They won't interfere each other and arbitrate in the same fashion. The work flow is as shown in Figure 1.8. The first stage is to select requests from the request pool. The

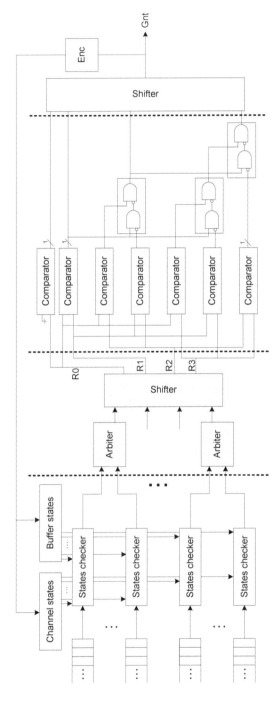

Figure 1.8   The control unit of the inter-chip network. The units are homogeneous and independent with each other, and thus only one unit is shown. The comparator in the figure will output "true" if there is no collision between two input requests.

request can be selected under two conditions: the intended link is free and there is available buffer at destination. The second stage is to reduce the collision possibility among the nodes. We set each data channel into several regions logically and assign at most one request to each region every time. In this way, the possibility of collision can be reduced within a region. The following stage is to check the collision among the selected requests. We adopt the shortest path algorithm to determine the path for each request so that for each request the possible path is fixed for the given sender-receiver pair. If there are overlaps among the paths for different request. We say that the related requests will collide and be chosen in round-robin to preserve fairness. The last stage is first to permutate the requests back to the original order as in input of the second stage. The grant information for the selected requests will be sent out to the nodes.

1.6 Evaluations

In this section, we evaluate and compare the performance and energy efficiency of SUPERB and electrical mesh. The reason we choose electrical mesh is that it is a popular and mature NoC in electrical domain. The simulations are done with a heterogenous multiprocessor system simulation platform [32]. This simulation platform considers not only interconnects, but also the processor, cache/memory system, which are indispensable but ignored by many network designs. By taking these subsystem into our design, we can get a close look and meaningful comparison on our design and the electrical mesh under realistic applications, Fast Fourier Transform (FFT), Low-Density Parity-Check (LDPC), Reed-Solomon decoder (RSD) and Machine Learning for financial market (MLf) provided by [32]. It is assumed to be 16 main memory channels for each system and they are distributed evenly at specified locations on each chip. Every processor has its own private 32 KB L1 instruction and data caches. Shared L2 caches are evenly distributed among the whole system so that each node has a 256 KB L2 cache bank. One memory port is allocated for every 16 cores. A directory based MOSI cache coherency protocol is used for the evaluation.

To compare SUPERB with electrical mesh under different network scales, we consider systems of 16-core, 64-core, 128-core and 256-core, which are corresponding to 2, 4, 4 and 4 chips with the same number of cores on each chip for both SUPERB and electrical mesh. We only consider chips less than 4 is because it is difficult to stack too many chips together though 3D techniques can be used. The intra-chip scale for different systems will vary according to the total number of cores on the system. For SUPERB optical interconnects, each data channel are set to have 4 waveguide. The number of channel groups for each intra-chip and inter-chip networks will be varied based on the number of communication nodes. For each waveguide, we assume to multiplex 16 wavelengths at 10 Gbps, resulting an aggregate bandwidth of 160 Gbps for each waveguide. In all designs, the electrical clock frequency is assumed to be 2.5 GHz and the link bit width for electrical mesh is assumed to be 64 so that the link bandwidth is 160 Gbps. To calculate the propagation delay of signals, the chip size is assumed to be $10\,mm \times 10\,mm$ and the optical group refractive index of the silicon waveguide is 4.2. The propagation length will be adjusted according to the network scale and sender-receiver pairs. We evaluate and compare SUPERB and electrical mesh (Emesh) on overall performance and energy efficiency.

1.6.1 Node Agent Power and Area

In SUPERB, thanks to channel partition, the node agent only needs to negoti-ate with neighboring agents so that the control scheme is relatively simple and the arbitration complexity is almost constant. As a result, the power and area overhead is kept in a reasonable range. In SUPERB, the intra-chip control module is located in the center of each chip, physically. The central arbiter is the composition of all node agents, which mainly accounts for the arbitration of the corresponding node requests in a relatively independent way. There is a centralized inter-chip controller to handle all inter-chip networks. Since only several chips involved in the inter-chip communication, the control is simple and the power and area overhead is relatively small compared with intra-chip networks. In the following evaluations, the synthesis is carried out for both intra-chip controllers and inter-chip controller at $45\,nm$ technology node assuming activity ratio to be 0.3. In Figure 1.9 and Figure 1.10, the power and area shown for both the whole control scheme and single node agent of different architecture scales. For all cases, the total power is kept less than $1\,W$. Figure 1.9 shows that dynamic power contributes more than 80 percentage. The total power increases mildly as the network scales up. The average power on each node agent becomes higher when the network scales are larger. It makes sense in that larger network scale will introduce extra energy consumption on information exchange and the number of links between agents will also increase. The area overhead of controller is scal-able, occupying less than 1% of the whole chip size. As we can see from Figure 1.10, the total area and average agent area rise almost linearly when

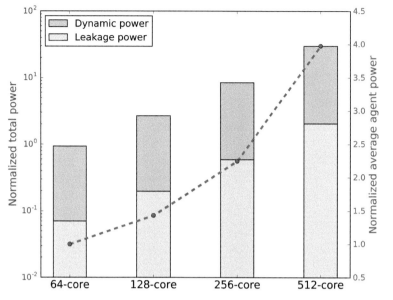

Figure 1.9 Power scalability of node agent.

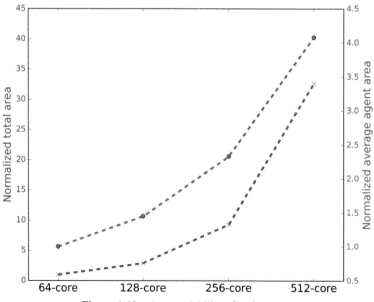

Figure 1.10 Area scalability of node agent.

the network scales up. These trends match what we discuss about SUPERB's control scheme and demonstrate the good scalability of the proposed design, which also contributes much to the performance and energy of the whole architecture.

1.6.2 Performance

Performance is one of the main design objective of the proposed architecture. To achieve a clear comparison on different items, we normalized the results for each network scale based on the minimum data in that scenario. Figure 1.11 shows the normalized system performance. It is obvious that SUPERB achieves higher performance than Emesh for all applications and network scales. The highest speedup is achieved by FFT on SUPERB compared to Emesh, as much as $1.6\times$. In most cases, SUPERB offers more performance as the network scale becomes larger. The explanation is that when mapping application tasks to more processing cores, the overhead of cache coherence control messages on use of shared resources and higher average packet latency will make the waiting longer, thus lower system performance for larger network scale.

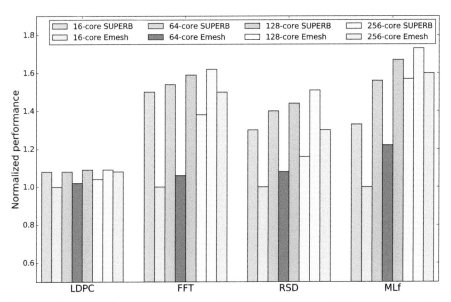

Figure 1.11   Normalized performance for SUPERB and Emesh on different network scales.

1.6.3 Energy Efficiency

Apart from the achieved better performance, SUPERB is also designed to be energy efficient. To show the power consumption of SUPERB, we adopt the nanophotonic power models in [12, 33]. There are various losses along the optical path and they are list in Table 1.1. For optical communication, EO/OE conversion is necessary and the conversion power is assumed to be $100\,fJ/bit$ as demonstrated in [34]. The sensitivity of the photodetector is assumed to be $10\ \mu W$ as in [10]. We also assume $1\ \mu W$ heating power per MR per Kelvin, and 20 K tuning range. The power for switching MRs are assumed to be $50\ \mu W$ per MR [33]. The power efficiency of off-chip laser is assumed to be 30% as in [33]. The power of node agent is synthesized with $45\,nm$ and taken into consideration. The on-chip lasers are used and can adjust the power consumption according to the path loss, thus can save extra energy. The huge advantage of SUPERB on energy efficiency can be found in Figure 1.12. Let's take FFT as an example. Emesh consumes almost $10\times$ more energy for transmitting one bit compared to SUPERB. The already serious power issue on modern computing system would suffer a lot from the electrical interconnects. While the optical network SUPERB achieves much better energy efficiency, it can relieve the power dissipation on chip and is indeed a promising candidate for the unified inter/intra-chip network.

Figure 1.13 shows the performance per energy consumption. It measures the offered performance under the same energy consumption. It is a indicative index for energy efficient and high performance computing system design. It shows that SUPERB achieves much higher performance per energy consumption than Emesh. It should be noticed that when the network scales up, the offered performance becomes higher compared to the corresponding Emesh, demonstrating optical interconnect's advantages for high performance computing systems containing many high-end processors.

Table 1.1 Optical loss

Component	Loss
Passing MR loss	0.05 dB
Filter Drop loss	1.5 dB
Waveguide Crossing loss	0.05 dB
Waveguide Propagation Loss	1 dB/cm
Coupler	1 dB
Bending	0.005 dB/90°

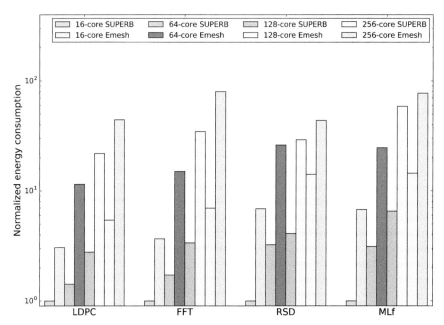

Figure 1.12   Normalized energy consumption of different applications for SUPERB and Emesh on different network scales.

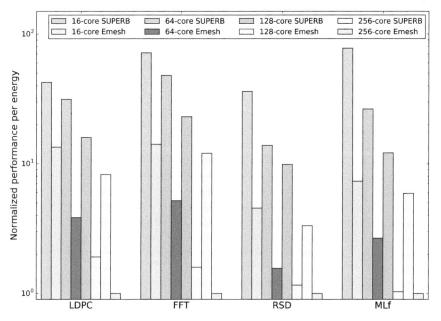

Figure 1.13   Comparison of normalized performance per energy consumption on different network scales.

1.7 Conclusion

The advances in nanophotonics have motivated us to exploit the benefits of optical interconnects for future manycore processor with a large number of cores. In this work, we propose an unified design of inter/intra-chip optical network called SUPERB which supports high-throughput and low-latency communication for the multi-chip system. The proposed network effectively explores the distinctive properties of optical interconnect to boost performance as well as reduce energy consumption. The evaluation and comparison shows that SUPERB achieves promising performance with good energy efficiency. There are still challenges on optical interconnects design. Thermal effects will dramatically weaken the optical link's reliability if not properly treated since most optical devices are temperature sensitive. Thermal variation, as well as process variation would be one of the major concerns about optical interconnect. It shows that optical interconnect is promising solution for the high performance computing systems with hundreds even thousands of manycore processors.

References

[1] H. Esmaeilzadeh, E. Blem, R. S. Amant, K. Sankaralingam, and D. Burger, "Dark silicon and the end of multicore scaling," in *2011 38th Annual International Symposium on Computer Architecture (ISCA)*. IEEE, 2011, pp. 365–376.

[2] D. Sekar, C. King, B. Dang, T. Spencer, H. Thacker, P. Joseph, M. Bakir, and J. Meindl, "A 3d-ic technology with integrated microchannel cooling," in *2008 International Interconnect Technology Conference*, June 2008, pp. 13–15.

[3] J. M. Perkins, T. L. Simpkins, C. Warde, and J. Clifton G. Fonstad, "Full Recess Integration of Small Diameter Low Threshold VCSELs within Si-CMOS ICs," *Opt. Express*, vol. 16, no. 18, pp. 13 955–13 960, 2008.

[4] A. V. Krishnamoorthy, K. W. Goossen, W. Jan, X. Zheng, R. Ho, G. Li, R. Rozier, F. Liu, D. Patil, J. Lexau *et al.*, "Progress in low-power switched optical interconnects," *IEEE Journal of Selected Topics in Quantum Electronics*, vol. 17, no. 2, pp. 357–376, 2011.

[5] G. Katti, M. Stucchi, K. De Meyer, and W. Dehaene, "Electrical modeling and characterization of through silicon via for three-dimensional ics," *IEEE Transactions on Electron Devices*, vol. 57, no. 1, pp. 256–262, 2010.

[6] D. Vantrease, R. Schreiber, M. Monchiero, M. McLaren, N. P. Jouppi, M. Fiorentino, A. Davis, N. Binkert, R. G. Beausoleil, and J. H. Ahn, "Corona: System Implications of Emerging Nanophotonic Technology," in *35th International Symposium on Computer Architecture*, 2008, pp. 153–164.

[7] I. O'Connor, "Optical solutions for system-level interconnect," in *Proceedings of the 2004 international workshop on System level interconnect prediction*. New York, USA: ACM, 2004, pp. 79–88.

[8] M. J. Cianchetti, J. C. Kerekes, and D. H. Albonesi, "Phastlane: a rapid transit optical routing network," in *Proceedings of the 36th Annual International Symposium on Computer Architecture*. New York, USA: ACM, 2009, pp. 441–450.

[9] A. Shacham, K. Bergman, and L. P. Carloni, "Photonic Networks-on-Chip for Future Generations of Chip Multiprocessors," *IEEE Trans. Comput.*, vol. 57, no. 9, pp. 1246–1260, 2008.

[10] Y. Pan, J. Kim, and G. Memik, "FlexiShare: Channel sharing for an energy-efficient nanophotonic crossbar," in *IEEE 16th International Symposium on High Performance Computer Architecture*, 2010, pp. 1–12.

[11] S. Bartolini and P. Grani, "A Simple On-chip Optical Interconnection for Improving Performance of Coherency Traffic in CMPs," *15th Euromicro Conference on Digital System Design (DSD)*, pp. 312–318, 2012.

[12] P. Koka, M. O. McCracken, H. Schwetman, X. Zheng, R. Ho, and A. V. Krishnamoorthy, "Silicon-photonic network architectures for scalable, power-efficient multi-chip systems," in *Proceedings of the 37th Annual International Symposium on Computer Architecture*. New York, USA: ACM, 2010, pp. 117–128.

[13] F. E. Doany, C. L. Schow, R. Budd, C. Baks, D. M. Kuchta, P. Pepeljugoski, J. A. Kash, F. Libsch, R. Dangel, F. Horst, and B. J. Offrein, "Chip-to-chip board-level optical data buses," in *Conference on Optical Fiber communication/National Fiber Optic Engineers Conference,2008.*, 2008, pp. 1–3.

[14] Y. Pan, P. Kumar, J. Kim, G. Memik, Y. Zhang, and A. Choudhary, "Firefly: illuminating future network-on-chip with nanophotonics," *In Proc. of the International Symposium on Computer Architecture*, vol. 37, no. 3, pp. 429–440, 2009.

[15] S. Pasricha and N. Dutt, "ORB: An on-chip optical ring bus communication architecture for multi-processor systems-on-chip,"

in *Design Automation Conference, Asia and South Pacific*, 2008, pp. 789–794.

[16] N. Kirman, M. Kirman, R. K. Dokania, J. F. Martinez, A. B. Apsel, M. A. Watkins, and D. H. Albonesi, "Leveraging Optical Technology in Future Bus-based Chip Multiprocessors," in *Proceedings of the 39th Annual IEEE/ACM International Symposium on Microarchitecture*. Washington, DC, USA: IEEE Computer Society, 2006, pp. 492–503.

[17] S. Koohi and S. Hessabi, "All-Optical Wavelength-Routed Architecture for a Power-Efficient Network on Chip," *IEEE Transactions on Computers*, vol. PP, no. 99, pp. 1–1, 2012.

[18] S. Bartolini, L. Lusnig, and E. Martinelli, "Olympic: A hierarchical all-optical photonic network for low-power chip multiprocessors," in *2013 Euromicro Conference on Digital System Design (DSD)*, Sept 2013, pp. 56–59.

[19] M. Browning, C. Li, P. Gratz, and S. Palermo, "Luminoc: A low-latency, high-bandwidth per watt, photonic network-on-chip," in *2013 ACM/IEEE International Workshop on System Level Interconnect Prediction (SLIP)*, June 2013, pp. 1–4.

[20] R. Morris and A. K. Kodi, "Exploring the Design of 64- and 256-core Power Efficient Nanophotonic Interconnect," *IEEE Journal of Selected Topics in Quantum Electronics*, vol. PP, no. 99, pp. 1–8, 2010.

[21] J. Carpenter and R. Melhem, "Deterministic multiplexing of noc on grid cmps," in *2013 IEEE 21st Annual Symposium on High-Performance Interconnects (HOTI)*, Aug 2013, pp. 1–8.

[22] K. H. Mo, Y. Ye, X. Wu, W. Zhang, W. Liu, and J. Xu, "A Hierarchical Hybrid Optical-Electronic Network-on-Chip," in *2010 IEEE Computer Society Annual Symposium on VLSI (ISVLSI)*, 2010, pp. 327–332.

[23] Y. Ye, J. Xu, X. Wu, W. Zhang, W. Liu, and M. Nikdast, "A Torus-Based Hierarchical Optical-Electronic Network-on-Chip for Multiprocessor System-on-Chip," *J. Emerg. Technol. Comput. Syst.*, vol. 8, no. 1, pp. 5:1–5:26, Feb. 2012.

[24] Z. Wang, J. Xu, X. Wu, Y. Ye, W. Zhang, M. Nikdast, X. Wang, and Z. Wang, "Floorplan Optimization of Fat-Tree Based Networks-on-Chip for Chip Multiprocessors," *IEEE Transactions on Computers*, vol. PP, no. 99, pp. 1–1, 2012.

[25] H. Li, H. Gu, and Y. Yang, "A hierarchical cluster-based optical network-on-chip," in *2010 2nd International Conference on Future*

Computer and Communication (ICFCC), vol. 2, May 2010, pp. V2-823–V2-827.

[26] J. Kim, W. Dally, S. Scott, and D. Abts, "Cost-efficient dragonfly topology for large-scale systems," in *Conference on Optical Fiber Communication – incudes post deadline papers, 2009. OFC 2009.*, March 2009, pp. 1–3.

[27] S. Pasricha and S. Bahirat, "OPAL: A multi-layer hybrid photonic NoC for 3D ICs," in *Design Automation Conference (ASP-DAC), 2011 16th Asia and South Pacific*, jan. 2011, pp. 345–350.

[28] S. Le Beux, J. Trajkovic, I. O'Connor, G. Nicolescu, G. Bois, and P. Paulin, "Optical Ring Network-on-Chip (ORNoC): Architecture and design methodology," in *Design, Automation Test in Europe Conference Exhibition*, 2011, pp. 1–6.

[29] Y. Xu, J. Yang, and R. Melhem, "Channel borrowing: an energy-efficient nanophotonic crossbar architecture with light-weight arbitration," in *Proceedings of the 26th ACM International Conference on Supercomputing*. ACM, 2012, pp. 133–142.

[30] R. W. Morris, A. K. Kodi, A. Louri, and R. D. Whaley, "Three-dimensional stacked nanophotonic network-on-chip architecture with minimal reconfiguration," *IEEE Transactions on Computers*, vol. 63, no. 1, pp. 243–255, 2014.

[31] I. Datta, D. Datta, and P. Pande, "BER-based power budget evaluation for optical interconnect topologies in NoCs," in *2012 IEEE International Symposium on Circuits and Systems (ISCAS)*, 2012, pp. 2429–2432.

[32] R. K. V. Maeda, P. Yang, X. Wu, Z. Wang, J. Xu, Z. Wang, H. Li, L. H. K. Duong, and Z. Wang, "Jade: a heterogeneous multiprocessor system simulation platform using recorded and statistical application models," in *Proceedings of the 1st International Workshop on Advanced Interconnect Solutions and Technologies for Emerging Computing Systems (AISTECS)*, January 2016.

[33] A. Joshi, C. Batten, Y.-J. Kwon, S. Beamer, I. Shamim, K. Asanovic, and V. Stojanovic, "Silicon-photonic clos networks for global on-chip communication," in *Networks-on-Chip, 3rd ACM/IEEE International Symposium on*, 2009, pp. 124–133.

[34] A. V. Krishnamoorthy, R. Ho, X. Zheng, H. Schwetman, J. Lexau, P. Koka, G. Li, I. Shubin, and J. E. Cunningham, "Computer Systems Based on Silicon Photonic Interconnects," *Proceedings of the IEEE*, vol. 97, no. 7, pp. 1337–1361, 2009.

2

Design and Optimization of Vertical Interconnections for Multilayer Optical Networks-on-Chip

Alberto Parini[1] and Gaetano Bellanca[2]

[1]FOTON Laboratory, CNRS UMR 6082, University of Rennes 1, ENSSAT, F-22305 Lannion, France
[2]Department of Engineering, University of Ferrara, I-44122 Ferrara, Italy

Abstract

This chapter presents an overview of material platforms and vertical interconnection schemes enabling the implementation of multilayer optical networks-on-chip. Issues related to the design of staked photonic circuits, with particular regard to optical isolation, will be highlighted and discussed. Among the various interconnection solutions outlined in the text, two significant ones will be more detailed: the former exploiting directional couplers with inverse adiabatic tapers; the latter relying on multimode interference devices. Results showing the feasibility and the optimization of these approaches are presented and discussed.

2.1 Introduction

In the last years, optical networks-on-chip (ONOCs) have been widespreadly proposed as possible solutions to cope with problems related to the increasing bandwidth demand of multiple-core processor architectures able to guarantee, at the same time, low latency and reduced power consumption [1, 2, 3]. Optical networks at chip level have therefore been the object of increased attention, and several implementations have been theorized and investigated [4].

A first classification of ONOCs can be made with respect to their topological layout. According to the definition given by A.W. Poon in [4],

41

a distinction can be made between cross-grid and banyan networks. The cross-grid class includes topologies whose layout is organized on a matrix-like mesh as, for example, the GWOR router [5], the Snake router [6] or the Cornell-Columbia 4×4 nonblocking SOI router [7]. Banyan class includes topologies such as the lambda-router [8] or the WDM cross connect switch proposed by Soref and Little [9]. A further category can be introduced to list ring-based topologies, as the Chameleon [10] or the Corona ones [11]. ONOCs can be also differentiated between active and passive, depending on the physical mechanism exploited for signal routing. In active networks, the routing paths linking different sources and destinations can be dynamically modified during operation by properly on/off tuning the optical response of the switching elements. This can be done with a shift of resonance combs in microring based solutions, as in [7], or by acting on the transmission state of Mach-Zehnder interferometers, as in [12]. For passive networks, on the contrary, physical and logical paths between sources and destinations are predefined at the design stage, and cannot be dynamically modified [13]. Active networks therefore allow a dynamic "on the fly" allocation of the routing paths, but require extra tuning circuitry (of thermo-optics or electro-optics nature). Passive solutions, by avoiding the presence of tuning mechanisms, reduce the technological complexity of the circuit, but impose stricter constraints at system levels due to the lack of dynamic flexibility. Nevertheless, passive networks may eventually require some extra-tuning devices to compensate unavoidable fabrication issues.

Within the ONOCs domain, a strong effort is devoted to system-level investigations, in order to identify topologies where the largest number of computational cores are allowed to communicate with the lowest energy consumption. However, it must be observed that, irrespective to the adopted layout, as the number of interconnected cores increases, the complexity of the network infrastructure grows as well, resulting in longer (hence more dissipative) optical links and in the unavoidable appearance of planar intersections between waveguides. Crossings and junctions are particularly critical structures, since they give rise to two combined detrimental effects: firstly, they add a further scattering loss factor caused by the breaking in the continuity of the waveguides; secondly, they generate crosstalk effects by diverting a fraction of the signal carried by the main waveguide into the traversing one. The critical impact of planar crossing on optical networks-on-chip performance is highlighted by the number of different technological solutions proposed so far for they optimization: from intersection smoothed by means of adiabatic tapers [14], to more complex configurations exploiting auto-focusing effects in multimode interference (MMI) devices [15].

Limitations arising from a planar approach in the design of photonic circuits can be eased by introducing a more challenging technological paradigm, in which also the third (vertical) dimension is exploited to create multilayer structures. The availability of multiple optical planes clearly expands the ONOCs design space toward multiple directions. From the topological side, superposed optical levels can be exploited to implement optical bridges to avoid single or multiple waveguide crossings and, more generally, to allow an easier placing and routing of the different elements composing the photonic circuit. From the functional side, the capability to stack heterogeneous materials with complementary features enables the creation of new devices otherwise unachievable. A typical example in this sense is the integration of $III-V$ active layers over a silicon substrate for the implementation of sources [16], amplifiers and detectors [17, 18, 19]. To fully exploit the potentials of hybrid electronics/photonics platforms, multilayer structures must be necessarily fabricated following CMOS compatible proceedings, to keep the integrity of electrical devices and circuits that may coexist on the same chip die. Intrinsic potentialities of multilayer architectures can be fully exploited only if the different optical levels are allowed to communicate efficiently. Therefore, interconnection structures enabling vertical transfer of optical power constitute the founding functional blocks of 3D photonics, and the assessment of their performance is a key issue for the development of this domain.

The aim of this chapter is to provide an overview of technological platforms and vertical interconnection schemes recently proposed for the implementation of multilayer photonics. The text is organized in two main parts, each one with different sections. In the first part, which is dedicated to an overview of materials and configurations allowing efficient vertical optical communications, Section 2.2 presents a selection of materials suitable for planar photonic circuits fabrications that, at the same time, can be stacked to create a variety of 3D photonic platforms. These platforms are then outlined in Section 2.3 and their characteristics discussed. Section 2.4 concludes the first part of the chapter by presenting a set of configurations enabling vertical optical interconnections. Section 2.5 opens the second part, by assessing the issue of optical isolation between superposed circuitry layers. Afterwards, two interconnection schemes are outlined in more detail: directional couplers with inverse adiabatic tapers, in Section 2.6, and vertical links exploiting multimode interference (MMI) devices, in Section 2.7, respectively. Concluding remarks are reported at the end.

2.2 Materials for Photonic Integrated Circuit Fabrication

Thanks to its dominance in microelectronics and to its transparency from $\approx 1.1 \, \mu m$ to the far-infrared region, silicon (Si) is the reference material for photonic integrated circuits (PICs) fabrication. With respect to optical properties, Si shows an high refractive index ($n = 3.45$) in the telecommunication wavelength range, thus enabling the fabrication of extremely compact structures when the guiding cores are surrounded by low-index cladding materials, such as silicon dioxide ($n = 1.45$). Beside silicon, nitrogen compounds such as silicon nitride and aluminium nitride are emerging for photonic circuits integration, these compounds presenting low insertion loss, compatibility with CMOS manufacturing processes and capability to be vertically deposed to form stacked architectures.

Crystalline silicon ($c - Si$) is the allotropic form of this element in which the entire bulk is organized as a perfect lattice of atoms. The technological platform enabling photonic fabrication on ($c - Si$) is the silicon-on-insulator (SOI) configuration, in which a thin film of crystalline silicon is stacked over a thick buffer of silicon dioxide (SiO_2). On SOI wafers, photonic structures are drawn in the crystalline layer while the buffer acts as a mechanical support. After fabrication, SOI circuits are usually covered by a passivation coating of a material (polymer or again silicon dioxide) whose refractive index matches the ones of the substrate. This top coating provides a mechanical protection and also mitigates scattering losses caused by waveguide roughness at the interface. Typical propagation loss of crystalline silicon waveguides are around $2.5 \, dB/cm$, with record values as small as $0.92 \, dB/cm$ and $1.7 \, dB/cm$ reported in references [20] and [21], respectively. SOI waveguides are usually dimensioned to operate in a single mode transverse electric (TE) polarization state, resulting in cross sections standardized around $500 \, nm \times 250 \, nm$ (width $\times$ height). Crystalline silicon shows excellent properties in terms of propagation losses. Nevertheless, this material suffers from a major technological drawback, as it cannot be chemically deposited over an host substrate by conserving its crystalline regularity. This clearly restricts the fabrication of ($c - Si$) photonic circuitry to a single layer. To overcome this major limitation, a big effort is underway toward the development of optical materials capable of being deposited, like polycrystalline silicon ($p - Si$), hydrogenated amorphous silicon ($a - Si$) or nitride compounds.

Amorphous silicon ($a - Si$) is characterized by the absence of long range order in its atomic lattice. This lack of regularity, with a consequent high density of point defects, degrades the optical quality of this material, making it unsuitable for photonics. Nevertheless, addition of hydrogen to amorphous silicon reduces the defect density and optical losses, thus resulting in a compound exploitable for photonic fabrications. In addition to this, hydrogenated amorphous silicon can be chemically deposited over preexisting layers, thus allowing the composition of 3D platforms. TE single mode waveguides fabricated in hydrogenated amorphous silicon present propagation loss ranging between $3\,dB/cm$ and $5\,dB/cm$ [22], which are slightly higher than SOI-based waveguides. However, time stability of these loss values for $a - Si$ is still an open issue [22]. Moreover, electrical carrier mobility of hydrogenated amorphous silicon is considerably lower with respect the one of crystalline silicon [23], thus making challenging the implementation of electro-optically driven switching devices.

Polycrystalline silicon ($p-Si$) is characterized by a mixed structure in which crystalline grains are separated by thin disordered grain boundaries. Polycrystalline silicon can be deposited above other substrates, but its inhomogeneity has a negative impact in terms of increased insertion loss with respect to amorphous and crystalline silicon counterparts ($6.5\,dB/cm$ reported in reference [24]). On the other side, the irregularity of the lattice structure has a beneficial impact on the electrical carrier mobility [23]. Therefore, this material can be eventually exploited to implement active functionalities.

Silicon nitride (SiN) is a compound of silicon and nitrogen which is gaining increasing interest for photonic circuits fabrications. Although its relatively low refractive index ($n = 2$) entails larger bending radius for curves and rings, reduced propagation loss ($1\,dB/cm$ reported in [25]) makes this material attractive for applications where energetic constraints are dominant over the topological (i.e. large footprint) ones. In addition to this, silicon nitride can be chemically deposited over silicon dioxide, thus making it exploitable for 3D stacked architectures.

Aluminium nitride (AlN) is a compound of aluminium and nitrogen which has been recently proposed in [26] for the fabrication of a multilayer photonic platform, in conjunction with amorphous silicon. As well as silicon nitride, aluminium nitride presents a relatively low refractive index ($n = 2$) and limited propagation loss ($1.4\,dB/cm$ at $1550\,nm$ reported in [26]). A remarkable property of aluminium nitride is its reduced

Table 2.1   Characteristic parameters of the guiding materials presented in the text

Material	Refractive Index	Waveguide Cross Section ($w \times h$) [nm]	Propagation Loss [dB/cm]
Crystalline silicon	3.45	500×250	≈ 2.5
Amorphous silicon	3.6	500×250	$3 \div 5$
Polycrystalline silicon	3.5	500×250	≈ 6.5
Silicon nitride	2.0	1000×400	$1 \div 2$
Aluminium nitride	2.0	1000×400	≈ 1.4

Thermo-Optic Coefficient (TOC), making this compound suitable for fabrication of temperature-insensitive circuitry. Table 2.1 summarizes the main characteristic parameters of the outlined materials, namely: the refractive index around $1550\,nm$, the typical cross section (width $\times$ height) of a TE single mode waveguide in the $1500\,nm \div 1600\,nm$ wavelength range, and the waveguide propagation loss in dB/cm.

2.3 Technological Platforms for Multilayer Photonics

Crystalline silicon under the form of SOI wafers is the main technological substrate for photonic integrated circuits, but, as deposition techniques able to retain its regular lattice structure are currently not available, a number of approaches leveraging on combinations of alternative materials have been explored for the implementation of multilayer photonic architectures. A possible (but clearly not unique) classification of multilayer platforms can be introduced between silicon-based and heterogeneous ones. Within the silicon-based class, one can list the platforms in which all the guiding layers composing the stacked structure are made of either crystalline, polycrystalline or hydrogenated amorphous silicon. Within the heterogeneous class, on the contrary, one can list the platforms where beside silicon, other materials are exploited to fabricate the guiding structures, such as silicon nitride or aluminium nitride. Figure 2.1 presents a selection of both homogeneous and heterogeneous multilayer platforms which have been recently proposed in literature for the fabrication of 3D photonic circuitry, and whose main features are detailed hereinafter.

(a) Amorphous silicon over crystalline silicon ($a - Si/SiO_2/c - Si$)**:** In this platform, the bottom guiding layer is made of SOI crystalline silicon ($c - Si$) while the top one is obtained through a chemical deposition of a thin hydrogenated amorphous silicon ($a - Si$) layer. Optical isolation between the two layers is provided by a thick interposed layer of silicon dioxide (SiO_2).

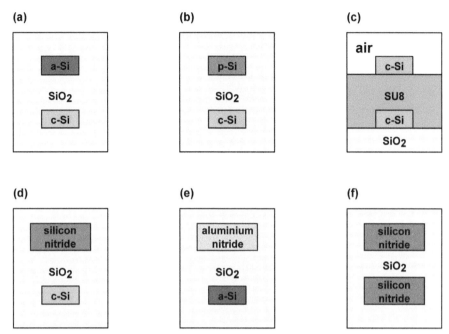

Figure 2.1 Schematic of the technological platforms for multilayer photonics described in the text. Each insets represents a vertical cross section of two superposed waveguides fabricated in the guiding layers, surrounded by the isolating substrate. (a): Amorphous silicon over crystalline silicon. (b): Polycrystalline silicon over crystalline silicon. (c): Adhesively bonded crystalline silicon nano-membranes. (d): Silicon nitride over crystalline silicon. (e): Aluminium nitride over amorphous silicon. (f): Silicon nitride over silicon nitride.

This platform has been presented and exploited in [27] to fabricate and assess a vertical coupler with inverse adiabatic tapers. Propagation losses of both layers have been estimated around $5.5\,dB/cm$.

(b) Polycrystalline silicon over crystalline silicon ($p - Si/SiO_2/c - Si$)**:** In this configuration, the bottom guiding layer in made in SOI crystalline silicon while the top one is added through a deposition of a polycrystalline silicon layer. An example of this platform can be found in reference [23], in which a low loss waveguide designed in the bottom layer is vertically coupled to a microring resonator fabricated in the top polycrystalline one. Thanks to its high electronic mobility, the polycrystalline is proposed for the implementation of active devices, whereas the bottom crystalline one can host low-loss interconnections.

(c) Bonded crystalline silicon nanomembranes $(c - Si/SU8/c - Si)$**:** This platform consists of two crystalline silicon nanomembranes, both obtained from standard SOI wafers, which are mechanically bonded one above the other by exploiting the adhesive properties of a $SU8$ polymer layer. On this platform, a 1×32 H tree optical network for clock distribution [28] and a grating-assisted vertical coupler [29] have been successfully fabricated and characterized. Authors of [28] measured a propagation loss of $4.3\,dB/cm$ for structures fabricated on the bonded top layer, against a value in the order of $2.5 - 3\,dB/cm$ for classical crystalline SOI waveguides.

(d) Silicon nitride over crystalline silicon $(SiN/SiO_2/c - Si)$**:** Here, the top guiding layer is fabricated through deposition of silicon nitride above a standard SOI substrate, with an interposed separating layer of silicon dioxide. In reference [30] this combination of materials is exploited for the fabrication of an optimized vertical tapered coupler providing an insertion loss between two waveguides separated by a $100\,nm$ silicon dioxide layer of $0.2\,dB$. Moreover, in the same reference, also a thermally tuned microring resonator hosted in the nitride layer is fabricated and characterized. Further examples of this platform can be found in references [31] and [32], in which the top silicon nitride layer is exploited to create a two levels optical bridge over one or more underlying SOI waveguides, in order to avoid a sequence of planar crossings.

(e) Aluminium nitride over amorphous silicon $(AlN/SiO_2/a - Si)$**:** In reference [26], a further nitride-compound/silicon hybrid configuration is proposed, in which the bottom guiding layer is made with deposited amorphous silicon, whereas the top one is in aluminium nitride (AlN). The bottom amorphous silicon layer enables the realization of compact low loss guiding structures, while the top aluminium nitride one, thanks to its reduced Thermo-Optic Coefficient (TOC), can host temperature-insensitive devices. Authors of [26] reports propagation loss of $3.8\,dB/cm$ for waveguides patterned in the bottom amorphous silicon layer and of $1.4\,dB/cm$ for waveguides in the top aluminium nitride one. Moreover, in the same reference, a variety of functional devices ranging from MMI to microring resonators are implemented on both layers, making this platform very interesting for 3D photonics.

(f) Silicon nitride over silicon nitride $(SiN/SiO_2/SiN)$**:** In this homogeneous configuration, bottom and top guiding layers are both made with deposed silicon nitride, while the optical isolation is provided by an interposed separating layer of silicon dioxide. This platform is adopted in reference [25] for the fabrication of a two levels microring resonator.

2.4 Overview of Vertical Interconnection Schemes

A number of different schemes enabling vertical optical interconnection have been proposed and assessed so far. In this section, a description of the operating principles of the most common ones will be provided, together with a selection of references where more details on these implementations can be possibly found.

(i) Vertical direction couplers: The simplest solution allowing an off-plane (vertical) transfer of optical power is the one depicted in Figure 2.2(i), which consists of two parallel superposed waveguides. This configuration is a classical directional coupler where the power exchange occurs vertically. Coupling efficiency is maximized when the modes propagating in the bottom and top waveguides are phase matched, i.e. they both have the same effective index at the considered working wavelength. This matching condition can be achieved only by properly designing the geometries of the two waveguides, making this solution extremely sensitive to fabrication tolerances and misalignments. It must be also observed that the power exchange between waveguides follows a periodic evolution; therefore, the interaction length has to be carefully tuned to avoid back-coupling effects.

(ii) Vertical directional couplers with inverse adiabatic tapers: This second configuration, which will be treated extensively later in Section 2.6, is presented in Figure 2.2(ii). It consists of a vertical directional coupler, as the one previously outlined, in which two inverse overlapping tapers are added to waveguide terminations. The variable cross section of the tapers introduces a progressive adiabatic variation of the modes effective indices along the propagating direction, thus enabling an automatic phase-matching condition even for waveguides fabricated on different materials. In addition to this, tapering also relaxes the fabrication constraints, thus increasing the robustness of this scheme with respect to technological fabrication tolerances.

(iii) Multimode Interference (MMI) based vertical links: Thanks to their excellent efficiency, large bandwidth and ease of fabrication, Multi-Mode Interference (MMI) devices are extensively applied in integrated optics to implement splitting and combining functions. The operation of these devices relies on the self-imaging principle, according to whom a field distribution (mode) on the input port(s) of the MMI is, totally or partially, transferred to the output port(s) if the length of the multimode section is suitability chosen. MMIs are generally exploited in a planar configuration, where input and output ports lie on the same physical level.

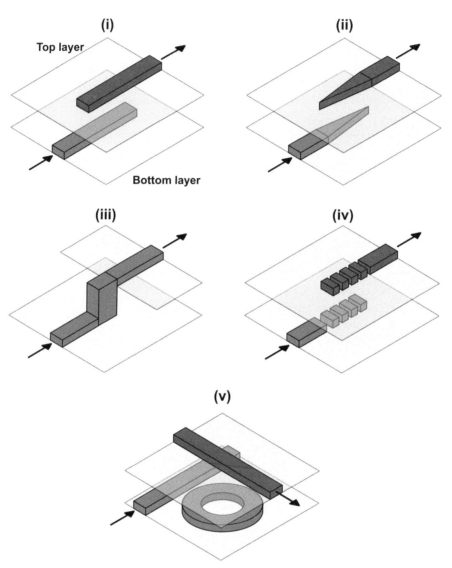

Figure 2.2 Representation of different technological solutions enabling optical interconnection between superposed layers of photonic circuitry. Different colors identify waveguides fabricated on heterogeneous materials. (i): Vertical directional coupler. (ii): Vertical directional coupler with inverse adiabatic tapers. (iii): MultiMode Interference (MMI) based vertical link. (iv): Grating-assisted vertical coupler. (v): Microring-assisted vertical coupler.

However, MMI structures can also be created following a vertical direction, as schematically shown in Figure 2.2(iii). Demonstration of MMI-based vertical links can be found in polymeric materials [33] and also on SOI platforms [34]. The governing equations and the design rules for this type of interconnections will be treated in more detail in the last sections of the chapter.

(iv) **Grating-assisted vertical couplers:** Transfer of optical power between superposed layers can be achieved also through diffractive elements, such as gratings. A schematic of this solution is presented in Figure 2.2(iv). Here, a Bragg structure etched in the bottom waveguide produces a out of plane (vertical) diffraction of the optical field. A second grating, etched in the top waveguide, couples and reconverts the diffracted radiation to a guided mode. Practical implementations of grating-assisted vertical couplers can be found in several previously described 3D platforms. In [35], a grating-assisted link is fabricated in the silicon nitride over crystalline silicon platform, to couple a bottom SOI waveguide to a top SiN microring resonator. A further implementation of a grating-assisted vertical coupler can be found in [29], where the technological platform adopted to fabricate the structure is the one based on two bonded crystalline silicon nanomembranes. Thanks to their intrinsic wavelength-dependent response, vertical interconnections comprising diffraction gratings can also perform signal switching and multiplexing within multilayer networks. An example can be found in [36], where four grating-assisted vertical couplers [37] are interconnected together to compose a 3D Generic Wavelength-routed Optical Router (GWOR) topology.

(v) **Microring-assisted vertical couplers:** Thanks to their selective behavior with respect to frequency, microring resonators are the key routing fabrics in wavelength-routed optical networks-on-chip. Moreover, microrings can also be exploited to vertically couple optical power, provided to arrange input and output bus waveguides on two different stacked levels, as illustrated in Figure 2.2(v). In this configuration, coupling between the ring and the bottom bus waveguide takes place on the horizontal plane, whereas it occurs vertically to the top bus waveguide. Unlike the previously outlined solution, winch where intrinsically wide band, it must be noted that in this configuration the vertical coupling is effective only if the wavelength of the signal is resonant with the ring. Two level microring-assisted coupling has been demonstrated on different platforms, including the polycrystalline silicon over crystalline silicon [23] and the silicon nitride over silicon nitride one [25].

2.5 Isolation between Layers in Vertical Stacked Structures

In the first part of the chapter, the main features of materials and inter-connection schemes allowing 3D optical stacking have been presented and discussed. The present section opens the second part of the chapter by treating a fundamental issue related to 3D photonics, namely the optical isolation among superposed guiding layers. As for the in-plane case, also off-plane undesired interactions can give rise to a number of detrimental effects, such as crosstalk and additional propagation losses, which can seriously affect the communication network performance. To understand the origin of this issue, it must be observed that two or more dielectric waveguides, laying in a close proximity, mutually interacts and exchange power. This interaction can be positively exploited to implement switching and coupling functionalities but, if not properly accounted and controlled, it can be equally a source of signal impairments. A straightforward way to mitigate any undesired interaction is to increase the reciprocal waveguide distance, at the cost of an increased final device footprint. Usually, mutual isolation issues affect waveguides and devices laying on the same plane but, in the case of 3D stacked architectures, isolation must be provided also in the vertical direction by raising the thickness of the interposed material between the different planes of the optical circuitry. Unfortunately, as the vertical gap between layers increases, the implementation of efficient and, at the same time, spatially compact schemes for vertical power transfer becomes more and more problematic. To quantify the impact of vertical distance between layers on the mutual waveguide interaction, the test-case simulation scenario presented in Figure 2.3 can be exploited, in which two waveguides located on two superposed optical levels crosses orthogonally, thus interfering with each other. In an ideal case, the two waveguides should not have any reciprocal interference and their modes should propagate undisturbed. However, in the real case, each waveguide acts as a perturbing structure for the other with a consequent detrimental influence on propagation. The strength of this undesired interference depends on two factors: (1) the distance g separating the two optical layers and (2) the distribution of the guided modes around each waveguide core at the working wavelength. As stated before, as the distance g grows, the mutual interaction decreases progressively and, at the same time, for a fixed value of g, a mode tightly confined around the guiding core will be less affected by any inhomogeneity located in the surrounding environment. As mode confinement is correlated to the effective waveguide index which, in turns,

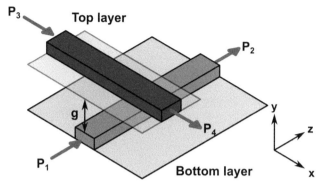

Figure 2.3 Orthogonal crossing between two waveguides laying on different optical levels. The vertical gap g is defined as the distance separating the upper side of the bottom layer waveguide from the lower side of the top layer waveguide. The value of g therefore corresponds to the thickness of the interposed isolating material.

depends on the difference between the refractive indices of the waveguide core and of the surrounding substrate, platforms comprising lower index guiding materials (as the ones with nitrides, for example) will in general require ticker isolating layers.

Mutual interaction effects can be quantified in terms of an additional power insertion loss factor affecting the propagation on each waveguide. Taking as a reference the scheme presented in Figure 2.3, two quantities can be defined:

$$IL_{BOTTOM} = -10 \cdot log_{10} \left(P_2/P_1 \right) \tag{2.1}$$
$$IL_{TOP} = -10 \cdot log_{10} \left(P_4/P_3 \right). \tag{2.2}$$

The IL_{BOTTOM} parameter accounts for the additional loss engendered on the bottom waveguide by the presence of the top one, while, similarly, IL_{TOP} represents the increased loss factor on the top waveguide due to the presence of the bottom one. For homogeneous layer stacking, the insertion loss in the two layers will be the same. On the contrary for heterogeneous stacked layers, as the silicon nitride over crystalline silicon previously presented in Figure 2.1(d), insertion losses of the two planes will be different. More specifically, for each technological platform a worst case scenario exists, which corresponds to the field propagation in the waveguide with smaller effective index. In this case, as the mode is less confined around the waveguide core, it is more subject to scattering losses due to perturbations determined by the presence of the other waveguide. The curves presented in Figure 2.4 show

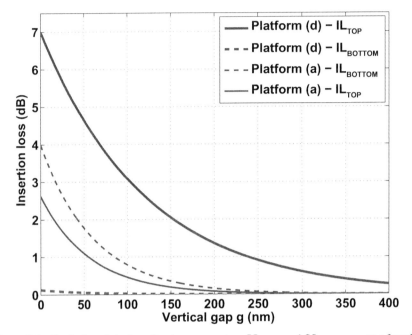

Figure 2.4 Evolution of the insertion loss parameters IL_{TOP} and IL_{BOTTOM} as a function of the vertical gap g between layers, for two orthogonal crossings implemented on platforms (a) and (d), respectively. On platform (d), the most unfavorable case corresponds to a propagation in the top silicon nitride waveguide (thick solid red curve) which, due to its low refractive index and wide mode area, is particularly perturbed by the presence of the bottom crystalline silicon one. Conversely, on this same platform, the propagation in the bottom crystalline silicon waveguide (thick red dashed curve) is almost unaffected by the top nitride one. For platform (a), the most unfavorable case is when the propagation takes place in the bottom crystalline silicon waveguide (thin dashed blue curve), this due to its slightly lower refractive index with respect to the top amorphous silicon one. The working wavelength is $\lambda_0 = 1550\,nm$.

the evolution of the insertion loss parameters IL_{TOP} and IL_{BOTTOM} with respect to the vertical gap g, for two orthogonal crossing implemented in platforms (a) and (d), respectively. Among the various platforms, the (a) and (d) ones constitute two opposite cases, whose results can be easily extended to similar configurations. Platform (a) is representative of solutions where both guiding layers are derived from high refractive index materials (crystalline and amorphous silicon), while platform (d) represents an heterogeneous solutions in which one of the guiding layers (silicon nitride) has lower index with respect to the other (crystalline silicon). For both (a) and (d) platforms, the worst case condition is when the field propagates in the layer with the lower index, and consequently the mode is widespread around the waveguide core. For platform (a), taking as a reference the physical parameters provided in Ref. [27] where this platform is exploited to fabricate inverse tapers, the most unfavorable case corresponds to a propagation in the bottom crystalline waveguide, which has a slightly lower bulk index ($n = 3.45$) with respect to the top amorphous one ($n = 3.6$). For this platform, waveguides in both layers are $480\,nm$ wide and $220\,nm$ tall, in order to have TE single mode operation around $1550\,nm$. For Platform (d), the worst case scenario is when propagation takes place in the top silicon nitride layer, which has a considerably lower bulk index ($n = 2$) with respect to the underlying crystalline silicon one ($n = 3.45$). Also for this second configuration, the two crossing waveguides are assumed to be TE single mode, with a cross-section (width $\times$ height) of $480\,nm \times 220\,nm$ for the bottom and of $1000\,nm \times 400\,nm$ for the top one, respectively. Note that, as a consequence of the lower refractive index of silicon nitride, the waveguides designed on this material should present wider cross sections, to guarantee guided propagation of the optical field.

This loss evaluation procedure is a simple example of the issues that need to be assessed when designing a 3D stacked photonic architecture. This scenario has been chosen as it is representative of situations where interference on waveguide propagation is undesired and cannot be avoided unless tremendously complicate the topology of the circuits. Moreover, it shows how, once fixed the maximum tolerable amount of losses, the gap among the layers guaranteeing the required performance is directly derived. The same approach could be used also to investigate other configurations as, for example, waveguides running parallel on two levels. Once determined the required (minimum) vertical distance between layers providing the targeted optical isolation, it is then possible to focus on the design of the vertical coupling structures, as will be outlined in next sections.

2.6 Design Space Exploration of Inverse Tapered Couplers

The design of a vertical interconnection scheme is subject to conflicting constraints. In fact, once fixed the vertical distance g between layers of the stacked architecture, the key objective is to implement a configuration providing the highest power transfer efficiency within the minimum spatial footprint. A vertical coupler with inverse adiabatic tapers is a typical case where this trade-off must be assessed. To achieve an efficient coupling between layers while keeping, at the same time, a required vertical distance between them, tapers with a considerable length may be needed, with a consequent negative impact on the final spatial occupation of the whole interconnection scheme. Therefore, a viable trade off must be identified. To this end, the aim of this section is to outline some of the interdependencies between geometric parameters and power transfer efficiency which characterize this vertical link solution, and to suggest some consequent optimization criteria.

By referring to Figure 2.5(A), the geometrical parameters taken into account within the design procedure are the bottom Lt_{BOTTOM} and top Lt_{TOP} taper lengths, together with their reciprocal shift S along the longitudinal direction. A value of $S = 0$ corresponds to a complete superposition of the tapers, i.e. the tip of top one starts to widen while the tip of the bottom one starts to shrink. The operating principle of this scheme can be understood by observing the FDTD (Finite Difference in Time Domain) simulated field propagation presented in Figure 2.5(B), where a cut on a vertical plane passing in the symmetry axis of the structure is presented. Here, the mode propagating in the bottom waveguide spread progressively in the substrate

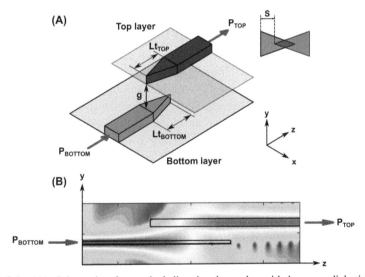

Figure 2.5 (A): Schematic of a vertical directional coupler with inverse adiabatic tapers enabling optical interconnections between two layers separated by a vertical gap g. Bottom and top tapers have length Lt_{BOTTOM} and Lt_{TOP}, respectively, with an eventual relative longitudinal shift S. (B): Vertical lateral cross-section of the electric field ($|E|^2$) propagating on a coupler implemented on a heterogeneous platform (platform (d)), showing the progressive transition of the optical power from the bottom to the top waveguide.

once the cross section of the bottom taper starts to shrink (causing a drop of the effective index). At the same time, widening of the cross section of the top taper allows collecting the mode inside the upper waveguide. Coupling takes place around the spatial coordinate where the effective indices of bottom and top waveguides match, tanks to this adiabatic tapering. In Figure 2.5(B), an hybrid coupler comprising different materials in the two layers is represented; this explains the different height of the bottom and top waveguides.

The transmission efficiency of the interconnection can be quantified by means of a power budget evaluation between bottom and top layers (up-link case):

$$IL_{LINK} = -10 \cdot log_{10}\left(P_{TOP}/P_{BOTTOM}\right). \tag{2.3}$$

A similar parameter can be also defined for the power transfer in the opposite direction (down-link case). For homogeneous platforms, as the (a) one, thanks to the reciprocity of the configuration, up-link and down-link insertion loss overlap. In the case of heterogeneous platforms, as the (d) one considered here, simulations have shown that the down-link case is affected by an additional power loss factor with respect to the up-link which is limited within $0.5\,dB$ in the $1.55\,nm$.

A first dependence that needs to be assessed within the design procedure of such a configuration is the one between IL_{LINK} and the lengths of the taper, for a fixed value of the vertical gap g. This study is presented in Figure 2.6 for two couplers implemented in platforms (a) and (d), respectively (up-link case). Here, the gap g between the layers is set to $g = 400\,nm$, this distance providing for both platforms losses smaller than $0.5\,dB$ in the case of a two level crossing, as previously shown in Figure 2.4. For sake of simplicity, a symmetric configuration with bottom and top tapers of the same length $(Lt = Lt_{TOP} = Lt_{BOTTOM})$ is considered, with a complete longitudinal superposition (i.e. $S = 0$).

From the curves presented in Figure 2.6 it can be observed how, by progressively increasing the length of the tapers up to $100\,\mu m$, the insertion loss value can be reduced to $0.2\,dB$ or less on both platforms. However, for the specific case of platform (d), a gap of $g = 400\,nm$ between layers might not be sufficient to limit the insertion loss in presence of an orthogonal crossings between waveguides. In fact, by observing the solid red curve in Figure 2.4, one can verify that a crossing with a gap $g = 400\,nm$ on platform (d) introduces an additional insertion loss of about $0.3\,dB$. This is a quite high value that can be detrimental, especially in the case of complex circuit topologies where more than tens of waveguide crossings can be expected on

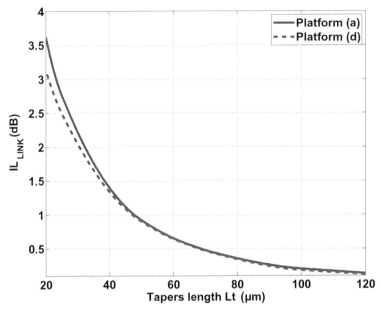

Figure 2.6 Insertion loss IL_{LINK} parameter as a function of the tapers length $Lt = Lt_{TOP} = Lt_{BOTTOM}$ for two inverse tapered couplers implemented in platforms (a) (blue solid line) and (d) (red dashed line), respectively. Here the vertical gap between the layers is set to $g = 400\,nm$, and the working wavelength is $\lambda_0 = 1550\,nm$. For a given gap value, an increase of the tapers lengths improves the power transfer efficiency at the expenses of a greater link spatial occupation.

each path. In such a situation, an increase of the gap g between layers is therefore mandatory.

To assess the constraints existing between the parameters of platform (d) [1], it is thus interesting to evaluate the dependence between the coupler insertion loss parameter IL_{LINK} and the vertical gap g between the stacked layers, when the tapers length is kept fixed at a given value. This study is presented in Figure 2.7 for a coupler with $Lt = Lt_{TOP} = Lt_{BOTTOM} = 100\,\mu m$ and a gap g progressively increasing from $200\,nm$ to $1000\,nm$. Unfortunately for a vertical gap $g = 1000\,nm$, which provides a loss level in orthogonal crossings lower than $0.05\,dB$, this configuration appears to have an insertion loss exceeding $9\,dB$, which

[1]The evaluations made for this platform, which is particularly interesting due to the asymmetry between layers, can be easily extended to the other stacked configurations.

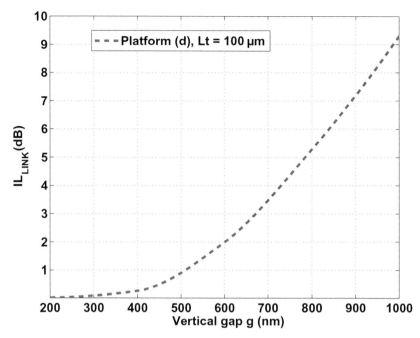

Figure 2.7   Evolution of the insertion loss parameter IL_{LINK} (up-link) as a function of the gap g, for a vertical tapered coupler implemented on platform (d) in which the length of the two tapers is fixed to $Lt = 100\,\mu m$ with no reciprocal relative shift ($S = 0$). As the distance between the layers is progressively increased, the insertion loss value grows with a quadratic trend. The working wavelength is $\lambda_0 = 1550\,nm$.

is unacceptable for practical applications. Increase in performance for this scheme can be achieved by acting on a further geometrical degree of freedom, namely the reciprocal longitudinal shift S between the taper tips. Figure 2.8 presents the evolution of the up-link insertion loss parameter IL_{LINK} as the shift S is increased from $0\,\mu m$ (corresponding to a complete tips superposition) to $80\,\mu m$. As S increases, the insertion loss value drops to the optimized minimum of $2.3\,dB$ for $S = 39\,\mu m$, before raising again. This behavior highlights the impact of the shifting parameter S on the optimization procedure of the coupler, and demonstrate the possibility to reduce the insertion loss of the coupler for gap g allowing, at the same time, a weak interference between waveguides located on different optical layers. Adiabatic tapers not only provide several degrees of freedom on which to act in order to reduce insertion loss, but also improve the spectral response of the coupler by widening and flattening the useful bandwidth. Figure 2.9 shows the up-link

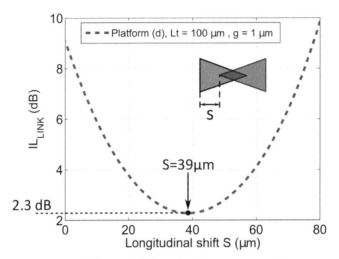

Figure 2.8 Evolution of the up-link insertion loss parameter IL_{LINK} as a function of the taper longitudinal shift S, for a gap g between layers of $1000\,nm$. The optimized shift providing the minimum insertion loss value of $2.3\,dB$ is $S = 39\,\mu m$. The the working wavelength is $\lambda_0 = 1550\,nm$.

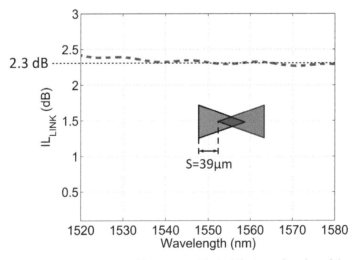

Figure 2.9 Up-link insertion loss IL_{LINK} (red dashed line) as a function of the wavelength for a tapered coupler fabricated on platform (d) with vertical gap $g = 1000\,nm$, tapers length $Lt = Lt_{TOP} = Lt_{BOTTOM} = 100\,\mu m$ and optimized reciprocal longitudinal shift $S = 39\,\mu m$. The insertion loss value of $2.3\,dB$ provided by this configuration at $\lambda_0 = 1550\,nm$ is guaranteed almost constant on a wide wavelength band.

insertion loss parameter IL_{LINK} evaluated in the range $1520\,nm - 1580\,nm$, for the optimized configuration with $S = 39\,\mu m$. As one can note, the value of $2.3\,dB$ provided by this coupler at $\lambda_0 = 1550\,nm$ is kept almost constant on the entire considered band. This means that, with this scheme, the channels of a WDM (Wavelength Division Multiplexing) frequency comb can be shifted between superposed layers without suffering from detrimental frequency-dependent power penalties.

2.7 Design of MMI-based Vertical Links

In this section, a further possible scheme enabling the interconnection of two waveguides laying on different levels, namely the MMI-based link, will be reviewed and discussed. A Multi-Mode Interference device (MMI) is basically a large core waveguide designed to support the propagation of several modes that beat and interfere with each other along the propagating direction. The optical signal is injected into the multimode section, and recovered out, from a number of access waveguides located at its beginning and at its end. These components are generally referred as $N \times M$ MMI, where N and M stand for the number of input and output waveguides, respectively. MMI devices are usually fabricated in a planar configuration, with the input and the output waveguides lying on the same horizontal level. However, an MMI can also operate vertically, if input and output waveguides are located on two different stacked layers, with the length and the height of the central multimode section suitably chosen. A schematic representation of an MMI-based vertical link is presented in Figure 2.10. The structure consists of an high-index guiding core (n_{core}) surrounded by a lower-index cladding (n_{clad}). Due to its monolithic layout, an MMI-based link may be utilized only to connect layers presenting the same refractive index (i.e. the same guiding material). This interconnection scheme is therefore intrinsically different with respect to the tapered coupler solution previously outlined, which, on the contrary, may eventually enable the interconnection of layers fabricated on heterogeneous materials. With respect to practical implementations, MMI-based links have been fabricated and characterized on several technological platforms, from some early examples on polymeric materials [33], to more recent implementation on SOI platform [34].

The geometric layout of an MMI-based link is described by a set of governing parameters, namely the height H and the length L of the multimode central section and the dimensions $w \times h$ of the input and output access waveguides. As for any vertical interconnection scheme, the design of this

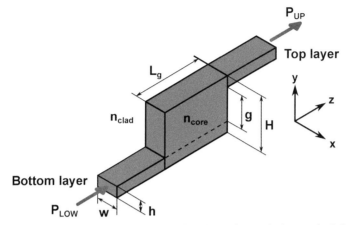

Figure 2.10   3D pictorial of a MMI-based vertical link. The vertical gap g is defined as the distance between the upper surface of the bottom waveguide and the lower surface of the top one.

solution starts form the definition of the gap g between the two layers to be linked. Once fixed this parameter, the height of the multimode section is consequently bound to be: $H = g + 2h$.

According to the self-imaging principle [38], in an MMI structure the transverse field distribution excited on the input side of the multimode section is reproduced, in single or multiple copies, at periodic intervals along the propagation direction (here the z axis). These self-images of the input field distribution appear periodically at distances L_p satisfying the following conditions:

$$L_p = p\,(3L_\pi) \quad p = 0, 1, 2, \cdots \tag{2.4}$$

for single images, and:

$$L_\pi = \frac{p}{2}\,(3L_\pi) \quad p = 1, 3, 5, \cdots \tag{2.5}$$

for multiple images.

L_π is the beating length between the two lowest-order modes of the multimode section which can be evaluated through the following relation:

$$L_\pi = \frac{\pi}{\beta_0 - \beta_1} \cong \frac{4n_{core}H^2}{3\lambda_0} \cong \frac{4n_{core}(g + 2h)^2}{3\lambda_0}, \tag{2.6}$$

where β_0 and β_1 are the propagation constants of the first and second order mode, respectively, n_{core} is the bulk refractive index of the material constituting the multimode section and λ_0 is the operating wavelength.

The first single non-trivial mirrored image which is exploitable to implement the vertical interconnection, providing also to the minimum length (and footprint) of the interconnection, is the one appearing at:

$$z = L_g = 3L_\pi. \qquad (2.7)$$

The operating principle of this interconnection scheme can be understood by observing the FDTD simulated field evolution presented in Figure 2.11(A), where a multimode section is fed, from the bottom-left corner, by a single-mode input waveguide. After a propagation distance $L_g = 3L_\pi$, the transversal field profile injected from the left bottom corner is mirrored on the right top one. By placing, on this position, a waveguide with the same parameters as the one used in the input section (see Figure 2.11(B)), efficient power transfer between the bottom and the top layer is obtained. The MMI structure presented in Figure 2.11 is supposed to be fabricated by deposition in a material with core refractive index $n_{core} = 3.45$ and completely surrounded by a cladding substrate with $n_{clad} = 1.45$. The two interconnected layers are separated by a vertical gap g of $400\,nm$, and the input and output waveguides have cross-section of $500\,nm \times 250\,nm$ ($w \times h$). Therefore, the height of the multimode section results $H = (g + 2h) = 900\,nm$. Beating L_π and optimal multimode section lengths L_g can be evaluated trough Equations (2.6) and (2.7), respectively, once the propagating constants β_0 and β_1 of the two lowest order TE modes of the multimode section are known [39]. For this case, assuming to route signals in the telecoms window around $\lambda_0 = 1550\,nm$, the multimode section length which produces a complete power transfer between the two layers is $L_g = 8\,\mu m$.

The transmission efficiency of an MMI interconnection scheme can be quantified, as usual, through a power budget between the top and bottom layers:

$$IL_{MMI} = -10 \cdot log_{10}\left(P_{TOP}/P_{BOTTOM}\right). \qquad (2.8)$$

In this case, as a result of the complete symmetric structure of the interconnection, up-link and down-link insertion losses coincide. Figure 2.12 shows a set of IL_{MMI} curves evaluated in the wavelength range $1520\,nm$ to $1580\,nm$ for a structure similar to the one depicted in Figure 2.11 and for increasing values of the gap g. As the distance between layers grows, the average loss value increases progressively, and the transmission curves start to show large ripples that may reach considerable high values. For gaps above $600\,nm$ these fluctuations can seriously compromise the correct operation of

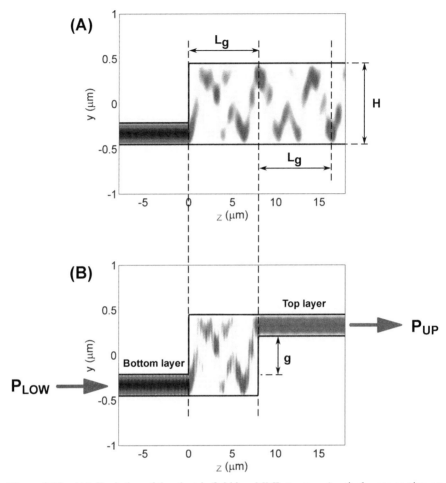

Figure 2.11   (A): Evolution of the electric field in a MMI structure (vertical cross section on the $y-z$ plane) showing the first and the second self-imaged copies of the input field appearing at coordinates $z = L_g$ and $z = 2L_g$, respectively. (B): Field transfer from the bottom to the top layer achieved by terminating the multimode section with an output waveguide having the same characteristics as the input one. Here the gap value g is $400\,nm$ and the consequent length of the multimode section is $L_g = 8\,\mu m$.

the device. Beside this, a further drawback of MMI-based schemes stems from the quadratic dependence of the beating length L_π, and consequently of the self-imaging distance L_g, with respect to the vertical gap g between the layers to be interconnected, as depicted graphically in the inset of Figure 2.11.

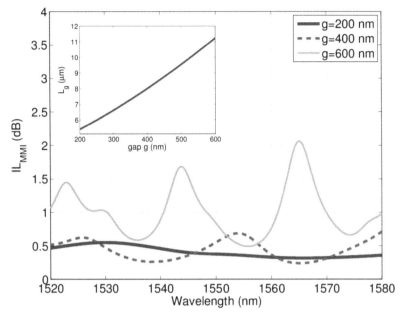

Figure 2.12   Insertion loss parameter IL_{MMI} of an MMI-based interconnection scheme in the wavelength range $1520\,nm - 1580\,nm$ for vertical gap $g = 200\,nm$ (thick solid blue line), $g = 400\,nm$ (red dashed line) and $g = 600\,nm$ (thin solid green line), respectively. The inset shows the length of the multimode section L_g as a function of the gap g.

In fact, by observing the approximated expression of L_π in Equation (2.6), one can note that the vertical gap g between the layers to be interconnected appears squared at the numerator. As this gap is increased, the first self-imaged distance $L_g = 3L_\pi$ exploitable to implement the link grows with a quadratic trend, with a clear detriment on the final footprint of the device.

2.8 Concluding Remarks

In this chapter, some key concepts regarding the challenging domain of multilayer optical architectures have been reviewed and discussed. In the first part, a general description of technological platforms and vertical inter-connection schemes enabling 3D photonics has been provided, with the aim to give the reader an overall picture of this wide and rapidly evolving domain. In the second part, after a discussion on the fundamental topic of optical isolation between layers, the design procedures of two of the presented schemes, namely the vertical couplers with adiabatic tapers and the

MMI-based links, have been detailed with the help of 3D-FDTD numerical simulations.

As outlined throughout the text, vertical stacking of multiple optical planes is mainly foreseen as a way to reduce the occurrence of waveguide intersections and, more generally, to ease the deployment of the components by spreading the whole circuit over different levels.

The choice of a specific interconnection scheme for the implementation of a multilayer photonic network depends on a number of interdependent factors, among which: the type the technological platform to be adopted for the circuit (homogeneous or heterogeneous one), the maximum footprint allowable for each vertical link, the required optical isolation level between layers (minimum gap) and, not least, the technological control over the fabrication steps. The most straightforward and conceptually simple solution enabling power transfer in vertical direction is the one realized with two superposed parallel waveguides. Within such a configuration, a total power transfer is guaranteed when top and bottom waveguide modes are in a perfect phase matching condition. In case of homogeneous layers, such a condition is automatically achieved by coupling two waveguides with the same geometry. Conversely, when the layers are made of different materials, phase matching is achievable only by carefully tuning the waveguide cross-sections until finding a coincidence between top and bottom mode effective indices. A practical implementation of this scheme clearly requires a high degree of control over the fabrication steps. Unfortunately, unavoidable tolerances and imperfections can quickly compromise the correct operation of the coupler, and reduce its efficiency. An evolution of the vertical coupler scheme is obtained by introducing waveguide tapering. As explained in Section 2.6, by adding two zones of adiabatic variation in the cross sections of the superposed waveguides, the phase matching condition is easily reached, thus enabling the interconnection of heterogeneous layers and also easing the technological constraints. Furthermore, the tapered coupler scheme can provide a transmission characteristic which is wavelength-independent over a wide band. Numerical evaluations presented in Section 2.6 have shown that, by means of a basic optimization procedure on the tapers relative shift, insertion losses as low as $2\,dB$ can be easily achieved even for a vertical gap of $1000\,nm$. Tanks to all these positive features (simple layout, resilience to technological imperfections, high coupling efficiency), adiabatic tapers are therefore largely adopted in integrated photonics and must be considered as the first technological option in the design of multilayer networks on-chip.

A further solution, which is suitable for the interconnection of homogeneous layers, is the one relying on MMI structures. MMI-based solutions can provide high power transfer efficiencies over a limited footprint. As shown in Section 2.7, silicon-based implementations of such a scheme can provide average insertion losses as low as $0.5\,dB$ for gaps up to $400\,nm$. Unfortunately, as the vertical distance to overcome increases, the transmission curves tends to become heavily wavelength-dependent, with spectral windows presenting by high loss ripples.

Grating-assisted and microring-assisted schemes can provide, beside power coupling, also switching and signal-processing functionalities. In grating-assisted configurations, in fact, vertical transmission is enabled only for wavelength which are in a specific relation (Bragg resonance condition) with respect to geometric pattern on one of both superposed waveguides. Thanks to this feature, such resonant schemes can find application as 3D add-drop multiplexers. Microring-assisted interconnections extend the concept of photonic switching element (PSE) to the third dimension. Therefore, their potential contexts of usage for on-chip signal-processing and networking are countless.

From the design side, it must be observed that all the modeling tools usually adopted at the system-level for complex design of planar PSEs (i.e. the tools based on Transmission or Scattering Matrix) can be directly extended to the 3D vertical case, provided to correctly evaluate the interactions taking place between different planes.

To conclude, it is interesting to observe how photonic integrated circuits are following a path similar to that formerly taken by electronic integrated circuits (IC). In the IC domain, in fact, the exponential increase in the component density has made it inevitable the development of technological solutions enabling interconnections over multiple levels, which nowadays are industrial standards. In this sense, 3D optical architectures could also boost the emerging field of automatic place-and-route tools for photonic circuitry, whose potential has been so far restrained to planar layouts. While topological benefits deriving from 3D optical stacking are the most apparent, it must be also observed how the capability to combine in the same structure materials with complementary features enables the implementation of functionalities otherwise unachievable with a single material acting alone as, for example, silicon-compatible laser sources. Heterogeneous stacking is therefore one of the main research lines over which the scientific community is directed, with particular regard to the exploration of platforms combining layers made of active materials, suitable to host switching structures, with layers providing low propagation loss on which design the interconnection structures. Advances in 3D photonics are therefore expected to benefit several application domains in the years to come, not only limited to on-chip interconnections, but also including sensing, photovoltaics, nonlinear and diffractive optics.

Acknowledgments

The research activity of Alberto Parini is supported by the *Labex CominLabs* (French National Research Agency program Investing for the Future, ANR-10-LABX-07-01) through the *3D Optical ManyCores* project. The research activity of Gaetano Bellanca is supported by the *WiNOT - PRIN 2015* Project of *MIUR*, and by *FAR 2016* funds of the *University of Ferrara*.

References

[1] A. Shacham, K. Bergman, and L.P. Carloni. Photonic networks-on-chip for future generations of chip multiprocessors. *Computers, IEEE Transactions on*, 57(9):1246–1260, Sept 2008.

[2] Howard Wang, Michele Petracca, Aleksandr Biberman, Benjamin G. Lee, Luca P. Carloni, and Keren Bergman. Nanophotonic optical interconnection network architecture for on-chip and off-chip communications. In *Optical Fiber Communication Conference/National Fiber Optic Engineers Conference*, page JThA92. Optical Society of America, 2008.

[3] Keren Bergman, Luca P. Carloni, Aleksandr Biberman, Johnnie Chan, and Gilbert Hendry. *Photonic Network-on-Chip Design*. Springer-Verlag New York, 2014.

[4] A.W. Poon, Xianshu Luo, Fang Xu, and Hui Chen. Cascaded microresonator-based matrix switch for silicon on-chip optical interconnection. *Proceedings of the IEEE*, 97(7):1216–1238, July 2009.

[5] Xianfang Tan, Mei Yang, Lei Zhang, Yingtao Jiang, and Jianyi Yang. A generic optical router design for photonic network-on-chips. *Lightwave Technology, Journal of*, 30(3):368–376, Feb 2012.

[6] Luca Ramini, Paolo Grani, Sandro Bartolini, and Davide Bertozzi. Contrasting wavelength-routed optical noc topologies for power-efficient 3D-stacked multicore processors using physical-layer analysis. In *Design, Automation Test in Europe Conference Exhibition (DATE), 2013*, pages 1589–1594, March 2013.

[7] Nicolás Sherwood-Droz, Howard Wang, Long Chen, Benjamin G. Lee, Aleksandr Biberman, Keren Bergman, and Michal Lipson. Optical 4x4 hitless silicon router for optical networks-on-chip (noc). *Opt. Express*, 16(20):15915–15922, Sep 2008.

[8] I O'Connor, M Brière, E Drouard, A Kazmierczak, F Tissafi-Drissi, Navarro D, F Mieyeville, Joni Dambre, D Stroobandt, J Fedeli, Z Lisik, and F Gaffiot. Towards reconfigurable optical networks on chip. In *Proceedings of ReCoSOC'05 (Reconfigurable Communication Centric SoCs)*, pages 121–128, 2005.

[9] Richard A. Soref and B.E. Little. Proposed n-wavelength m-fiber wdm crossconnect switch using active microring resonators. *Photonics Technology Letters, IEEE*, 10(8):1121–1123, Aug 1998.

[10] Sébastien Le Beux, Hui Li, Ian O'Connor, Kazem Cheshmi, Xuchen Liu, Jelena Trajkovic, and Gabriela Nicolescu. CHAMELEON: Channel Efficient Optical Network-on-Chip. In *IEEE International Conference on Design Automation and Test in Europe (DATE)*, Dresden, Germany, March 2014.

[11] D. Vantrease, R. Schreiber, M. Monchiero, M. McLaren, N.P. Jouppi, M. Fiorentino, A. Davis, N. Binkert, R.G. Beausoleil, and J.H. Ahn. Corona: System implications of emerging nanophotonic technology. In *Computer Architecture, 2008. ISCA '08. 35th International Symposium on*, pages 153–164, June 2008.

[12] Min Yang, William M. J. Green, Solomon Assefa, Joris Van Campenhout, Benjamin G. Lee, Christopher V. Jahnes, Fuad E. Doany, Clint L. Schow, Jeffrey A. Kash, and Yurii A. Vlasov. Non-blocking 4x4 electro-optic silicon switch for on-chip photonic networks. *Opt. Express*, 19(1):47–54, Jan 2011.

[13] A. Parini, G. Bellanca, A. Annoni, F. Morichetti, A. Melloni, M.J. Strain, M. Sorel, M. Gay, C. Pareige, L. Bramerie, and M. Thual. Ber evaluation of a passive soi wdm router. *Photonics Technology Letters, IEEE*, 25(23):2285–2288, Dec 2013.

[14] Tatsuhiko Fukazawa, Tomohisa Hirano, Fumiaki Ohno, and Toshihiko Baba. Low loss intersection of si photonic wire waveguides. *Japanese Journal of Applied Physics*, 43(2R):646, 2004.

[15] Hui Chen and A.W. Poon. Low-loss multimode-interference-based crossings for silicon wire waveguides. *Photonics Technology Letters, IEEE*, 18(21):2260–2262, Nov 2006.

[16] Jing Pu, Kim Peng Lim, Doris Keh Ting Ng, Vivek Krishnamurthy, Chee Wei Lee, Kun Tang, Anthony Yew Seng Kay, Ter Hoe Loh, and

Qian Wang. Heterogeneously integrated iii-v laser on thin soi with compact optical vertical interconnect access. *Opt. Lett.*, 40(7):1378–1381, Apr 2015.

[17] Guang-Hua Duan, Ségolène Olivier, Stéphane Malhouitre, Alain Accard, Peter Kaspar, Guilhem de Valicourt, Guillaume Levaufre, Nils Girard, Alban Le Liepvre, Alexandre Shen, Dalila Make, François Lelarge, Christophe Jany, Karen Ribaud, Franck Mallecot, Philippe Charbonnier, Harry Gariah, Christophe Kopp, and Jean-Louis Gentner. New advances on heterogeneous integration of iii–v on silicon. *J. Lightwave Technol.*, 33(5):976–983, Mar 2015.

[18] Shahram Keyvaninia, Gunther Roelkens, Dries Van Thourhout, Christophe Jany, Marco Lamponi, Alban Le Liepvre, Francois Lelarge, Dalila Make, Guang-Hua Duan, Damien Bordel, and Jean-Marc Fedeli. Demonstration of a heterogeneously integrated iii-v/soi single wavelength tunable laser. *Opt. Express*, 21(3):3784–3792, Feb 2013.

[19] Gunther Roelkens, Joost Brouckaert, Dirk Taillaert, Pieter Dumon, Wim Bogaerts, Dries Van Thourhout, Roel Baets, Richard Nötzel, and Meint Smit. Integration of inp/ingaasp photodetectors onto silicon-on-insulator waveguide circuits. *Opt. Express*, 13(25):10102–10108, Dec 2005.

[20] M. Gnan, S. Thoms, D.S. Macintyre, R.M. De La Rue, and M. Sorel. Fabrication of low-loss photonic wires in silicon-on-insulator using hydrogen silsesquioxane electron-beam resist. *Electronics Letters*, 44(2):115–116, January 2008.

[21] Fengnian Xia, Lidija , Sekaric, and Yurii Vlasov. Ultracompact optical buffers on a silicon chip. *Nature Photonics*, 1(1):65–71, 2007.

[22] Shiyang Zhu, G. Q. Lo, and D. L. Kwong. Low-loss amorphous silicon wire waveguide for integrated photonics: effect of fabrication process and the thermal stability. *Opt. Express*, 18(24):25283–25291, Nov 2010.

[23] Kyle Preston, Bradley Schmidt, and Michal Lipson. Polysilicon photonic resonators for large-scale 3D integration of optical networks. *Opt. Express*, 15(25):17283–17290, Dec 2007.

[24] Q. Fang, J. F. Song, S. H. Tao, M. B. Yu, G. Q. Lo, and D. L. Kwong. Low loss (6.45 db/cm) sub-micron polycrystalline silicon waveguide integrated with efficient sion waveguide coupler. *Opt. Express*, 16(9): 6425–6432, Apr 2008.

[25] Nicolás Sherwood-Droz and Michal Lipson. Scalable 3D dense integration of photonics on bulk silicon. *Opt. Express*, 19(18):17758–17765, Aug 2011.

[26] S. Zhu and Guo-Qiang Lo. Vertically-stacked multilayer photonics on bulk silicon toward three-dimensional integration. *Lightwave Technology, Journal of*, PP(99):1–1, 2015.

[27] Rong Sun, Mark Beals, Andrew Pomerene, Jing Cheng, Ching yin Hong, Lionel Kimerling, and Jurgen Michel. Impedance matching vertical optical waveguide couplers for dense high index contrast circuits. *Opt. Express*, 16(16):11682–11690, Aug 2008.

[28] Yang Zhang, Xiaochuan Xu, D. Kwong, J. Covey, A. Hosseini, and R.T. Chen. 0.88-thz optical clock distribution on adhesively bonded silicon nanomembrane. *Photonics Technology Letters, IEEE*, 26(23):2376–2379, Dec 2014.

[29] Yang Zhang, David Kwong, Xiaochuan Xu, Amir Hosseini, Sang Y. Yang, John A. Rogers, and Ray T. Chen. On-chip intra- and inter-layer grating couplers for three-dimensional integration of silicon photonics. *Applied Physics Letters*, 102(21), 2013.

[30] Ying Huang, Junfeng Song, Xianshu Luo, Tsung-Yang Liow, and Guo-Qiang Lo. Cmos compatible monolithic multi-layer si3n4-on-soi platform for low-loss high performance silicon photonics dense integration. *Opt. Express*, 22(18):21859–21865, Sep 2014.

[31] Adam M. Jones, Christopher T. DeRose, Anthony L. Lentine, Douglas C. Trotter, Andrew L. Starbuck, and Robert A. Norwood. Ultra-low crosstalk, cmos compatible waveguide crossings for densely integrated photonic interconnection networks. *Opt. Express*, 21(10):12002–12013, May 2013.

[32] A.M. Jones, C.T. DeRose, A.L. Lentine, D.C. Trotter, A. Starbuck, and R.A. Norwood. Layer separation optimization in cmos compatible multilayer optical networks. In *Optical Interconnects Conference, 2013 IEEE*, pages 62–63, May 2013.

[33] Jong-Moo Lee, Joon-Tae Ahn, Doo-Hee Cho, Hong-Seok Seo, Jung-Jin Ju, Myung-Hyun Lee, and Kyong-Hon Kim. Polymeric double layered waveguides vertically coupled through stepped mmi coupler. In *Conference on Lasers and Electro-Optics/Quantum Electronics and Laser Science Conference*, page CWA41. Optical Society of America, 2003.

[34] Chris J. Brooks, Andrew P. Knights, and Paul E. Jessop. Vertically-integrated multimode interferometer coupler for 3d photonic circuits in soi. *Opt. Express*, 19(4):2916–2921, Feb 2011.

[35] Majid Sodagar, Reza Pourabolghasem, Ali A. Eftekhar, and Ali Adibi. High-efficiency and wideband interlayer grating couplers in multilayer

si/sio2/sin platform for 3D integration of optical functionalities. *Opt. Express*, 22(14):16767–16777, Jul 2014.

[36] Giovanna Caló and Vincenzo Petruzzelli. Generic wavelength-routed optical router (gwor) based on grating-assisted vertical couplers for multilayer optical networks. *Optics Communications*, 366:99 – 106, 2016.

[37] Giovanna Caló and Vincenzo Petruzzelli. Grating-assisted vertical couplers for signal routing in multilayer integrated optical networks. *Optics Communications*, 386:6 – 13, 2017.

[38] L.B. Soldano and E.C.M. Pennings. Optical multi-mode interference devices based on self-imaging: principles and applications. *Lightwave Technology, Journal of*, 13(4):615–627, Apr 1995.

[39] Alberto Parini, Giovanna Caló, Gaetano Bellanca, and Vincenzo Petruzzelli. Vertical link solutions for multilayer optical-networks-on-chip topologies. *Optical and Quantum Electronics*, 46(3):385–396, 2014.

[40] Keren Bergman, Luca P. Carloni, Aleksandr Biberman, Johnnie Chan, and Gilbert Hendry. *Photonic Network-on-Chip Design*. Springer-Verlag New York, 2014.

3

Optical Interconnection Networks: The Need for Low-Latency Controllers

Felipe Gohring de Magalhães[1,2], Mahdi Nikdast[2], Fabiano Hessel[1], Odile Liboiron-Ladouceur[3] and Gabriela Nicolescu[2]

[1]Pontifical Catholic University of Rio Grande do Sul – GSE, Porto Alegre, Brazil
[2]Polytechnique Montréal, Montréal, Canada
[3]McGill University, Montréal, Canada

Abstract

Design trends for the next-generation multiprocessor systems point to the integration of a large number of processing cores, requiring high-performance interconnects. One solution to improve the communication infrastructure in such systems is to use networks-on-chip (NoCs) as they present considerable improvement in the system bandwidth and scalability. Nevertheless, as the number of integrated cores continues to increase, metallic interconnects in NoCs become a performance bottleneck due to their higher latency, lower bandwidth, and higher power consumption when the system scales. Optical interconnection networks (OINs) are one of the most promising paradigm in this design context, addressing the aforementioned issues with the metallic interconnects. A critical requirement to realize high-performance OINs is to develop low-latency controllers for such systems. Control techniques, such as those based on circuit-switching, can impose a high latency while performing the correct network routing, and hence a better solution must be found in order to fully utilize OINs. This chapter overviews some of the common control techniques employed in OINs. Moreover, a low-latency control approach is presented, which relies on a centralized core and pre-calculated routes, reducing the overall system communication latency through granting different requests within one clock cycle.

3.1 Introduction

Nowadays, computational systems present a rising number of features, leading to a significant growth in applications' design complexity. Also, systems are implemented based on integrating multiple processing components. In multiprocessor systems, one of the main design concerns lies in how the communications among the components are performed. Bus-based systems present a well-known solution with a reasonable bandwidth and ease of implementation. However, as the number of integrated processing components continues to rise, buses become less likely to be employed due to the growing design complexity of the system [1]. In traditional bus-based architectures, the communication can become a bottleneck, compromising the system operation [2].

Networks-on-chip (NoCs) tend to provide a better communication performance [3], where the communication is performed by routers that forward packets over the network. In addition to the gain in the communication capability, NoCs usually have an improved energy efficiency with a high level of re-usability [4]. However, as the number of possible integrated cores continues to further increase in the system, metallic interconnects in NoCs become a bottleneck due to their high power consumption, limited bandwidth, long latency and poor scalability. The International Technology Roadmap for Semiconductors (ITRS) [5] pointed out the need for a new technology to overcome such restrictions. In this design context, optical interconnection networks (OINs) are currently considered to be one of the most promising paradigms [6], addressing the aforementioned issues in NoCs to realize high-bandwidth and low-power interconnects in multiprocessor systems [7]. OINs are already a reality for long-distance communications [8], and their employment for short-distance communications, such as inter-chip communications, have already been proved to be feasible [9, 10]. For example, [11] presented an OIN with a low power consumption, a low insertion loss (7.9 dB for an 8×8 network), and a power penalty of less than 1 dB. This work brings forward OINs as an attractive candidate for high-performance multiprocessor architectures.

The performance and efficiency of OINs are highly constrained by their controllers. The control part has a significant impact on the overall performance of OINs, as the gain obtained on the fast optical path can be completely shattered by a high latency controller. Although low-latency controllers have been proposed for electrical networks, they rely on structures that are not available in OINs (e.g., using buffers to temporarily store the incoming traffic). Therefore, new controlling techniques must be proposed for OINs. Previous works demonstrated architectures with controllers that impose either a long setup time or are too complex, thus challenging their employment in practice [12, 13]. Most OINs are built relying on two optical components: Mach-Zehnder interferometers (MZIs) and Microring Resonators (MRs). Both components are used as switching elements, which are grouped to form different network architectures.

In this chapter, we present an overview of the most common techniques for the control of OINs. Also, a low-latency controller is designed which takes benefits of pre-calculated routes stored in look-up tables (LUT), and hence it is called the Look-Up Table Centralized Controller (LUCC). The LUCC is designed to reduce the latency required to configure the network and solve request conflicts. It employs different techniques, such as the shortest path first and matrix reduction algorithms (introduced later in this chapter), to reduce the controller overhead in time and guarantee the best possible network routing. This way, the LUCC is able to process requests and configure the network within one clock cycle. The rest of the chapter is organized as follows. Next Section presents an overview of different control techniques usually applied to OINs. Following, Section 3.3 introduces the developed low-latency controller, called LUCC, followed by Section 3.4 that presents the obtained results when applying the LUCC in different OINs. Section 3.5 overviews the state-of-the-art among related works, positioning our approach within existing works. Finally, Section 3.6 draws the conclusion.

3.2 Control Strategies

When it comes to optical interconnection networks, the design of the controller of the system is just as important as choosing the right topology. Unlike NoCs, optical networks do not rely on buffers to temporarily store data in every switch. If buffering is required, then several conversions between the optical and electrical domains are required (i.e., optical-electrical and electrical-optical conversions), leading to high and undesired costs. The controller is usually based on one of the following techniques:

- **Time sharing**: where time windows are set for each IP to transmit its information. Within each time window, a set of IPs are granted to send their data, while the others are stalled until the time window ends;

- **Dynamically path setting**: presents a behavior similar to those found in circuit-switching NoCs. The path is determined at the beginning of the transmission. In other words, the controller should compute all the inputs and choose all the paths, either by assigning wavelengths in passive networks or by tuning the components in active networks, and;
- **Wavelength division**: which is based on wavelength division multiplexing (WDM), where multiple wavelengths can be simultaneously transmitted into an optical waveguide (counterpart of electrical wires in OINs). This technique divides the total available bandwidth into a series of non-overlapping sub-channels, assigning a different wavelength to each transmitting IP. This technique is usually employed in passive networks, but not exclusively.

3.2.1 Time Sharing

Different network topologies for OINs are possible, each one with its pros and cons. Some topologies, such as mesh or ring, rely on distributed nodes to perform parallel communication at the cost of more area and power hungry structures. Also, there are topologies in which only one single transmission at a time is permitted, and hence all the other requesting nodes should wait. In such a case, the control unit should manage the requests in a manner that no request waits forever to be granted access (i.e., deadlock free). Figure 3.1 presents a scenario in which seven IPs share a single communication media where a centralized controller is employed. In this case, conflicts happen when more than one IP tries to simultaneously access the shared media.

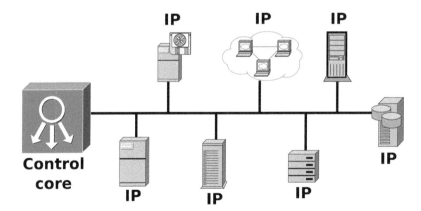

Figure 3.1   An example of a system with seven IPs sharing the same communication media.

Addressing a conflicting situation such as the one in Figure 3.1, a common approach is to use a time sharing/division multiplexing (TDM) technique [14], providing several recurrent time windows (slots), one for each IP. The time slots might have the same duration depending on priorities attributed to the IPs. Also, the amount of information transmitted over the media within each time slot may either be the same for all the IPs or vary according to their priorities.

The simplest implementation for TDM is based on the round robin (RR) algorithm [15]. It implements a first-in-first-out (FIFO) queue, where each position holds a single IP identification (ID). For every new time slot, the IP with the ID found on the next entry of the FIFO will have its access granted. This kind of control imposes a fair latency overhead on the system, and hence is usually implemented using a single centralized control unit, which is more suitable for systems with a low design complexity. Very often, centralized control units are fully integrated into the network (as one black box). In these cases, the IPs that access the network cannot differentiate between the network and the control unit. Figure 3.2 illustrates an example of a centralized control unit and an OIN, where both the control and interconnection network are presented as one unified unit.

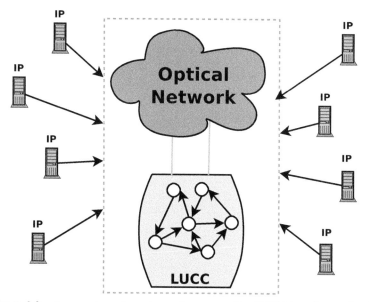

Figure 3.2 An overview of a system using a centralized TDM-based control unit.

3.2.2 Circuit-Switching

Another approach for controlling OINs concerns the circuit-switching technique. This technique is particularly suitable for large networks, composed of a large number of inputs/outputs (I/Os) and network nodes. In this case, the controller has access to all the network nodes and configure those involved in each communication. The main appeal of circuit-switching is its utilization in three-dimensional (3D) stacked integrated-on-chip systems, in which each layer of the chip holds one parcel of the entire architecture [16].

Considering a hybrid 3D stacked optical-electronic architecture, each optical network node (i.e., optical router) is directly connected to an electrical network one (i.e., electronic router), which is placed just above/under the optical node using a through silicon via (TSV) [17] (see Figure 3.3). The IPs are connected to the electrical network nodes, which receives different communication requests. Once a routing request is received, the electrical nodes configure an optical path (i.e., from the source to the destination core) on the optical network. Once the optical route is configured, the message starts to be sent through the optical network.

Figure 3.3 shows an overview of the 3D stacked networks and the connections. Figure 3.3(a) presents the top system view. Figure 3.3(b) shows a lateral systems view, where it is possible to see the TSV connections between the electrical nodes and the optical nodes. Figure 3.3(c) illustrates the optical

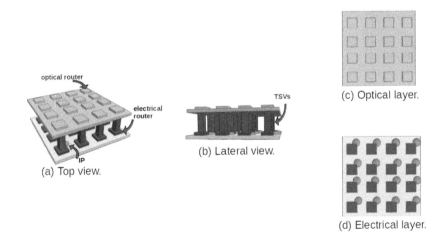

Figure 3.3   An overview of a system using a distributed circuit-switched control unit.

network layer. Finally, Figure 3.3(d) indicates the electrical layer including the IPs connected to the electronic routers.

Considering the circuit-switching technique, the time it takes to configure all the nodes can impose a high latency overhead. Still, this technique remains useful for systems in which large-size messages are exploited as the high cost to close the routing path can then be compensated by the gain to transmit bulky messages over the optical network.

3.2.3 Wavelength Division

The last introduced technique applied to control units for OINs is the wavelength division multiplexing [18]. This technique allows the simultaneous transmission of multiple optical wavelengths in a single optical waveguide. WDM is similar to TDM, but instead of using different time slots, WDM employs different optical wavelengths. When using WDM, the available bandwidth of the optical channel is divided into several sub-channels, where each sub-channel supports a single optical wavelength.

Figure 3.4 depicts an example of a centralized controller for a WDM-based OIN including four requesting IPs, each of which is assigned with a distinct wavelength (i.e., λ_1 to λ_4). The WDM technique can be employed for two different scenarios: **(i)**: different IPs request for access simultaneously, as presented in Figure 3.4, and; **(ii)**: for the cases when only one IP is requesting access and all the wavelengths are free. In both cases, the controller might assign different optical wavelengths to the same IP for each communication

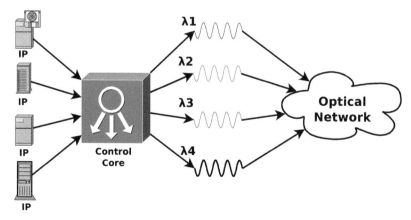

Figure 3.4   An overview of a system using a centralized WDM-based control unit.

request, so the transmission might occur in a parallel stream. The potential to have parallel transmission is one of the main benefits when using optical networks. On the opposite to electrical wires, optical waveguides are capable of handling more than one transmission at a time, given that each transmission occurs using a distinct wavelength. This way, WDM controllers can take benefits of this capacity and configure the network to transmit data in parallel, even for shared waveguides.

3.3 Low-Latency Controlling Solution for OINs

The performance and efficiency of OINs are constrained by their controllers. Long setup time of circuit-switching techniques makes them not efficient for OINs, while the physical design and implementation of TDM and WDM techniques for OINs can become very complex in practice. In this section, we present the design and development of the Look-Up Table Centralized Controller (LUCC) for OINs, which is based on pre-calculated routes stored in lookup tables, realizing low-latency communications in such systems.

Controllers in OINs have two major roles, including addressing different requests to access the network and also configuring the network. The requests are basically from the input nodes demanding to send data through the network. Each request should be computed in order to look for possible conflicts and guarantee a routing path availability. Considering the network configuration, an optical path is setup between a source and a destination node, through which a message is sent.

The design of LUCC consists of four different parts, including the connectivity matrix definition, the conflict resolution, the path configuration, and the runtime calculation (i.e., dynamic setup). The matrix definition is a technique used to describe all the network connections, facilitating applying different methods for possible simplifications (e.g., the matrix reduction method [19]). The conflicts resolution is responsible for detecting same-destination conflicts (i.e., when several IPs simultaneously request to communicate with the same IP core), and solving them using a given algorithm. The path configuration is a memory block used to store static data accessed by the controller during the runtime, which can be built using the output from the matrix definition. This memory is used mainly to reduce computation time, thus reducing the overall control latency. The runtime calculation is a block responsible for real-time calculations (e.g., computations related to path assignments and readings of memory addresses). Figure 3.5

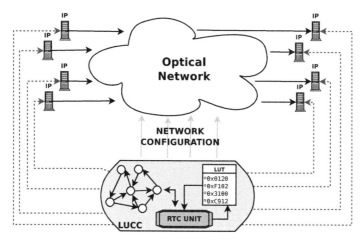

Figure 3.5 LUCC design overview.

illustrates the designed controller overview. The matrix definition and the path configuration are performed at the design time and result in the LUT which holds the network paths, while the conflict resolution and real-time calculations are performed at runtime.

We start by defining the connectivity matrix for all the I/O ports in the network. It is also possible to consider methods to reduce/simplify the matrix [19], thus reducing the controller overhead. In particular, such reduction methods can be used to remove the redundant and/or unused network nodes in the matrix when not all the network nodes are connected to each other (depending on the topology). Table 3.1 indicates the connectivity matrix $\mathcal{M}$ for one specific 8×8 network, in which all the nodes can communicate with all the others, but themselves. In the matrix, each paired index *i,j* represents a communication from input *i* to output *j*, such as:

$$\forall_{ij} \in \mathcal{M}, \; if \; \mathcal{M}_i \; communicates \; with \; \mathcal{M}_j \Rightarrow \mathcal{M}_{ij} = 1,$$

resulting in the connectivity matrix $\mathcal{M}$.

After defining the connectivity matrix, the second step is to compute all the input requests as fast as possible. In order to minimize the controller overhead, we target one clock cycle latency in the design of LUCC, resulting in a latency in a range of nanoseconds. Algorithm 1 describes the logic for the conflicts resolution and dynamic setup, determining the best available path for each input request. In the algorithm, it is possible to see different possible scenarios, such as when several input requests are received and

Table 3.1   Connectivity matrix for an 8×8 network

Node	0	1	2	3	4	5	6	7
0	0	1	1	1	1	1	1	1
1	1	0	1	1	1	1	1	1
2	1	1	0	1	1	1	1	1
3	1	1	1	0	1	1	1	1
4	1	1	1	1	0	1	1	1
5	1	1	1	1	1	0	1	1
6	1	1	1	1	1	1	0	1
7	1	1	1	1	1	1	1	0

conflicts are found. The algorithm works by checking all the input requests (*while $i <$ numberOfInputs*) and looking for conflicts for each one of them (*checkConflicts(i)*). When no conflict is found and the target is not busy (*noOpenComm(i)*), the requesting node is granted access. If a conflict is found (*conflict()*), the controller checks to see if the IP with its ID on the next slot of the round robin queue is the requesting node

---

Algorithm 1 Conflicts Resolution and Granting Control

---

while $i <$ *numberOfInputs* **do**
   checkConflicts(i);
   if *requestReceived(i) and noOpenComm(i)* **then**
      if *destIsAvail() and noDestConf()* **then**
         acknowledge(i) $\leftarrow$ 1;
         writeEnable(i) $\leftarrow$ 1;
      else if *destIsAvail() and conflict()* **then**
         if *roundRobinControl()* **then**
             acknowledge(i) $\leftarrow$ 1;
             writeEnable(i) $\leftarrow$ 1;
         end if
         roundRobinSpin();
      end if
   else if *endOfCommunication(i)* **then**
      acknowledge(i) $\leftarrow$ 0;
      writeEnable(i) $\leftarrow$ 0;
      endOfComm(i) $\leftarrow$ 1;
   else
      endOfComm(i) $\leftarrow$ 0;
   end if
end while
pathDefinition();

($roundRobinControl()$), and it grants access if that is the case. For the cases where the the round robin queue is used, a new node is set as the next one in the round robin queue ($roundRobinSpin()$). Lastly, the controller defines the network configuration ($pathDefinition()$) by accessing the LUT.

Finally, the last step for the controller is to configure the path that the message should take from the source to the destination node. For every input request, more than one optical path might be selected, generating a huge number of path possibilities for the controller to compute. One possible approach to address this issue is to use a shortest path first (SPF) algorithm, like the Dijkstra algorithm [20]. The SPF algorithm looks for the minimal distance between the source and the destination in a network. The distance between the source and destination is counted as the number of network nodes through which the message should pass. It works by considering the value of each edge, which connects any two nodes, and then looking for the minimal sum of edge values between the source and the destination. The edge weight can be, for example, the distance or power consumption depending on the routing technique objective and is defined by the designer. If power consumption is to be considered, each edge can express the power dissipated between each pair of nodes. As it would be very complex to calculate a new path for every new input request, a pre-calculated path allocation lookup table is generated and stored in memory (i.e., static memory) by implementing the SPF algorithm and running it during design time.

Applying the SPF algorithm, we first represent the network as a series of graphs. The OIN architecture in our work consists of several optical switches integrated to form the network. As a result, the graph representation starts from MZI-based optical switches. Figure 3.6 presents the graph abstraction of a 2×2 MZI-based optical switch, where inputs and outputs are represented as a graph node and connections are represented as edges.

Employing several 2×2 switching blocks, it is possible to form a 4×4 optical switch shown in Figure 3.7. As can be seen, six 2×2 optical switches are interconnected, forming a larger optical switch with four inputs and four

Figure 3.6   Graph representation of a 2×2 MZI-based optical switch.

outputs. At this point, the resulting graph is already complex to compute during runtime as the number of possible source-target combinations is already large. Consequently, the possible hardware implementation, although feasible, would be costly.

Employing the 2×2 and 4×4 optical switches and connecting them accordingly, one can construct different OINs. Figure 3.8 shows the graph representation of one possible OIN topology, the 8×8 Beneš network, in which eight 2×2 and two 4×4 optical switches are employed. As can be seen, the complexity increases with the addition of extra nodes and I/O ports, justifying the choice for a pre-computed path lookup table. The path allocation lookup tables have to be generated for every new topology. This is due the fact that for different topologies various possible paths can be found. It is worth mentioning that this computation can be performed once and it is off-line, reducing the final design complexity.

In addition to the control algorithm, another important aspect to guarantee the LUCC employment in practice is its memory contents generation and optimization. For a small number of I/O ports, the memory size is negligible. However, as the number of ports increases, the memory consumption considerably grows up to a point that using the controller is prohibitive. Addressing this issue, a reduction technique is employed. It works by looking for all duplicated entries in the table and removing them. Besides that, the remaining entries are compressed and reallocated. Applying this reduction technique, it is possible to reduce up to 70% of the memory size. The allocating paths algorithm is presented in Algorithm 2. The algorithm works by generating the input request combinations ($inputsCombinations()$) and then running the SPF algorithm ($spf\_algo()$) for each combination and eventually paths creation. Finally, the resulting memory array is optimized ($allocOptmization()$ and $allocReduction()$) as discussed.

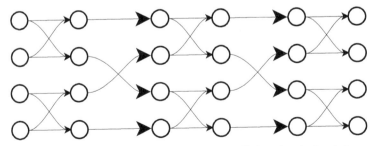

Figure 3.7   Graph representation of a 4×4 MZI-based optical switch.

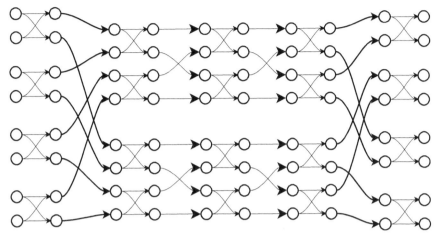

Figure 3.8 Graph representation of an 8×8 Beneš network based on employing several 2×2 and 4×4 optical switches.

Regarding the control core and its conflict resolution algorithm, different approaches are possible. Minimizing the latency overhead, we consider using the iSLIP algorithm [21] as an inspiration for the LUCC. This algorithm has been proved to work efficiently for small to medium size networks and it runs through 'n' iterations, where each iteration can be divided into three steps:

- **Request:** all requesting input ports signal their intention to have access to the network and are stored in virtual queues, one for each port;
- **Grant:** each output port that has received a request should determine which input port can be granted access. For the situations where only one input port is requesting access, this step is straightforward. For the cases where two or more requests are targeting the same output, a round robin algorithm is used to decide which input port can have its access granted first; and,
- **Accept:** the input ports consider all the received grant messages and choose from them based on the round robin algorithm. An input port notifies the targeted output that its access is granted, and once it receives an acknowledgement, it then moves to the next request.

The iSLIP runs through multiple iterations. As a result, the number of input-to-output matches are improved. Previous work showed that running the iSLIP beyond four iterations does not yield significantly greater matches [21]. Similar to the iSLIP, the LUCC relies on three steps but it does not

---

Algorithm 2 LUT Generation and Reduction

$inputsCombinations()$;
for all $inputs$ **do**
   $spf\_algo()$;
end for
$allocOptmization()$;   ▷ exclude duplicated table entries / reallocate values
while $allocSize \geq threshold$ **do**[1]
   $allocReduction()$;
end while

---

perform recursive iterations or utilizes virtual queues at the input ports, reducing the implemented core complexity and focusing on a fast and efficient conflict resolution. As the recursive iterations in the original iSLIP implementation are designed to improve the throughput, the LUCC throughput might be inferior for some specific scenarios compared to the original iSLIP implementation.

The LUCC conflict resolution works based on a matrix method in which for every new request round all the source-target pairs are mapped to a matrix $\mathcal{R}$ of requests, and then each column j is checked for any possible conflicts. For instance, the matrix for a 3×3 optical switch, where all the inputs are requesting to communicate with the output two ($\langle 0 \rightarrow 2, 1 \rightarrow 2, 2 \rightarrow 2 \rangle$), is:

$$\mathcal{R} = \begin{bmatrix} 0 & 0 & 1 \\ 0 & 0 & 1 \\ 0 & 0 & 1 \end{bmatrix}.$$

Considering the same switch, but under a different scenario such as $\langle 0 \rightarrow 2, 1 \rightarrow 0, 2 \rightarrow 1 \rangle$, the request matrix changes to:

$$\mathcal{R} = \begin{bmatrix} 0 & 0 & 1 \\ 1 & 0 & 0 \\ 0 & 1 & 0 \end{bmatrix}.$$

The matrices are created based on the IDs of the requesting input port and the requested output port. For example, using the same switch as discussed above, $\mathcal{R}(i,j) = 1$ if the input port i request access output port j in the switch, such as:

$$\forall_{ij}, if\ request(i) = j \Rightarrow \mathcal{R}_{ij} = 1.$$

---

[1]Here, a threshold size might be defined as any limitation on the final memory size. For example, memory utilization on FPGAs can be defined as the upper threshold size.

Note that the matrix can be accessed directly (as a hardware register) and no extra processing is needed, accelerating the conflict detection.

Once the matrix is generated, all the columns of the matrix (each column is associated with an output port) are verified to find any possible conflicts, where a conflict is defined as any situation in which two or more inputs are requesting the same output. This can be defined as:

$$\forall j \in \mathcal{R}, \quad \neg XOR(j) \wedge OR(j) \implies conflict(j) = 1.$$

When a conflict is found, a FIFO queue is implemented based on the round robin algorithm to decide which port will be granted access. Since the round robin algorithm uses a FIFO for each input, each request is treated individually as one particular process that is triggered by the conflict bit controlled by the matrix method.

Lastly, the control algorithm implements the signalizing process of the LUCC. This algorithm is responsible for checking the status signals, such as the conflicts and requests, and decide which requesting port will have its access granted based on these signals. Each request port generates one control block that implements the control algorithm, which generates one individual hardware process. In other words, during runtime, each request port has one particular piece of hardware handling its request individually, thus expediting the execution time.

It is worth mentioning that the conflict resolution and the round robin algorithm execute independently of the system clock and they are triggered by the matrix $\mathcal{R}$ and conflict signals. Both blocks are fully implemented using only combinatorial logic, thus leaving the control block to be constrained by the system clock. This helps reduce the controller latency as less hardware is clock dependent. Figure 3.9 overviews the decision flow chart of the LUCC, where it is possible to see the steps taken by the controller from the request received until the end of the communication.

3.4 Results

Different traffic patterns, such as complement and all-to-all, are considered to assess the performance of LUCC under various request conditions. The complement traffic pattern is used to verify the longest paths in the network. All-to-all traffic pattern comprises all possible communication combinations in the network, as each IP in the system requests access to all the nodes. Figure 3.10 depicts both traffic patterns using the 8×8 Beneš network, in

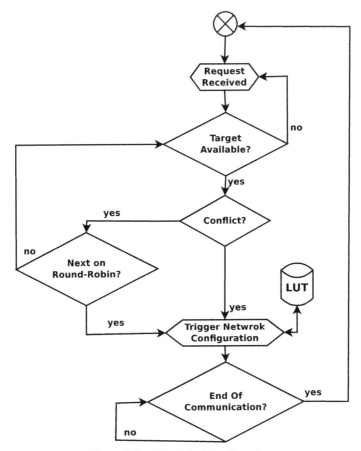

Figure 3.9 LUCC decision flow chart.

which the IPs zero and seven (*IP 0 and IP 7*) are used as an example. Considering the figure and as an example, in the case of the *all-to-all* traffic pattern, *IP 0* requests access to all the nodes in the network, while it requests access to the other extreme of the network (e.g., *IP 7*) under the *complement* traffic pattern (*IP 7* also requests access to its other extreme, which is *IP 0*).

An effort is made to verify the one clock cycle latency of the LUCC under different scenarios, as well as exploring the correct network configuration. We first perform simulations and then the same scenarios were applied on FPGA prototypes. For both cases (simulation and prototyping), a list of signals are employed to illustrate the results, presented as arrays of values organized in a little-endian fashion. As a result, the rightmost and leftmost values of the

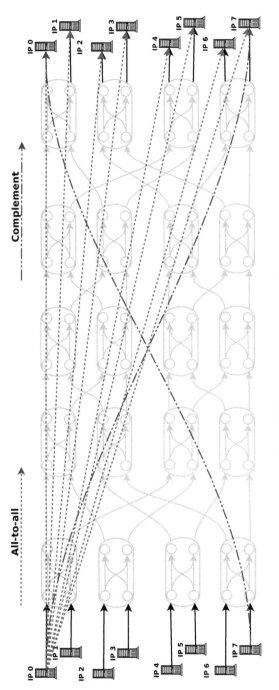

Figure 3.10 Traffic patterns exemplification.

array hold the information of *IP 0* and *IP 7*, respectively. The most important used signals are:

- **target:** it holds the target for each requesting port. If the value for a position equals -1, it means that the IP on that position is not requesting access to the network. Otherwise, it shows the desired destination. For instance, the array $target = 0 -1 -1 -1 -1 -1 3 -1$ indicates that the *IP 7*, whose position in the array is on the leftmost, is targeting *IP 0* and *IP 1* is targeting *IP 3*. Also, the other IPs are not requesting any access;
- **request:** it is used by the IPs to signal their intention to access the network, where *1* shows a request for access and *0* means idle (i.e., no request). For instance, the array *request = 00000001* indicates that the *IP 0*, whose position in the array is on the rightmost, is requesting access to the network, while all the other IPs are idle, and;
- **ack:** it is used by the LUCC to signal IPs when the access is granted, where *1* means that the access is permitted and *0* means that it is not. For instance, the array *ack = 10010000* shows that the *IP 7* and *IP 4* are permitted to access the network, while all the others are not.

Figure 3.11 depicts the one cycle response time of LUCC, where the controller runs at 500 MHz (i.e., period of 2 ns). Two scenarios are illustrated in the figure, including one with conflicts and one without conflicts, and considering a small system to improve the results visibility. Considering the first scenario, marked with orange boxes in the figure, four simultaneous input requests are generated (targeting outputs 3, 2, 1 and 0). As can be noticed, the access is granted after one clock cycle (i.e., in the figure, **ack** signals turn to **1** only one clock cycle after **request** signals become **1**). Also, the orange arrow shows the time it takes for the LUCC to acknowledge the access, which is one clock cycle in this case.

Considering the second communication scenario, marked with yellow boxes in Figure 3.11, four simultaneous input requests are again generated, such that inputs 0, 1 and 3 are targeting output 2 and input 2 is targeting output 0. Note that more than one input is targeting the output 2, and hence a conflict occurs. It is possible to see the resolution of the conflicted request at a time, and even for the cases in which a conflict is found, the controller latency is not affected for computing the requests. First, among the conflicting request ports, *IP 0* is granted access (i.e., *ack = ***1*). Next, *IP 1* has its access granted (i.e., *ack = **1**), and finally *IP 3* is granted access (i.e., *ack = 1****). In this situation, the receiver side takes two clock cycles to set its

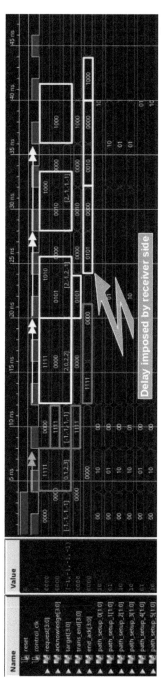

Figure 3.11 LUCC simulation diagram running at 500 MHz, which shows: input requests (*request*), LUCC grants (*acknowledge*), destination outputs (*target*), end of transmissions (*trans_end*), acknowledges for transmissions ending (*end_ack*) and the bits to set up the MZI switches (*path_setup**). In each array of bits, the bits on the rightmost and leftmost represent the signal for the I/O zero and I/O three, respectively.

connection, thus blocking its port. This fact delays the granting response time of LUCC, as the destination is blocked until the receiver side computes the end of the transmission. This can be noticed in the figure by considering **trans_end** and **end_ack** signals. The simulation results indicate the fast response time of LUCC whose latency is constrained by the frequency of the platform, on which the control unit is able to solve requests in one clock cycle.

Indicating the feasibility of LUCC and its functionality using FPGA technology, we synthesized LUCC for a Xilinx FPGA (Xilinx II-Pro) [22]. The signal readings for prototype evaluation were performed using the Chipscope Pro Analyzer, also from Xilinx. The FPGA execution results are indicated in Figure 3.12. As can be seen, it is possible to see four ports are requesting (**RX_***) access and having their requests granted (**ACK_***) after one clock cycle. Also, the figure shows the end of communication signals (**TAIL_*** and **TAIL_ACK_***), illustrating the protocol used at the end of a transmission. The protocol works based on handshaking: the sender signals that it will end the transmission by setting TAIL signal to '1' and then the controller acknowledges the end of transmission by setting TAIL_ACK to '1'. According to Figure 3.12, the LUCC latency is based on the employed technology and it is able to compute requests in one clock cycle. For the specific scenario in this figure, the FPGA execution frequency was configured to be 50 MHz, resulting in a period of 20 ns. Therefore, the LUCC latency is 20 ns for the considered scenario and using the proposed FPGA.

Next, the design was prototyped for an Altera's FPGA [23], where the signal readings were performed using Altera's SignalTap Logic Analyzer Tool [24]. For this step, not only was the LUCC prototyped but fast transceivers were prototyped as well. The fast transceivers are hardware components provided by Altera and used for fast communications. It is composed of different blocks, such as serializers and synchronizers. They are used in order to access the full communication capabilities of the OIN. The main goal of this step is to show the LUCC interacting with IPs, in this case traffic injectors. Here, the LUCC receives requests from input blocks that inject traffic into the fast transceivers only after having their access granted. The output of the fast transceivers was virtually connected to the desired target. The FPGA used in this case was the Stratix IV 330T, configured to an execution frequency of 100 MHz, and hence with a period equals 10 ns.

Figure 3.13 presents the obtained readings, where it is possible to see the request, ack, tail and tail_ack signals with the same functionalities as

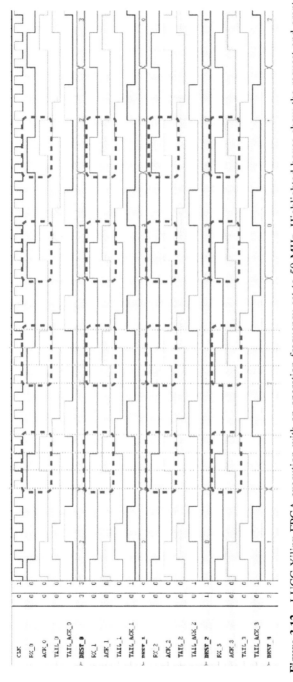

Figure 3.12 LUCC Xilinx FPGA execution with an operation frequency set to 50 MHz. Highlighted boxes show the request and grant moments, happening in one clock cycle.

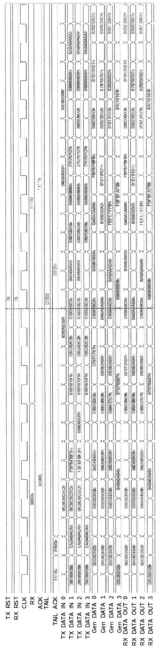

Figure 3.13   LUCC prototyped on an Altera's FPGA executing with an operation frequency set to 100 MHz.

the anterior case. In addition to these signals, three signal groups are further presented in this figure: *Gen_DATA_*, *TX_DATA_** and *RX_DATA_**. These signals correspond to the traffic generated in the inputs (*Gen_DATA*), the fast transceivers input (*RX_DATA*), and the fast transceivers output (*TX_DATA*).

Finally, we integrated the LUCC with a fabricated optical switch to verify its behavior in a realistic and dynamic scenario. Considering the co-design, a 4×4 optical switch was controlled by the LUCC running on an FPGA. The FPGA is used as it enables the fast prototyping and evaluation of the design. Furthermore, the FPGA board contains needed interfaces to interact with the optical switch. Also, in addition to the LUCC design, the FPGA was equipped with a request generation block. The lab-setup overview is illustrated in Figure 3.14[1], in which the main blocks are presented and the inputs **0** and **1** as well as the outputs **5** and **6** are indicated. Also, the 2×2 switch in the bottom-left of the switch (i.e., MZI 3) is not being controlled because of the component malfunction[2].

Figure 3.15 presents a microscopic picture of the silicon photonic (SiPh) switch. It is a 4×4 optical switch based on the Spanke-Beneš topology [26] that includes five 2×2 integrated MZIs that are directly controlled by the LUCC. Carrier injection tuning method was employed to bias one arm of the MZI for high-speed and efficient switching. This method involves varying the power applied to the MZI arm, which affects the signal transmitted in the biased waveguide. The SiPh switch was fabricated by the IME foundry and the measured switching voltage and switching time are $0.18\ Vmm$ and 6 ns, respectively.

Figure 3.16 presents the extracted results, where the FPGA readings are shown. The following configuration executions are presented: (**i**) input *0* is targeting output *5*, (**ii**) once again, input *0* is targeting output *5*, and (**iii**) inputs *0* and *1* are targeting output *5*, which results in a conflict. The figure illustrates the FPGA readings for each scenario, where it is possible to see the requests, acks, tail, tail acks, MZIs configuration and target signals. This experiment indicates the correct performance of the LUCC while controlling a prototyped optical switch. Considering the outputs and comparing them with the expected outcome, it is possible to observe that the LUCC is able to

---

[1]The setup was at the Photonic Systems Group at McGill University, Montréal, Canada.

[2]Optical components are very sensitive to fabrication process variations [25]. In this case, the MZI 3 was affected through the fabrication process. Consequently, we decided not to rely on this component for evaluation purposes.

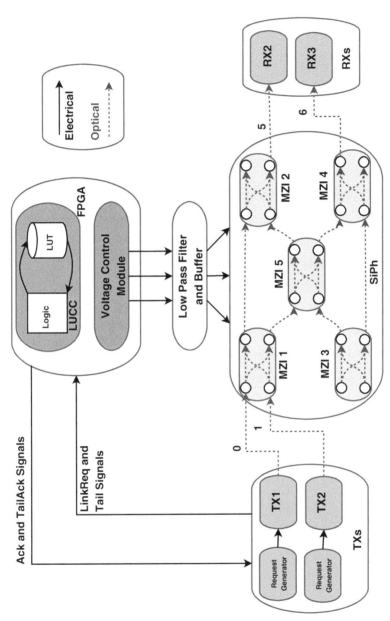

Figure 3.14   Lab-setup overview for integrating the LUCC with a 4×4 optical switch.

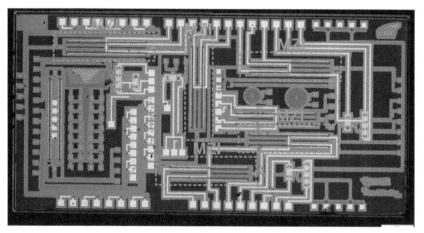

Figure 3.15 Microscopic picture of the 4×4 optical switch based on MZIs.

successfully perform all the connections, dealing with the requests, conflicts and network configuration within one clock cycle.

3.5 State of the Art

Most state-of-the-art works opt for designing a network controller that is applicable to a specific topology or type of architecture, aiming to optimize at most the performance of that specific network. Nevertheless, the controller of an OIN should work as a standalone unit that does not directly relate to the topology or architecture of the controlled network.

The work in [27] presents the design of a 5×5 MR-based optical router, which uses a mixed active-passive design approach. In this router, when the communication is from the left-side of the switch towards the right-side of the switch (i.e., west to east), only passive components are used, thus reducing the power consumption. The paper claims that the router is designed to be used as the basic block in large-scale networks. The control unit for this router is based on circuit-switching, where package switching [28] is applied on the electrical layer to set the paths on the optical layer. The main drawback of this approach lies in the fact that circuit-switching techniques are time demanding and as the networks scales their response time might get prohibitive.

A new architecture for optical networks-on-chip (ONoCs) along with the controlling scheme are presented in [29]. The system is claimed to consume

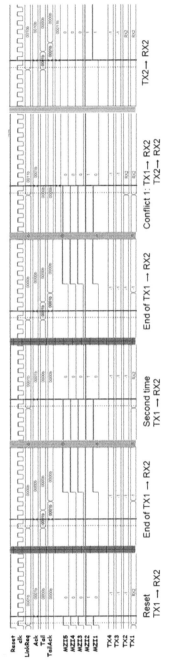

Figure 3.16 Extracted data from lab experiments.

less power through using a dynamic resource provisioning. In order to guarantee a message delivery, a given time is provided before turning the injector laser off, which is defined according to the number of hops the message will pass through and is dynamically calculated by the controller. The latency of the presented method is around 3.5 ns in an 8×8 network in which resonators and peripherals run at 5 GHz. The main drawback for the solution proposed in this work is that using circuit-switching for path allocation comes at a high cost. Even for a fairly small network size (eight IP nodes), the control latency is high, i.e., at a 5 GHz frequency operation, the period is 0.2 ns, meaning that it takes around 17 clock cycles for every request to be computed.

The work presented in [30] brings the design of an ONoC as an extension of other previous works like λ-Router [31]. The control unit employed in this work is based on wavelength routing, in which each I/O is assigned to one specific wavelength and might communicate at any time without arbitration. Although this kind of structure solves any kind of contention, the overhead imposed by this solution is very high as '*n*' buffers must be used for every I/O port, where *n = number of IPs*.

A routing technique based on wavelength selection integrated with spatial routing is introduced in [32]. The technique is based on the unused spectrum that exists between the resonances of a broadband MR switch by interleaving additional wavelength channels in the unused spectral space. Circuit-switching technique is used along with WDM. Also, each router includes several receivers and modulators. For a small message (e.g., 100 kbit messages) at a 100 Gbps transmission rate, the reported latency is around 100 ns. The latency is relatively high acknowledging the fact that this architecture is fully tailored [3].

An optical router is presented in [33] that can be used as the basic block in constructing on-chip optical networks. Four types of MR-based 4×4 switches are introduced. The goal of this structure is to use the minimum possible number of resources. The control unit applied to such network is based on WDM, where a specific transmission frequency is selected for each basic block. This leads to a more complex controlling solution since the cost for computing the required frequencies rises with the increase in the number of basic blocks.

An electrical-optical mixed approach is presented in [34], where a butterfly-based architecture is introduced, in which each node is composed

---

[3]Fully tailored points out to the fact that the entire system is customized and the components are designed based on each other, thus not having a generic behavior.

of an optical switch and an electrical switch. Optical switches are composed of MRs and are in charge of transmitting data based on circuit-switching, while electrical switches are in charge of closing the path by using package switching techniques. The main contribution of this solution is the fact that when the control unit is not able to close the optical path and after a set of trials, the message can be transmitted using the electrical layer directly. As for the other cases where circuit-switching is used, the tuning time that is necessary to close the optical path is the main issue.

An asynchronous and variable-length packet switching technique is presented in [35]. This technique uses a two layered approach to dynamically set the message path over the network. The main drawback of the presented technique is that the gain of using the optical path might be tainted depending on the delay time imposed by the controller. Also, the delay required for applying this technique imposes modifications on the network layer, which is often not desirable.

A multi-cast scheduling control solution is proposed in [36], focusing on input-queued switches based on an weight based arbiter (WBA). The proposed technique is based on TDM and aged-based weight calculations. The goal of this design is to achieve a time slot of roughly 51 ns for a data rate of 40 Gbps. The algorithm works by attributing weights to all the input cells on the beginning of every time slot. In cases of weight equality, the controller randomly chooses one of the conflicting inputs. Results show the latency of approximately 20 ns for a 64×64 network. The main issue with this approach lies in the fact that it is only applicable to the network considered in [36].

Finally, [37] presents a controlling solution for contention handling based on optical-buffering, by introducing a three stage buffering method. This method uses electronically controlled wavelength routing switches in combination with optical delay lines to temporarily store data [38, 39, 40]. The optical-buffer-based scheme was tested on an experimental setup composed only of a laser and a modulator, which injecting packets at a data rate of 2.5 Gbps. Although this technique is promising, as it directly deals with optical signals, it is still in an early stage of employment. Also, optical buffering is costly.

As discussed before, most of related works present network controllers that are designed/tailored for a specific topology or type of architecture, hence optimizing the performance of the architectures for which they have designed. However, the application of topology-constrained controllers is limited as their employment is hard-limited by the very one topology they are built for.

Also, a good number of works are based on circuit-switching techniques, thus adding a control latency that could shatter the benefits of using OINs, besides adding extra power consumption for each electrical node added to the control layer (i.e., electrical layer).

Compared with the aforementioned contributions, our work stands out thanks to the fact that our proposed control unit is suitable for a large variety of network topologies: the topologies based on tunable switches or the ones that only need arbitration. Moreover, as opposed to the controllers based on circuit-switching, the proposed controller unit latency does not affect the overall system performance. Still, for the devices where multiple frequencies are allowed, designers can slightly modify the controller so it comports with WDM routing. Table 3.2 presents the comparison between the controlling schemes presented previously and the LUCC. We consider the following points in the table:

- **Topology-Constrained:** this property specifies if the control unit is constrained by the controlled topology or if it has a generic behavior. Three different categories are used: (1) hard-yes, for the cases that are highly constrained to a specific topology; (2) yes, for the cases that are directly related to a specific topology, but are applicable to a few others, and; (3) no, for the cases where the controller can be used in any topology with none or small modifications;
- **Routing Algorithm:** it is the algorithm used by the controller in the network;
- **Strategy:** it specifies if the control is centralized, distributed or abstract, and;

Table 3.2 Comparing control solutions

Reference	Topology-Constrained	Routing Algorithm	Strategy	Latency
[27]	Hard Yes	Circuit-switching	Distributed	High
[29]	Yes	Circuit-switching	Distributed	High
[30]	Hard Yes	WDM	Distributed	Low
[32]	No	Circuit-switching & WDM	Distributed	High
[33]	Hard Yes	WDM	Centralized	Low
[34]	No	Circuit-switching	Distributed	High
[35]	Yes	WDM	Distributed	Low
[36]	Hard Yes	TDM	Centralized	Low
[37]	Hard Yes	WDM	Distributed	Low
LUCC	No	TDM[1]	Centralized	Low

[1]The implemented control is TDM-based, not using strict time slots.

- **Latency:** it is the time overhead due to the controller. For the cases where no precise value is provided, an estimation is made based on the presented data and routing techniques. This information is normalized and simply presented as LOW or HIGH in the table.

3.6 Conclusion

Optical interconnection networks are considered to be a solution for high bandwidth communications. However, the controller for such networks is still left unattended, not being fully improved. This chapter presents an overview of the most common techniques employed when controlling OINs. Furthermore, the design and validation of a centralized controller for OINs, LUCC, are presented. Results indicate a fast response time when employing LUCC in OINs: it takes only one clock cycle delay for every request to be computed. This type of controller is suitable for small to medium size topologies through using the developed methods of off-line computing and compression. Furthermore, while MZI-based switches are considered in this work, the proposed approach can be extended to any type of switches (MZI-based or microrings) and OIN topologies.

References

[1] T. Le and M. Khalid. NoC prototyping on FPGAs: A case study using an image processing benchmark. In *2009 IEEE International Conference on Electro/Information Technology*, pages 441–445, June 2009.

[2] C. Hilton and B. Nelson. PNoC: a flexible circuit-switched NoC for FPGA-based systems. *IEEE Proceedings in Computers and Digital Techniques*, 153(3):181–188, May 2006.

[3] S. Tota, M.R. Casu, M.R. Roch and M. Zamboni. A multiprocessor based packet-switch: performance analysis of the communication infrastructure. In *IEEE Workshop on Signal Processing Systems Design and Implementation, 2005*, pages 172–177, nov 2005.

[4] L. Benini and G. De Micheli. Powering networks on chips: energy-efficient and reliable interconnect design for SoCs. In *Proceedings of the 14th International Symposium on Systems Synthesis*, ISSS '01, pages 33–38, New York, NY, USA, 2001. ACM.

[5] ITRS. International technology roadmap for semiconductors, http://www.itrs.net/ - last access on 12/2016.

[6] D. A. B. Miller. Physical reasons for optical interconnection. In *International Journal of Optoelectronics, volume 11 - no. 3*, pages 155–168. International Journal of Optoelectronics, 1997.

[7] I. O'Connor and F. Gaffiot. On-chip optical interconnect for low-power. In *Ultra Low-Power Electronics and Design*, pages 21–39. 2004.

[8] A. F. Benner, M. Ignatowski, J. A. Kash, D. M. Kuchta and M. B. Ritter. Exploitation of optical interconnects in future server architectures. *IBM Journal of Research and Development*, July 2005.

[9] A. V. Rylyakov, C. L. Schow, B. G. Lee, W. M. J. Green, S. Assefa, F. E. Doany, M. Yang, J. Van Campenhout, C. V. Jahnes, J. A. Kash and Y. A. Vlasov. Silicon Photonic Switches Hybrid-Integrated With CMOS Drivers. *IEEE Journal of Solid-State Circuits*, 47(1):345–354, Jan 2012.

[10] B. G. Lee, A. V. Rylyakov, W. M. J. Green, S. Assefa, C. W. Baks, R. Rimolo-Donadio, D. M. Kuchta, M. H. Khater, T. Barwicz, C. Reinholm, E. Kiewra, S. M. Shank, C. L. Schow and Y. A. Vlasov. Monolithic Silicon Integration of Scaled Photonic Switch Fabrics, CMOS Logic, and Device Driver Circuits. *Journal of Lightwave Technology*, 32(4):743–751, Feb 2014.

[11] M. S. Hai, P. Liao, M. M. Shafiei, O. Liboiron-Ladouceur. MZI-based Non-blocking SOI Switches. In *Asia Communications and Photonics Conference 2014*, page ATh3A.147. Optical Society of America, 2014.

[12] A. Biberman, B. G. Lee, N. Sherwood-Droz, M. Lipson and K. Bergman. Broadband Operation of Nanophotonic Router for Silicon Photonic Networks-on-Chip. *IEEE Photonics Technology Letters*, 22(12):926–928, June 2010.

[13] Z. Li, M. Mohamed, X. Chen, H. Zhou, A. Mickelson, L. Shang and M. Vachharajani. Iris: A Hybrid Nanophotonic Network Design for High-performance and Low-power On-chip Communication. *J. Emerg. Technol. Comput. Syst.*, 7(2):8:1–8:22, July 2011.

[14] W.P. Boothroyd and E.M. Creamer. A time division multiplexing system. *Transactions of the American Institute of Electrical Engineers*, 68(1):92–97, July 1949.

[15] R. V. Rasmussen and M. A. Trick. Round robin scheduling - a survey. Technical report, European Journal of Operational Research, 2006.

[16] J. Carson. The emergence of stacked 3D silicon and its impact on microelectronics systems integration. In *Proceedings of the Eighth Annual IEEE International Conference on Innovative Systems in Silicon, 1996*, pages 1–8, Oct 1996.

[17] M.G. Smith and S. Emanuel. Methods of making thru-connections in semiconductor wafers, September 26 1967. US Patent 3,343,256.

[18] C. White. *Data communications and computer networks : a business user's approach.* Thomson Course Technology, Boston, Mass, 2007.

[19] S. Le Beux, I. O'Connor, G. Nicolescu, G. Bois and P. Paulin. Reduction Methods for Adapting Optical Network on Chip Topologies to 3D Architectures. *Microprocess. Microsyst.*, 37(1):87–98, February 2013.

[20] N. Jasika, N. Alispahic, A. Elma, K. Ilvana, L. Elma and N. Nosovic. Dijkstra's shortest path algorithm serial and parallel execution performance analysis. In *Proceedings of the 35th International Convention MIPRO*, May 2012.

[21] N. McKeown. The iSLIP scheduling algorithm for input-queued switches. *IEEE/ACM Transactions on Networking*, 7(2):188–201, Apr 1999.

[22] XILINX. www.xilinx.com, Last access: http://www.xilinx.com. December 2016, 2007.

[23] Altera Stratix IV - http://tinyurl.com/neqmmj3 - last access on 07/2016.

[24] Signaltap II Embedded Logic Analyzer - http://tinyurl.com/nhx69nj - last access on 07/2016.

[25] M. Nikdast, G. Nicolescu, J. Trajkovic, and O. Liboiron-Ladouceur. Chip-scale silicon photonic interconnects: A formal study on fabrication non-uniformity. *Journal of Lightwave Technology*, 34(16):3682–3695, August 2016.

[26] R. A. Spanke and V. E. Benes. N-stage planar optical permutation network. *Appl. Opt.*, 26(7):1226–1229, Apr 1987.

[27] H. Jia, Y. Zhao, L. Zhang, Q. Chen, J. Ding, X. Fu and L. Yang. Five-Port Optical Router Based on Microring Switches for Photonic Networks-on-Chip. *IEEE Photonics Technology Letters*, 25(5):492–495, March 2013.

[28] L. Kleinrock. *Information Flow in Large Communication Nets.* Rle quarterly progress report, Massachusetts Institute of Technology, April 1962.

[29] Z. Li and T. Li. ESPN: A case for energy-star photonic on-chip network. In *2013 IEEE International Symposium on Low Power Electronics and Design (ISLPED)*, pages 377–382, Sept 2013.

[30] H.A. Khouzani, S. Koohi and S. Hessabi. Fully contention-free optical NoC based on wavelenght routing. In *2012 16th CSI International Symposium on Computer Architecture and Digital Systems (CADS)*, pages 81–86, May 2012.

[31] M. Briere, B. Girodias, Y. Bouchebaba, G. Nicolescu, F. Mieyeville, F. Gaffiot and I. O'Connor. System Level Assessment of an Optical NoC in an MPSoC Platform. In *Design, Automation Test in Europe Conference Exhibition, 2007. DATE '07*, pages 1–6, April 2007.

[32] J. Chan and K. Bergman. Photonic interconnection network architectures using wavelength-selective spatial routing for chip-scale communications. *IEEE/OSA Journal of Optical Communications and Networking*, 4(3):189–201, March 2012.

[33] X. Tan, M. Yang, L. Zhang, Y. Jiang and J. Yang. A Generic Optical Router Design for Photonic Network-on-Chips. *Journal of Lightwave Technology*, 30(3):368–376, Feb 2012.

[34] J. Wang, B. Li, Q. Feng and W. Dou. A Highly Scalable Butterfly-Based Photonic Network-on-Chip. In *2012 IEEE 12th International Conference on Computer and Information Technology (CIT)*, pages 33–37, Oct 2012.

[35] H. Yang, V. Akella, C. N. Chuah and S. J. B. Yoo. Design of Novel Optical Router Controller and Arbiter Capable of Asynchronous, Variable length Packet Switching. In *International Conference on Photonics in Switching, 2006. PS '06.*, pages 1–3, Oct 2006.

[36] M. Shoaib. Selectively Weighted Multicast Scheduling Designs For Input-Queued Switches. In *2007 IEEE International Symposium on Signal Processing and Information Technology*, pages 92–97, Dec 2007.

[37] Y. Liu, M. T. Hill, H. de Waardt, G. D. Khoe and H. J. S. Dorren. All-optical buffering using laser neural networks. *IEEE Photonics Technology Letters*, 15(4):596–598, April 2003.

[38] D.K. Hunter, M.C. Chia and I. Andonovic. Buffering in optical packet switches. *Journal of Lightwave Technology*, 16(12):2081–2094, Dec 1998.

[39] M. Renaud, C. Janz, P. Gambini and C. Guillemot. Transparent optical packet switching: The European ACTS KEOPS project approach. In *IEEE Lasers and Electro-Optics Society 1999 12th Annual Meeting*, 1999.

[40] T. Sakamoto, K. Noguchi, R. Sato, A. Okada, Y. Sakai and M. Matsuoka. Variable optical delay circuit using wavelength converters. *Electronics Letters*, 37(7):454–455, Mar 2001.

4

Interconnects and Data System Throughput

Keyon Janani, Sakshi Singh, Moustafa Mohamed and Alan Mickelson

Electrical, Computer and Energy Engineering, University of Colorado at Boulder, Boulder 80309-0425 CO, USA

Abstract

The role of optics and optical devices in data centers is analyzed and discussed. Reasoning behind the present use of optics for interconnections of distances exceeding a few meters, of copper for micron to meter distances and doped semiconductor for the shortest distances is elucidated. The present day data center interconnection architecture is described. Discussion of latency, insertion loss, pulse distortion and dissipation of optical in comparison with electrical components indicates why the performance of optical components cannot be evaluated by present data center modeling techniques. The potential role of optics in future disaggregated architectures is the last topic of the chapter before discussion and conclusions.

4.1 Introduction

Optics has been the interconnection medium of choice for telephone networks since the early 1980's. Wire connections of the 1970's could not support repeater-less 2 km city trunk lines when the requirement for bandwidth reached forty-five Mbps. In 1975, it was demonstrated that multimode optical fiber could [30]. Single mode fiber in the long haul network now can support 8 terabits of bandwidth over 40 km employing 80 channel dense wavelength division multiplexing (DWDM) and four level phase modulation with polarization multiplexing at 25 Gbps (for 100 Gbps/channel). An undersea cable with eighty fibers can then support 640 Tbps of data. But multilevel coding schemes and DWDM are expensive. The problem we address here is one that addresses massive bandwidth requirements in a more cost conscious manner. Here we are concerned with increasing bandwidth but decreasing link lengths. The problem is to determine if and/or how optical waveguide technology could prevail for ever shorter data interconnections.

Optical fiber technology is a primary enabler of today's internet. The cloud is internet for hire, a potpourri of leasable resources of internet centric companies available to users who find maintaining large computers and data banks unprofitable. The unit of the cloud is the corporate cloud server, an array of long haul fiber interconnected data centers where the data center is a physical resource of interconnected racks stuffed with servers. The telecommunications problem has been to connect ever longer distances with ever larger bandwidths. The problem with data communications is to enable ever greater greater data transfer between ever more tightly spaced yet more powerful (bandwidth hungry) processing electronics.

Conductive electrical interconnections have a length bandwidth product that is significantly smaller than that of optical links. Optical links will always have significantly greater technology overhead, more complex transceivers, than electrical links. Processing power, though, increases relentlessly, even if at sub-Moore rates. By necessity, optical interconnection technology is working down the hierarchy of the data center from the (north-south) internet connection to the (east-west) interconnection of server racks and even into the rack. A recent yet all encompassing consideration affecting the march of optical data technology is that the cloud has come to be a significant (1.8% of total) electricity user in the United States (US) [55] and worldwide yet cloud computing continues to grow at an exponential rate [56]. Although interconnections draw little or no power, the interconnection architecture must be mindful of the power drawn by electronics.

The primary question to be addressed here is whether optics can improve the energy efficiency of a data center (throughput per watt-hour) and if so, how. In past generations of electronics, efficiency has always been achieved through miniaturization. In the case of the data center, unless new improvements in power efficiency can be found, further miniaturization will be stymied by heat dissipation requirements. It is not likely that optical innovations at the rack to rack interconnection level can have measurable effect on data center energy efficiency. Transceivers, the primary source of optical dissipation, dissipate two or three watts whereas racks dissipate in excess of 10 kWs, limited by the heat that can be removed from a rack by cost effective means. Optics can only enable power savings through enabling new architectures, that is, enabling new arrangements of resources that allow for cooler operation.

At present, optics is contributing to architecture by allowing higher and higher data rates to be communicated between servers. In fact, the bottleneck is now the electrical switches and how many fibers in they can connect with how many fibers out. Those electrical switches, though, contain memory and other electronics that allow powerful energy saving methods to be implemented including interconnection virtualization and software definition of the interconnection paths. All optical switching at the inter rack level may cause more problems than such a technology could solve. New architectures within the rack, though, may be enabled by ever smaller and finer optical interconnection technology. Large improvement to efficiency will be achieved in the disaggregation of intra-rack resources [29]. If optical technologists can address a number of bottleneck problems here, optics will contribute greatly in this next step of internet development.

The chapter is organized as follows. In the next section, general models for electrical and optical interconnections are considered. This section serves as a basis for calculations interspersed throughout the development. Discussion then turns to the characteristics of the optical transceivers that are necessary to initiate and terminate optical links. As the length scale of the interconnection shrinks, packing density and size become more stringent transceiver requirements. The discussion then turns to characteristics and needs of data centers before discussing modeling tools for power consumption and throughput of these centers. We will see that although the tools are myriad, the questions asked by the tools do not well address the questions we would have answered. The following section goes into the process of disaggregation within a data center rack. It is here that we see a promise for optical implementation if a number of issues can be addressed. A discussion of the chapter's results then proceeds a section of conclusions.

4.2 Interconnections

Although comprehensive reviews that evaluate optics as an interconnection technology appear every few years [26, 44, 45], there are salient features of the electrical and optical interconnections that are important to the present exposition. Electrical interconnection technology is quite resilient to replacement by optical technology in any application. The case for replacement must be irrefutable to be effective. Some basic considerations can make the case for or against replacement more understandable.

4.2.1 Electrical and Optical Interconnections

An information processing system will consist of discrete functional blocks and interconnections that connect those blocks. Regardless the size of the blocks, a point to point interconnection will appear as in Figure 4.1. Here, the input and output streams are assumed to be electrical. The network box could purely electrical, for example, consisting of a metal conductor connecting the inner conductor of the input port with the inner conductor of an electrical output port. The network could also consist of a carefully matched network that amplifies the input signal to drive a laser that in turn illuminates the core of an optical fiber whose output is focused into an optical detector whose electrical output is carefully matched to a transimpedance amplifier that is in turn matched to an output coax line that plugs into a data out line. The development of the present section is general and only assumes that some replica of the data in signal appears at the data out port.

In digital systems, the Data In stream $d_i(t)$ and Data Out stream $d_o(t)$ will generally consist of analog signals that transition between roughly fixed signal levels for periods of time that are roughly multiples of a bit period, T_b. The tops of these levels are distorted with noise, the transitions are not sharp and generally will overshoot and ring. The signal input to the second (output) block, $d_o(t)$, will be an attenuated and spectrally distorted (generally smoothed) version of the signal $d_i(t)$ input to the interconnect. The best way to quantify the attenuation and distortion is spectrally. The signals are information bearing and, thereby, most simply considered as time-stationary, stochastic signals. The spectral density, $S(\omega)$, of a signal $d(t)$ can then be defined as the Fourier transform $\mathcal{F}$ of the autocorrelation $r_{dd}(\tau)$

$$S(\omega) = \int_{-\infty}^{\infty} r_{dd}(\tau) \exp\left[-i\omega\tau\right] d\tau \qquad (4.1)$$

Figure 4.1   A schematic depiction of an interconnection between blocks. The output of the first block that functions as input to the interconnection is labeled as $d_i(t)$ whereas the output of the interconnect that functions as the input to the second block is labeled as $d_o(t)$.

where the autocorrelation is the result of taking the expectation, E of a stationary information bearing process $d(t + \tau)d(t)$

$$r_{dd}(\tau) = E\left[d(t + \tau)d(t)\right]. \tag{4.2}$$

Any degradation due to propagation can be represented by a frequency dependent transmission function $T(\omega)$ such that the spectral density exiting the link $S_0(\omega)$ is given by

$$S_o(\omega) = T(\omega)S_i(\omega). \tag{4.3}$$

In general, for the linear response we are assuming, we can further break up the transmission function into a spectral attenuation $\alpha(\omega)$ and dispersion $D(\omega)$ such that

$$T(\omega) = D(\omega)\exp\left[-2\alpha(\omega)L\right] \tag{4.4}$$

where the L is the length of the point to point interconnect. Although the integrals are written with infinite upper limits, in practice the upper limit of the frequency is roughly equal to the inverse of the information rate or

$$\omega_b = \frac{2\pi}{T_b} = \frac{2\pi c}{n\lambda_b} \tag{4.5}$$

where the subscript b refers to bit, c (non-italic, italic c will be capacitance per unit length) is the speed of light and n defines the velocity of propagation on the interconnection line v_p by the relation c$/n$. The λ_b is the length of bit T_b in the interconnection medium and ω_b the frequency extent (first zero of the Fourier transform) of a single bit. Information bearing systems generally possess only the minimum bandwidth necessary to pass the required information stream due to cost constraints.

4.2.1.1 Electrical interconnections

The simplest data interconnection can be depicted as in Figure 4.1. If the connection is electrical, the simplest connection between data source and receiver is a wire. Such a connection will only effectively transmit data streams if the information rate is low and the wire short and pleasantly plump. We will soon quantify low, short and pleasantly plump in terms of the phase change in traversing the link, the resistance per unit length and the skin depth. For higher (than low) data rates and longer (than short) links, one needs to use a transmission line. Evidently, a wire is a representation of a transmission line in the low-frequency, short-link limit. A higher frequency transmission line consists of (a) signal wire(s) and ground plane extending along a propagation direction with fixed geometrical relation between one and another. A ladder circuit representation of such a transmission line would appear as as in Figure 4.2. The r, l, c (note that non-italic c is the speed of light in vacuum) and g are the per unit length resistance, inductance, capacitance, and conductance of the transmission line. The resistance R, inductance L, capacitance C, and conductance G of a length of Δz of line would then be given by $R = r\Delta z$ with comparable definitions for L, C and G.

An infinite ladder is hard to use for prediction. This infinite series of impedances (each of infinitesimal length in order that the composite line has a length Λ) of this ladder, however, can be summed [25, 47]. The result is that the infinite ladder can be replaced by an impedance Z where the line impedance is given by

$$Z = \sqrt{\frac{r + j\omega l}{g + j\omega c}} \tag{4.6}$$

where ω is as before the angular frequency and here j the square root of -1. The infinite ladder network can also be seen to support a propagating wave with a propagation constant, γ, given by

$$\gamma = \sqrt{(r + j\omega l)(g + j\omega c)} \tag{4.7}$$

The propagation constant, γ, appears in the expression for propagation of a voltage or current wave traveling down a finite length of the line in the expression

$$v(z, t) = \Re\{v_0(x, y)\exp(j\omega t)\exp(\gamma z)\} \tag{4.8}$$

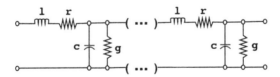

Figure 4.2   A transmission line model of electrical interconnects where the series resistance, R, series inductance L, shunt capacitance C and shunt conductance G are expressed as per unit length r, l, c and g respectively. The total length of the line, Λ, is taken to be the limit of $N\Delta z$ where $N \to \infty$ and $\Delta z = \Lambda/N$, Δz being the length of each r, l, c, g unit.

where we have assumed that z in the direction of propagation of an (x, y, z) coordinate system. The transverse dependence of the $v_0(x, y)$ is indicative of the location of signal and ground. The γ determines the loss and dispersion of the electrical line. The Z will determine the coupling (reflection and transmission) of the line when the line is attached to the data source and data receiver. The earlier power spectral density $S(\omega)$ is still defined as per Equation (4.1) where the $d(t)$ could be defined as $\sqrt{v(t)i(t)}$ in order to make the identification of $S(\omega)$ in terms of the measurable voltage and current.

The $D(\omega)$ and $\alpha(\omega)$ that determine the signal distortion are determined from the $\gamma(\omega)$ and $Z(\omega)$ that we have here denoted as functions of the angular frequency ω. γ is often expressed as

$$\gamma = \alpha + j\beta. \tag{4.9}$$

Here, it is clear that α is the amplitude attenuation. The solution of a propagation problem requires equating field components at boundaries. When the section of line is long enough that the wave sees a phase difference between input and output, impedance mismatch can lead to reflection. The dispersion function $D(\omega)$ then can be seen to be a result of any mismatch of the interconnection line with the data in and data out line. The $D(\omega)$ will also be a result of the frequency dependence of the impedance, $Z(\omega)$ and loss $\alpha(\omega)$.

We note here that when the conductance g and resistance r of the line are negligible, the line impedance, $Z(\omega)$, is real and equal to $\sqrt{l/c}$ and the propagation constant, $\gamma(\omega)$ is purely imaginary and equal to $j\sqrt{lc}$. The $Z(\omega)$ and $\gamma(\omega)$, may still vary with frequency but are lossless. In practice, the g is usually negligible but the r is always the limitation on the use of an electrical interconnect.

4.2.1.2 Optical interconnections

When we employ an optical interconnection between the data in and data out ports of our archetypal interconnection of Figure 4.1, the resulting block diagram appears as in Figure 4.3. This link is nowhere near as simple as the transmission line of Figure 4.2 which may physically consist of wire in fixed proximity to a ground plane. Even when the modulation and propagation characteristics of the optical line are much superior to the electrical, they must be so much better so as to definitively overcome the extra overhead of requiring source, driver, modulator, detector and amplifier. There is also an issue of size of the overhead. The components of the optical interconnection of Figure 4.3 that appear to the left of the channel are referred to as the

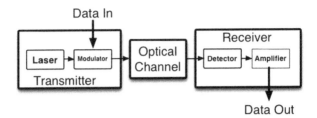

Figure 4.3 Block diagram representation of optical interconnects consisting of the transmitter (laser and optical modulator), channel (optical fiber) and receiver (photo-detector and trans-impedance amplifier).

transmitter and those to the right of the channel as a receiver. A box that contains both a transmitter and receiver (has both incoming and outgoing data ports as well as outgoing and incoming electrical ports) is referred to as an optical transceiver.

As with the electrical interconnection, the optical interconnection will both distort and attenuate a data signal that traverses the link from input to output. As with the electrical interconnection, the propagation can be represented as in Equation (4.4). A difference is that the ω in Equation (4.4) for the optical case needs to be considered as the deviation from the optical carrier ω_{opt}. For the electrical case, the information rate and the actual transmission rate are the same. This is often made explicit by referring to the optical frequency in terms of the wavelength $\lambda = c/f_{opt}$ where f_{opt} is the optical frequency and $\omega_{opt} = 2\pi f_{opt}$. A practical consequence of the optical frequency being on the order of 2×10^{14} Hz and the electrical (for a typical on-chip clock rate) being on the order of 4×10^9 Hz is that the attenuation of an optical link, $2\alpha(\omega)$, over the information bandwidth is constant. The optical dispersion in units of picoseconds (of packet spreading) per nanometer (of optical bandwidth) kilometer (of propagation distance) is over the optical bandwidth, which may be greater than the information bandwidth. Generally, though, the optical source bandwidth is so much less than the optical center frequency that the $D(\lambda)$ is also generally constant over the information over the source bandwidth.

4.3 Photonic Transceivers

An optical element containing the transmitter and receiver of Figure 4.3 is referred to as an optical transceiver. We see that the transceiver links the data in and data out ports of Figure 4.1 by electrical connections. The transceiver is linked to distant transceivers by an output optical waveguide and input optical waveguide. The data in port must be connected to a driver that in turn is connected to a modulator that impresses the data stream upon the optical stream. The input fiber port must be connected to a detector that in turn is connected to an amplifier (generally a trans-impedance amplifier as high speed optical detectors generate current without voltage) and the amplifier to the data out port. Evidently, the most important characteristics of the transceiver are its size, power throughput, bandwidth, and power consumption.

An optical transceiver can be small. The largest of electronic chips, for example, Intel Xeon processors that may contain tens of cores (up to 72) and billions of transistors (up to 4 billion) fit into LGA1366 (Socket B) packages until 2012 [62] that requires a chip dimension of 4.5 cm × 4.25 cm. Evidently, if such a chip can contain 72 cores, a single core chip with hundreds of millions of transistors can be smaller than 4.5 cm × 4.25 cm area divided by 72. Chips such as drivers and trans-impedance amplifiers that require ten of thousands of transistors can be made of areas of a mm^2 or less. Silicon photonic circuits require significant real estate per device but require many fewer devices. Transceivers presently in use for inter-rack interconnections satisfy the small form factor package (SFP) standard [63] that limits the outer dimension of the transceiver enclosure to be 5.6 cm × 1.34 cm × 0.85 cm or roughly 6 mm^3. Transceivers could be much smaller, though, especially if placed on boards instead of in standalone enclosures.

A complete optical transceiver that monolithically integrated all the CMOS drive and amplification electronics and silicon photonic optics (apart from optical sources) occupied 8 mm × 12 mm or 0.96 cm^2 was discussed in [16]. Techniques in which laser are bonded to the top of silicon photonic chips have now been demonstrated that allow the integration to expand in the third dimension without increasing the chip footprint [14]. In [16], however, the authors concluded that the minimum thickness of oxide (130 nm) necessary to produce tolerable waveguides was so thick as to have a severe effect on CMOS transistor performance. The authors concluded that such monolithic integration should be avoided. Luxtera, among others, had already been experimenting with bonding of electronic chips on photonic chips [27] before the work of [16]. Indeed, hetero-integration of electronics and SoI photonics has been demonstrated in various works [13, 59] and a recent roadmap [60] indicates that integration of source, SoI components and electronics should soon result in 1 cm^2 integrated transceivers that may even include processing power in addition to simply signal amplification and grooming.

The power consumption throughput of an optical transceiver is an important consideration. Most semiconductor lasers at the telecommunication windows of 1.3 μm and 1.55 μm radiate on the order a few mW out of a facet. Thermal noise dominates over shot noise in high speed receivers. The thermal noise floor, the value of the noise power in a detector load, is given by $4kT\Delta f$ where k is Boltzmann's constant, T the temperature, and Δf the inverse of the information rate. The minimum detectable optical power given this noise floor is

$$P_{min} = \frac{\sqrt{4kT\Delta f / R_L}}{\mathcal{R}} \tag{4.10}$$

For data rates of 50 Gbps, 50 Ω loads, and responsivities $\mathcal{R}$ on the order of 1 amp/watt, the minimum detectable signal is on the order of 2 μW. The bit error rate (BER) of a link is roughly proportional to 10^{-SNR} where SNR is the optical signal to noise ratio at the receiver. If the laser output is 2 mW, then a link must have loss less than roughly 20 dB in order to be viable. Once an SNR has dropped to some value it cannot be recovered. Optical amplification amplifies both signal and noise.

The first optical interconnections used for data communications were VCSEL transmitters radiating into multimode fibers. The first generation of such transmitters were rated at 1 Gbps but graduated to 10 Gbps and presently there are coarse wavelength division multiplexed (CWDM) versions that operate at 25 Gbps per channel over 100 m multimode fiber links. Indium phosphide (InP) distributed feedback lasers can be modulated at rates of 25 Gbps per channel. Higher speeds generally require external modulators. Silicon photonic (SiP) modulators that promise integration with multiplexers and demultiplexers are just coming into commercial use. SiP modulators have been demonstrated with 25 Gbps modulation rates with active lengths of as short as 100 microns [7] and switching powers of 200 fJ/bit [8]. 50 Gbps SiP modulators using simple on-off keying have also been demonstrated [22]. SiP slot modulators loaded with nonlinear polymers have been used in demonstrations of modulation rates as high as 100 GHz [6, 36]. Switching powers as low as 1.6 fJ/bit have been demonstrated [23]. Such power are sufficiently low that slot loaded SiP's have been directly modulated with FPGAs without need for amplification [24, 64].

The power consumed by a transceiver evidently varies with technology. A laser at a few mW output may draw as little as tens of mW. The power draw of a driver certainly depends on the power the modulator draws. Assuming one of the higher values of the last paragraph, 200 fJ/bit and a bitrate of 25 Gbps, the modulator draw is less than 100 mW. By including the drive overhead the draw may result in 200 mW. The trans-impedance amplifier after the detector will need to recover signal levels of 3 Volts looking into as low as a 50 Ω line may draw order of 200 mW. A common package these days is PSM4 that outputs 4×25 Gbps channels into 4 optical fibers. Four channels at roughly 500 mW per channel will lead to a draw of perhaps 2 W. Many of the multiple vendor service agreements require wavelength control of the sources. The thermo-electric coolers on these packages used to control wavelength can draw significant current. A completely packaged 100 Gbps four channel transceiver, though, still can be produced to dissipate no more than circa 3 watts.

4.4 Data Centers

4.4.1 Energy Use

Storage, retrieval and processing of data in the cloud (cloud use) is increasing. According to suppliers of data center equipment such as Cisco, cloud use will continue to increase. The global data center traffic (the number of bytes of data passing through all of the world's data centers) will increase at a rate of 25% per year [56] for the next five years. The power consumed by data centers has become an issue in recent years. In 2014, United States (US) data centers consumed roughly 70 GW-h of electricity, representing roughly 1.8% of US energy consumption [55]. This energy use is estimated to increase at a rate of roughly 4% per year [19, 55]. It should be noted, however, that the rate of increased energy use is less than the growth in traffic indicating an improvement in data center efficiency that has been the trend for some years. This improvement is, in part, due to interconnection virtualization and software definition of the interconnection. We will consider more on where within the center energy is dissipated in the next next paragraph.

4.4.2 Folded Clos (Leaf-Spine) Architecture

An individual data center is a collection of racks (generally of inner width of 19") vertically filled with trays that are integral numbers of rack units

(1 ru = 1.75") in height. The trays are fully populated with servers. The inter-connection of racks (the east-west traffic network of the center) is depicted in Figure 4.4. The north south connections are those that enter from above into the spines. The north traffic is that which enters and the south is that which exits (even if it exits through the top in the arrangement drawn as it could be drawn to exit through the bottom). The structure linking the racks is what is referred to as a Clos network [11, 18] in the archival literature or, more recently, as a leaf spine interconnection [4, 5]. Simple requests, for example look-up, generate north south traffic only. More complex requests, for example, requests to carry out some computations, will generally generate significant east west traffic that can exceed (or even greatly exceed depending on the amount of data generated and subsequently erased internally) the north south traffic. In today's data centers, east west traffic exceeds north south traffic.

An interesting question we can now address is where within the data center power is dissipated. Reference [19] presents an average distribution of dissipation for an archetypal warehouse scale data center. Cooling amounts to 50% of the power budget. This cooling is primarily the removal of heat from the racks whose operation expends roughly one quarter of the power

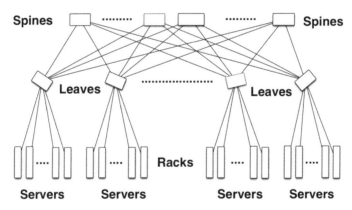

Figure 4.4   Depiction of an individual data center. Each of the racks that make up the data center contain N servers. These servers are connected through the top of the rack to a leaf switch. The leaves connect to spines. Requests from a network interface card of a server to another server traverse a leaf, a spine and then another leaf before finding the intended destination. The spines may be addressed from above by long lines from the web (cloud) or by shorter lines from a local connection center that collects the incoming traffic from the cloud and redistributes that traffic to the spines.

budget. A rack may contain 50 servers with 20 cores per CPU with each core dissipating circa 10 watts. This 10 kW per rack is considered to be about the maximum amount of heat that can be cost effectively removed from a standard rack volume (roughly 48 cubic feet = 2' × 4' × 6'). There is 3% listed for lighting, 11% for power conversion (power supplies) and 10% is listed as networking. These together amounts to roughly another one quarter of the total. A typical large scale data center may consist of 5,000 racks and, therefore, the center would dissipate roughly 200 MW total power.

The 10% that is attributed to networking refers primarily to connections to the internet (cloud). As we have seen previously, a transceiver dissipates circa 3 watts. The electrical switches of the Clos network have at most now 48 ports with 96 port versions to come. Depending on the details of the leaf-spine interconnect, there then may be as many as circa 100 transceivers per rack. 100 transceivers corresponds to 300 W, an unmeasurable quantity in comparison with the 10 kW per rack.

4.4.3 Hierarchy of Interconnection

Evidently, the data center is characterized by a hierarchy of interconnections. From the top of the figure down, we see connections from electrical switches that make up the islands of the spine layer, down to the switches of the leaf layer and on down to the top of rack switches. The racks contain servers that consist of chips and peripherals whose outputs are connected by a bus from which data is switched in and out as well as connected with metal traces and possibly InfiniBand type CPU to memory interconnection fabric. Traces and fabric carry the larger data transfers. The CPU chips consist of multiple cores where the cores consist of blocks that in turn consist of gates. The intra-chip interconnection consists of conductive traces connecting switches that may be gates or multiple gates serving as outputs of the functional blocks or of a core.

The hardwire connections (when they exist) between data centers of a company cloud are generally assumed to be from 20–80 km in length. Other long lines are internet and of traffic varying length. In a warehouse scale data center containing 10,000 racks, connections could be up to 2 km long, although the majority of all of the interconnections are more likely on the order of 300 m. The shortest of fiber connections are all less than 100 m. More discussion of chips is given a paragraph hence and discussion of the rack scale in the section of disaggregation.

4.4.4 Electrical Switching

All switching in all commercial information transmission system is done electrically. Much effort has gone into developing optical switches, especially for the telephone network. The problem that has proven unbeatable to the present is the 3R problem. The great success of the integrated circuit has been that at each individual transistor stage, the signal can be regenerated, reshaped and retimed, the three R's. Optical signal regeneration can be accomplished with optical amplifiers. The development of the erbium doped optical amplifier was an impetus for a massive effort into optical switching. Simultaneous reshaping and retiming is an elusive goal. Hence, the optics in the above rack system is used solely to link together optical transceivers that are plugged into electrical switches.

4.4.5 Electrical versus Optical

As we have mentioned previously, in 1975, fiber won out over electrical interconnection when the 2 km trunk lines needed to upgrade to rates of 45 Mbps. Multimode fiber in 1975 already exhibited bandwidth of 100 MH-km. Although electrical coax and twisted pair have improved since that time, the improvement in fiber (and fiber components) has been meteoric. Multimode optical fiber now can exhibit bandwidths on the order of 2 GHz-km. These are the fibers that are used in the 100 m 25 Gbps links from rack top to the lower level of the folded Clos (leaf). Single mode fiber exhibit bandwidths three or more orders of magnitude higher. Optical interconnections are the only ones above the rack top. Connections less than 100 m are multimode and others use single mode fiber. The standard SMF28 single mode fiber in use may have bandwidth on the order of 10 ps/(nm-km) near the dispersion minimum around 1.31 μm operating (carrier) wavelength. Optical fiber losses are 3 dB/km near 0.8 μm wavelength, 0.5 dB/km near 1.3 μm wavelength, and 0.2 dB/km near 1.55 μm wavelength.

At the lowest level of the hierarchy, the intra-chip, electrical interconnection is the medium of choice. Wherever optical interconnections are the medium of choice, those connections consist of optical waveguides connecting optical transceivers. Clearly, use of the optical transceivers we have previously discussed on-chip is not an option as each transceiver consists of at least one chip, but more generally, multiple chips. An optical solution at the intra-chip level would need be disruptive by nature. The exponential increase in transistor density that has embodied Moore's law has resulted in the present day multi-core chip architecture. Presently, CPU chips of 1 or

2 cm on a side are partitioned into cores that are hundreds of microns on a side and shrinking. The core is the largest unit of transistors that can be heat sinked. The transistor free regions between the cores allow for dissipation as well as for interconnection lines between the cores. The clock speed on a chip has also saturated at rates of circa 4 GHz, an 8 Gbps switching rate. Higher clock speeds cannot be heat sinked at the transistor densities that have been standardized. The minimum length then of a bit (Equation (4.5)) is circa 2 cm assuming the line is in silicon (index of refraction, $n = 3.4$). The length is longer in silicon dioxide (index of refraction, $n = 1.45$).

Gate to gate (local) interconnects are of sub-micron length whereas block to block lengths may be hundreds of microns. We note that impedance matching (Figure 4.2 and Equation (4.6)) of these connections are not necessary as they are much shorter than the bit length. These short interconnects than can consist of simple wires. The gate to gate consist often of polysilicon wires and the block to block (semi-global) of TiN or doped metal. The loss (Equation (4.7) and Equation (4.9)) and delay of these interconnections must be controlled.

As is evident from from Equation (4.6) and Equation (4.7), when g is negligible, r determines the loss and delay. The resistance per unit length of a conductor is given

$$r = \rho/A \tag{4.11}$$

where ρ is the resistivity of the material and A is an effective area. For low frequency, A is the physical area. For high frequencies, A is the square of the skin depth, δ, where

$$\delta = \sqrt{\frac{2\rho}{\omega\mu}} \sqrt{\sqrt{1 + (\rho\omega\epsilon)^2} + \rho\omega\epsilon} \approx \sqrt{\frac{2\rho}{\omega\mu}} \tag{4.12}$$

where μ is the material permeability, ϵ the material dielectric constant that is also equal to $n^2\epsilon_0$ where n is the index of refraction and ϵ_0 the permittivity of free space, and where the approximation is good when $\omega \ll 1/\rho\epsilon$, which is true for any good conductor for frequencies through the microwave. The value of the skin depth for copper at 4 GHz is close to 1 micron. For on-chip interconnections this is not a restriction as a one micron thick line is about the most practical on-chip.

As previously stated, copper is a usual interconnection metal at least for the longest (global) on-chip interconnections. The resistivity of copper is $\rho = 1.7 \times 10^{-8}\Omega$-m. Silver is slightly lower, gold slightly higher and aluminum higher still ($\rho = 2.7 \times 10^{-8}\Omega$-m), but copper is easier to use. Polysilicon and TiN that are used in the fabrication process have resistivities 3 orders of magnitude lower than copper. But even for areas of 0.01 μm^2 (smaller than can be fabricated at present) on poly silicon, the resistance over 1 mm is 1 Ω. The main point here is that for the local and semi-global interconnects, one can choose the physical area to match the material in order to keep loss sufficiently low.

For global interconnects, there is space for ground planes and much wider lines that can be of the lower resistivity copper. When r is negligible, for interconnections with a velocity of propagation equal to that of light in silicon dioxide, the ratio between l and c can be fixed from Equation (4.6) and Equation (4.7). For a 50 Ω, we find that the inductance per unit length, l, is roughly 2.5 nH/cm and the capacitance per unit length, c, is roughly 1 pF/cm. These values are not hard to obtain with the present design constraints. Keeping the lines of a sufficiently high cross-sectional area then results in matched and low enough loss lines to traverse the longest of the inter-core distances.

There are enough available degrees of freedom that on-chip interconnections are satisfactory for the foreseeable future. More will be said on intra-rack connections in the section on disaggregation.

4.5 Modeling of Optically Connected Data Centers

As we have seen, present day data centers are optically interconnected from rack top up. That is, the electrical switches that control the flow of messages between servers are interconnected with optical fibers. Those fibers emanate from and terminate on optical transceivers that in turn plug into the electrical switches. The transceivers consume a minimum of power on the scale of data center consumption, on the order of three watts per transceiver or six watts per the pair required for the fiber link. Although there may be many optical links (fiber plus transceivers) in a warehouse scale data center, at present the only real function of fiber links is to interconnect electrical subsystems just as would wire for shorter length interconnections. There is promise that optics will play a greater role in data center operation as is evidenced by the number of papers that discuss new architectures involving optics [3, 10, 12, 16, 17, 21, 35, 51, 50, 66]. These architectures involve all optical operations as well as improved transceivers.

In order to determine promising directions, modeling is certainly preferable to development and testing of all options. For adoption, a disruptive approach will need to show a definitive improvement in data center operation. A number of recent works [20, 48, 49] have focused on predicting the optical throughput (loss budget and crosstalk) of the optical subsystems. There are works [37, 38, 39, 46] that have used computational benchmarks to predict multi-core chip performance for optical on-chip architectures. What is really necessary in order to evaluate new directions in data center architecture, though, is a tool that can predict the data center throughput and energy efficiency in terms of the characteristics of the optical components.

4.5.1 Data Center Modeling

A multi-core chip is much simpler to model [37, 38, 39, 46] than a data center that may contain more than a million multi-core chips arranged into racks interconnected into a Clos network. The power usage issue [55], however, has stimulated much interest. Recent reviews on the topic point to causes of inefficiencies [1], methodologies that might lead to better control [42] and to a myriad of modeling techniques that can be used to aid in decision making processes [19]. The lack of energy proportionality in data centers, that is, the dependence of the throughput on the nature of the request rather than the rate of requests [1] complicates prediction and necessitates computational aids to decision making.

Despite complexity, data center simulators abound as is evident from a recent review [54]. These simulators range from the very computer intensive to statistical simulators based on queuing [40] such as Big House [41, 43]. Big House uses analytical solutions to simple queuing problems [28, 57] in order to build up solutions to general queuing problems with general distributions of input and output requests distributed among k processors. In the notation of Kendall [28, 32], Big House uses solutions to (M1/M2/1) problems to affect solution to (G1/G2/k) problems. The speed-up afforded by Big House allows full data centers with thousands of racks to be simulated in hours whereas the more intensive simulators must be applied to sample systems in order to reduce runtimes even to days.

Despite the large number of simulators and the differences in implementation, [54] discusses a common design framework. A salient aspect of the common framework is that that components (component models) are placed into blocks of network or application models. Generally, the data center designer is interested in throughput as a function of parameters that are easily controlled such as scheduling and load balancing, that is, programmable attributes. Seldom can the simulator simulate anywhere near the number of servers that are contained in an actual data center due to time constraints but the designer is interested in seeing how the system scales, how operation can be extended and how operation can be customized. Although the models used for the individual components that go into the individual blocks should affect the overall system throughput, simulators are not focused on this level. The problem with evaluating optical transceivers and their effect on throughput is masked by more practical considerations as can be described with the help of Figure 4.5. Optical transceivers appear only at the inputs and outputs of switches. The characteristics of the transceivers then must conform to the switch characteristics in terms of data rate. Optical transceivers are specified to operate at some bandwidth and some power draw. Assuming the transceivers meet this minimal specifications, other characteristics are masked. that is, the latency of the switch is much greater than that of the transceiver. The built-in signal reconstruction of the switch counteracts loss and signal distortion that a transceiver may induce. As will be discussed further below, the transceiver is nearly invisible to any form of data center simulator, at least any form of the simulators reviewed in [54] which is a very general review.

4.5.2 Optics in Data Center Simulation

It should be pointed out that scheduling and load balancing, in part, exist within the electrical switches that act as the nodes of the leaf-spine (Clos

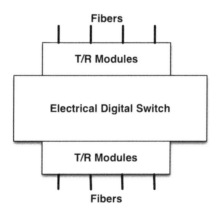

Figure 4.5 A schematic depiction of the location of the optical transceivers (labeled R/T's) inside of the data center. The most salient feature of the R/T's are that they are all located at the input and outputs of switches.

network) switching architecture. The data center manager does not have runtime control over the virtualization and software definition that also reside in the electrical switches. However, those more or less automatic functions are driving the 25% year growth in traffic with only a 4% increase in power, that is, a Moore's law-like increase in data efficiency. Many simulators such as Big House use data to generate load models and then use the simulator to calculate differences between new scheduling and load balancing schemes different from those used in the baseline. Different virtualization and software definition would require new benchmarks as the virtualization and software definition are built into the load models. Any use of data center simulators to model new designs of optical switches would require attention to degree of virtualization, software definition and programmable scheduling and load balancing allowed by the optical switching approach. These are details that are not usually attended to when considering optical switching architectures.

One can model different interconnection network configurations using actual optical transceiver characteristics to compare data center designs. Numerous works over the last decade have extolled the virtue of high radix switching [2, 33, 34] and why optics may offer high radix [3, 12]. High radix means that there are more ports in each switch. If each switch has more ports, there will be more non-blocking routes (paths) between destinations. More non-blocking routes between the endpoints reduce queues, that is, there are more paths over which to distribute the same queued traffic. When the center is processor bound, that is, when the processing of a command takes longer than the average time between reception of commands, there will be queues. High radix will then, in general, shorten the average wait time by equalizing the lengths of the difference queues over a larger number of queues that share the same total traffic. In the present data center interconnection, the optical transceivers are all attached to electrical switches. The radix of the interconnection is determined only by the radix of the electrical switch. If the radix of each optical transceiver is increased, the electrical switch still switches the same number of inputs and outputs. The others wait. No matter what the radix of the optical switches, only so many ports can be accommodated by the electrical switch.

The electrical switches also determine the latency of switching. The latency of an electrical switch is generally greater than 100 ns. An optical transceiver has no buffer. The optical signal simply propagates through the transceiver with perhaps a small time constant for the detector and/or transimpedance amplifier rise time. The transceiver simply acts as an extension of the fiber. The optical fiber determines the overall latency of the link. If we assume that fiber propagation induces a latency of roughly 5 nanoseconds per meter of propagation and interconnections are generally longer than 200 m (two of the 100 m multimode fibers), then the fiber latency is 120 ns minimum. Again, the optical transceiver, if functioning correctly, is essentially invisible when it comes to salient interconnection characteristics. The reason for this is at least partially historical. The architecture defines the role of the optics.

Although data center modeling tools exist, applying such tools to compare novel optical data center architectures is an arduous task that does not seem to have been profitably applied to the present. In part, the reason for this is that the effect of the optics in the interconnection seems to have a minimal, at best, effect on data center throughput when one considers that interconnection optics exists only in the interconnection network above the rack top.

4.6 Disaggregation

As we have seen, from the rack top up, the interconnection within a warehouse scale data center consists of optical fibers connecting to pluggable transceivers. The data centers themselves are connected together with long haul optical links, be they private long haul or part of the world wide web. The chips that reside in the servers are enabled by electrical interconnections. Now that on-chip clock rates are fixed, the smaller feature sizes of the coming transistor generations will not pose insurmountable scaling. But the other intra-rack connections, the board level ones that connect chips and chips with peripherals and all with the network interface card, are yet to be considered. Here the data rates will continue increase for the foreseeable future. We have only said so far that the intra-rack connections are electrical. More is to be said here about the interconnection as it is and to what it is migrating in the next decade.

A rack is generally a structure that with nineteen inch opening (total external width of 23.62 inches) across at the front, a depth that is often double this, and a height that consists of some number (forty-eight is common) of rack units. A rack unit is 1.75 inches and is the minimum depth of a tray that could be slipped into the rack. In general, the trays that slip into the rack are some number of rack units. For example, completely filling a forty-eight rack unit deep rack with three server across, three rack unit deep trays will result in a server with forty-eight servers. This is a typical number.

Within the rack, the servers are connected with coax/twisted pair (CAT) cable from their output ethernet card (network interface card (NIC)) to a rack top switch. The switch can interconnect servers in the rack or connect the server with a distant server addressed by the header that was inserted on the data stream by the ethernet card (network interface). The addresses in the data headers are generally virtual addresses. One of the schemes for address virtualization employs look-up tables. At each switch that an output signal traverses, there is a table that prioritizes the virtual address in terms of actual locations, sending the signal to the location that is closest, the most likely to be open (possess the shortest queue). In a software defined network, the tables are dynamically updated.

Within the server, communications are also electrical. The chips of the servers lie on a board and are connected to board interconnects through output pins. On board connections consist of metal traces, infiniband connections to memory as well a PCI express interface bus that connects all of the processing, memory chips and peripherals with the ethernet network interface card. On board traces can be used for high speed data transfer but suffer all of the problems of electrical interconnections. Distances on the boards range from centimeters to tens of centimeters at data rates that require impedance matching. As data rates increase, loss and dispersive delay increase correspondingly. Fix-ups can be found [31] but for ever higher rates will be more costly. Numerous proposals are on the table for optical backplanes to alleviate such problems. [9, 15, 53, 58, 61]. But for the moment, the problem (opportunity) we wish to discuss relates to a more architectural theme.

As we have stated above, the units of computational power in a present day data center is the server that comes complete with a network interface card. Connection of servers with each other in the center is generally an ethernet connection although connection with memory and between memories may often be handled by infiiniBand. As pointed out in [29], under typical data center use, the distribution of relative memory to CPU usage varies by more than 3 orders of magnitude. The implication is that separating the memory, CPU and other server functions from each other, using an optical scheme to allocate those resources could covert those 3 orders of magnitude of mismatch into improvement in energy efficiency. All of those resources are activated in the server but go unused despite their power dissipation. Optimal allocation then would have no power cost and further, would lead to a direct size reduction.

Some problems with disaggregation at present include the removal of the network interface card and that there is no software defined control over the switches within the rack, that is, the board level interconnections. Various authors are proposing techniques to achieve disaggregation within optical architectures [52, 65]. Activity in this area promises to greatly increase in coming years.

4.7 Discussion

Electrical interconnections display inherent limitations primarily due to material properties. Now that on-chip clock rates are fixed, the shortness of on-chip interconnections allow designers to control loss, dispersion and impedance by controlling conductor thickness and shape. At the board level, the data rates will continue to increase. From the rack up, fiber is the only viable choice for interconnection medium. Fiber requires no impedance matching and exhibits low loss and dispersion. Optical transceivers for rack to rack interconnects are compact, low loss and meet the modulation requirements. As electrical interconnections inside the rack become more stressed by increased data rates, the rate of conversion to optics in the rack will be determined more than all else by the progression of optical transceiver technology. The transceivers will become ever more compact, low power and higher bandwidth as optical replaces the electrical interconnections in the rack.

Present day data center modeling software is quite insensitive the characteristics of the optical transceivers employed in the interconnect fabric. Optics simply does not have a significant effect on power usage either directly (the optical transceivers use negligible power compared to the colossal number of transistors in a rack) or through enabling system level power control methodology. This may change when optics enters the rack.

4.8 Conclusion

Data center performance continues to improve. Power consumption increases but at slower rate than throughput, indicating that data center efficiency is improving. This improvement is the result of a number of effects. A pervasive cause is software definition of virtualization of the rack to rack interconnections. The optical transceivers and fibers that connect the racks have little effect on these ongoing improvements. To the present, optical links in the data center function only as a replacement for wire.

The next research frontier in the data center is the disaggregation of the rack. Already the intra-rack electrical interconnections are being stressed by increasing data rates. Optics could well serve a major role in disaggregation. It is early to predict what form this role will take.

References

[1] Dennis Abts, Michael R. Marty, Philip M. Wells, Peter Klausler, and Hong Liu. Energy proportional datacenter networks. *ACM SIGARCH Computer Architecture News – ISCA'10*, 38(3):338–347, June 2010.

[2] N. R. Adiga, M. A. Blumrich, D. Chen, P. Coteus, A. Gara, M. E. Giampapa, P. Heidelberger, S. Singh, B. D. Steinmacher-Burow, T. Takken, M. Tsao, and P. Vranas. Blue gene/l torus interconnection network. *IBM Journal of Research and Development*, 49(2.3):265–276, March 2005.

[3] Dan Alistarh, Hitesh Ballani, Paolo Costa, Adam Funnell, Joshua Benjamin, Philip Watts, and Benn Thomsen. A high-radix, low-latency optical switch for data centers. *ACM SIGCOMM Computer Communication Review – SIGCOMM'15*, 45(4):367–368, October 2015.

[4] Mohammad Alizadeh and Tom Edsall. On the data path performance of leaf-spine datacenter fabrics. In *Proceedings of the 21st Annual Symposium on High-Performance Interconnects*, pages 71–74, 2013.

[5] Mohammmad Alizadeh et al. Conga: Distributed congestion-aware load balancing for datacenters. In *Proceedings of SIGCOMM'14*, pages 503–514, August 17–22, 2014

[6] Luca Alloatti, Robert Palmer, Sebastien Diebold, Kai Philipp Pahl, Baoquan Chen, Raluca Dinu, Maryse Fournier, Jean-Marc Fideli, Thomas Zwick, Wolfgang Freude, Chritian Koos, and Juerg Leuthold. 100 GHz silicon-organic hybrid modulator. *Light Science and Applications*, 3, May 2014.

[7] T. Baba et al. 25-gbps operation of silicon p-i-n mach-zhender optical modulator with 100-micro-long phase shifter. In *Proceedings Conference on Laser and Electro-Optics (CLEO)*, 2012.

[8] Tom Baehr-Jones et al. Ultralow drive voltage silicon traveling-wave modulator. *Optics Express*, 20,11:12014–12020, 2012.

[9] N. Bamiedakis, A. Hashim, R. V. Penty, and I. H. White. A 40 gb/s optical bus for optical backplane interconnections. *Journal of Lightwave Technology*, 32(8):1526–1537, April 2014.

[10] R. G. Beausoleil, M. McLaren, and N. P. Jouppi. Photonic architectures for high-performance data centers. *IEEE Journal of Selected Topics in Quantum Electronics*, 19(2):3700109–3700109, March 2013.

[11] Vaclav Benes. *Mathematical Theory of Connecting Networks and Telephone Traffic*. Academic Press, 1965.

[12] N. Binkert, A. Davis, N. P. Jouppi, M. McLaren, N. Muralimanohar, R. Schreiber, and J. H. Ahn. The role of optics in future high radix switch design. In *Proceedings 38th Annual International Symposium on Computer Architecture (ISCA)*, pages 437–447, 2011.

[13] F. Boeuf et al. A multi-wavelength 3d-compatible silicon photonics platform on 300mm soi wafers for 25 gb/s applications. In *Transactions of the IEEE International Electron Devices Meetings – IEDM*, pages 1–13, 2013.

[14] John E. Bowers. Heterogenous iii-v/si photonic integration. In *Proceedings of the Conference of Lasers and Electro-optics-CLEO Science and Innovations 2016*, 2016.

[15] L. Brusberg, D. Manessis, C. Herbst, M. Neitz, B. Schild, M. Toepper, H. Schroeder, and T. Tekin. Single-mode board-level interconnects for silicon photonics. In *Proceedings of the Optical Fiber Communications Conference and Exhibition (OFC)*, March 2015.

[16] J. F. Buckwalter, X. Zheng, G. Li, K. Raj, and A. V. Krishnamoorthy. A monolithic 25-gb/s transceiver with photonic ring modulators and ge detectors in a 130-nm cmos soi process. *IEEE Journal of Solid-State Circuits*, 47(6):1309–1322, June 2012.

[17] Nicola Calabretta, Wang Miao, and Harm Dorren. High-performance flat data center network architecture based on scalable and flow-controlled optical switching system. *Proc. SPIE*, 9753:97530W–97530W–10, 2016.

[18] Charles Clos. A study of non-blocking switching networks. *Bell System Technical Journal*, 32(2):406–432, 1952.

[19] M. Dayarathna, Y. Wen, and R. Fan. Data center energy consumption modeling: A survey. *IEEE Communications Surveys Tutorials*, 18(1):732–794, 2016.

[20] Luan H. K. Duong, Zhehui Wang, Mahdi Nidkast, and Jiang Xu. Coherent and incoherent crosstalk noise analyses in interchip/intrachip optical interconnection networks. *IEEE transactions on Very Large Intergation*, 24,7:2475–2487, 2016.

[21] C. Kachris et al., editor. *Optical Interconnects for Future Data Center Networks*. Springer-Verlag, 2013.

[22] David J. Thomson et al. 50-gb/s silicon optical modulator. *IEEE Photonics Technology Letters*, 24,4:234–236, 2012.

[23] R. Palmer et al. High-speed silicon-organic hybrid (soh) modulator with 1.6 fj/bit and 180 pm/v in-device nonlinearity. In *Proceedings of the*

2013 European Conference on Optical Communications, page We.B.3, 2013.

[24] Stefan Wolf et al. 10 gbd soh modulator directly driven by an fpga without electrical amplification. In *Proceedings of the 2014 European Conference on Optical Communication (ECOC)*, page Mo.4.5.4, 2014.

[25] Richard Feynman, Robert Leighton, and Mathew Sands. *Feynman Lectures on Physics*. California Institute of Technology, 1963.

[26] Joseph W. Goodman, Frederick I. Leonberger, Sun-Yuan Kung, and Ravindra A. Athale. Optical interconnections for vlsi systems. *Proceedings of the IEEE*, 72(7), July 1984.

[27] Cary Gunn. Cmos photonics for high-speed interconnects. *IEEE Micro*, pages 58–66, March-April 2006.

[28] Varun Gupta, Mor Harchol-Balter, J. G. Dai, and Bert Zwart. On the inapproximability of m/g/k: why two moments of job size distribution are not enough. *Queueing Systems*, 64(1):5–48, 2010.

[29] Sangjin Han, Norbert Egi, Aurojit Panda, Sylvia Ratnasamy, Guangyu Shi, and Scott Shenker. Network support for resource disaggregation in next-generation datacenters. In *Proceedings of the Twelfth ACM Workshop on Hot Topics in Networks, HotNets'13*, pages 10:1–10:7, 2013.

[30] Jeff Hecht. *City of Light: The Story of Fiber Optics*. Oxford University Press, 1999.

[31] R. Ho, T. Ono, F. Liu, R. Hopkins, A. Chow, J. Schauer, and R. Drost. High-speed and low-energy capacitively-driven on-chip wires. In *2007 IEEE International Solid-State Circuits Conference. Digest of Technical Papers*, pages 412–612, Feb 2007.

[32] David G. Kendall. Stochastic processes occurring in the theory of queues and their analysis by the method of the imbedded Markov chain. *The Annals of Mathematical Statistics*, 24(3:338–354, 1953.

[33] J. Kim, J. Balfour, and W. Dally. Flattened butterfly topology for on-chip networks. In *40th Annual IEEE/ACM International Symposium on Microarchitecture*, pages 172–182, Dec 2007.

[34] John Kim, William J. Dally, and Dennis Abts. Flattened butterfly: A cost-efficient topology for high-radix networks. *ACM SIGARCH Computer Architecture News*, 35(2):126–137, 2007.

[35] P. Koka, M. O. McCracken, H. Schwetman, C. H. O. Chen, X. Zheng, R. Ho, K. Raj, and A. V. Krishnamoorthy. A micro-architectural analysis of switched photonic multi-chip interconnects. In *Computer*

Architecture (ISCA), 2012 39th Annual International Symposium on, pages 153–164, June 2012.

[36] J. Leuthold, C. Koos, W. Freude, L. Alloatti, R. Palmer, D. Korn, J. Pfeifle, M. Lauermann, R. Dinu, S. Wehrli, M. Jazbinsek, P. Gnter, M. Waldow, T. Wahlbrink, J. Bolten, H. Kurz, M. Fournier, J. M. Fedeli, H. Yu, and W. Bogaerts. Silicon-organic hybrid electro-optical devices. *IEEE Journal of Selected Topics in Quantum Electronics*, 19(6):114–126, Nov 2013.

[37] Z. Li, M. Mohamed, H. Zhou, L. Shang, A. Mickelson, D. Filipovic, M. Vachharajani, X. Chen, W. Park, and Y. Sun. Global on-chip coordination at light speed. *IEEE Design Test of Computers*, 27(4):54–67, July 2010.

[38] Zheng Li, Moustafa Mohamed, Xi Chen, Eric Dudley, Ke Meng, Li Shang, Alan R. Mickelson, Russ Joseph, Manish Vachharajani, Brian Schwartz, and Yihe Sun. Reliability modeling and management of nanophotonic on-chip networks. *IEEE transactions on Very Large Scale Integration (VLSI) Systems*, 20,1:98–111, 2012.

[39] Zheng Li, Moustafa Mohamed, Xi Chen, Hongyu Zhou, Alan Mickelson, Li Shang, and Manish Vachharajani. Iris: A hybrid nanophotonic network design for high-performance and low-power on-chip communication. *J. Emerg. Technol. Comput. Syst.*, 7(2):8:1–8:22, July 2011.

[40] D. Liao, K. Li, G. Sun, V. Anand, Y. Gong, and Z. Tan. Energy and performance management in large data centers: A queuing theory perspective. In *Proceedings of the International Conference on Computing, Networking and Communications (ICNC)*, pages 287–291, 2015.

[41] D. Meisner, J. Wu, and T. F. Wenisch. Bighouse: A simulation infrastructure for data center systems. In *Proceedings of the 2012 IEEE International Symposium on Performance Analysis of Systems and Software (ISPASS'12)*, pages 35–45, 2012.

[42] David Meisner, Christopher M. Sadler, Luiz Andre Barroso, Wolf-Dietrich Weber, and Thomas F. Wenisch. Power management of online data-intensive services. In *Proceedings of the 38th ACM International Symposium on Computer Architecture – ISCA'11*, page 12, 2011.

[43] David Meisner and Thomas F Wenisch. Stochastic queuing simulation for data center workloads. In *Exascale Evaluation and Research Techniques Workshop*, page 9, 2010.

[44] D. A. B. Miller. Optical interconnects to silicon. *IEEE Journal of Selected Topics in Quantum Electronics*, 6(6):1312–1317, Nov 2000.

[45] David A. B. Miller. Rationale and challenges for optical interconnects to electronic chips. *Proceedings of the IEEE*, 88, 6, June, 2000.

[46] Moustafa Mohamed, Zheng Li, Xi Chen, Li Shang, and Alan R Mickelson. Reliability aware design flow for silicon photonics on-chip interconnect. *IEEE Transactions on Very Large Scale Integration (VLSI) Systems*, 22,8:1763–1776, August, 2014.

[47] Jeremy T. Moody. Efficient methods for calculating equivalent resistance between nodes of a highly symmetric resistor network. Worcester Polytechnic Institute, 2013.

[48] M. Nikdast, G. Nicolescu, J. Trajkovic, and O. Liboiron-Ladouceur. Chip-scale silicon photonic interconnects: A formal study on fabrication non-uniformity. *Journal of Lightwave Technology*, 34(16):3682–3695, Aug 2016.

[49] M. Nikdast, J. Xu, L. H. K. Duong, X. Wu, X. Wang, Z. Wang, Z. Wang, P. Yang, Y. Ye, and Q. Hao. Crosstalk noise in wdm-based optical networks-on-chip: A formal study and comparison. *IEEE Transactions on Very Large Scale Integration (VLSI) Systems*, 23(11):2552–2565, Nov 2015.

[50] Lorenzo Pavesi and David J. Lockwood, editors. *Silicon Photonics III: Systems and Applications*. Springer-Verlag, 2016.

[51] S. Rumley, D. Nikolova, R. Hendry, Q. Li, D. Calhoun, and K. Bergman. Silicon photonics for exascale systems. *Journal of Lightwave Technology*, 33(3):547–562, Feb 2015.

[52] G. M. Saridis, S. Peng, Y. Yan, A. Aguado, B. Guo, M. Arslan, C. Jackson, W. Miao, N. Calabretta, F. Agraz, S. Spadaro, G. Bernini, N. Ciulli, G. Zervas, R. Nejabati, and D. Simeonidou. Lightness: A function-virtualizable software defined data center network with all-optical circuit/packet switching. *Journal of Lightwave Technology*, 34(7):1618–1627, April 2016.

[53] K. Schmidtke, F. Flens, and D. Mahgarefteh. Taking optics to the chip: From board-mounted optical assemblies to chip-level optical interconnects. In *Proceedings of the Optical Fiber Communications Conference and Exhibition (OFC)*, page 3, March 2014.

[54] Mohamed Abu Shar, Ali Kanso, Abdallah Shami, and Peter Ohlen. Building a cloud on earth: A study of cloud computing data simulators. *Computer Networks*, 108:78–96, 2016.

[55] Shehabi et al. *United States Data Center Energy Usage Report*. Ernest Orlando Lawrence Berkeley National Laboratory, 2016.

[56] Cisco Systems. Cisco global cloud index: Forecast and methodology, 2014-2019 white paper, 2015.

[57] János Sztrik. Basic queueing theory. *University of Debrecen, Faculty of Informatics*, 193, 2012.

[58] M. Tan, P. Rosenberg, G. Panotopoulos, M. McLaren, W. Sorin, S. Mathai, L. Kiyama, J. Straznicky, and D. Warren. Optopus: Optical backplane for data center switches. In *Proceedings of the Optical Fiber Communications Conference and Exhibition (OFC)*, page 3, 2014.

[59] E. Temporiti, G. Minoia, M. Repossi, D. Baldi, A Ghilioni, and F. Svelto. A 3d-integrated 25 gbps silicon photonics receiver in pic25g and 65 nm cmos technologies. In *Proceedings of the 40th European Solid State Circuits Conference – ESSCIR*, pages 1–3, 2014.

[60] David Thompson et al. Roadmap on silicon photonics. *Journal of Optics*, 18:20, June 24 2016.

[61] Zhehui Wang, Jiang Xu, Peng Yang, Xuan Wang, Zhe Wang, L. H. K. Duong, Zhifei Wang, Haoran Li, R. K. V. Maeda, Xiaowen Wu, Yaoyao Ye, and Qinfen Hao. Alleviate chip i/o pin constraints for multicore processors through optical interconnects. In *Proceedings of the 20th Asia and South Pacific Design Automation Conference*, pages 791–796, 2015.

[62] wikipedia. Socket b. https://en.wikipedia.org/wiki/LGA_1366, 2011.

[63] Wikipedia. Small form factor package. https://en.wikipedia.org/wiki/Small_form-factor_pluggable_transceiver, 2012.

[64] Stefan Wolf. Dac-less amplifier-less generation and transmission of qam signals using sub-volt silicon-organic hybrid modulators. *Journal of Lightwave Technology*, 33,7:1425–1432, 2015.

[65] Y. Yan, G. M. Saridis, Y. Shu, B. R. Rofoee, S. Yan, M. Arslan, T. Bradley, N. V. Wheeler, N. H. L. Wong, F. Poletti, M. N. Petrovich, D. J. Richardson, S. Poole, G. Zervas, and D. Simeonidou. All-optical programmable disaggregated data centre network realized by fpga-based switch and interface card. *Journal of Lightwave Technology*, 34(8):1925–1932, April 2016.

[66] W. Zhang, H. Wang, and K. Bergman. Next-generation optically-interconnected high-performance data centers. *Journal of Lightwave Technology*, 30,24:3836–3844, 2012.

PART II
Developing Design Automation Solutions and Enabling Design Exploration

5

Design Automation Beyond Its Electronic Roots: Toward a Synthesis Methodology for Wavelength-Routed Optical Networks-on-Chip

Marta Ortín-Obón[1], Andrea Peano[2], Mahdi Tala[2], Marco Balboni[2], Luca Ramini[2], Maddalena Nonato[2], Víctor Viñals-Yufera[1] and Davide Bertozzi[2]

[1]Departamento de Informática e Ingeniería de Sistemas, University of Zaragoza, Spain
[2]Dipartimento di Ingegneria, University of Ferrara, Italy

Abstract

While the information and computing revolution is often credited to Moore's Law scaling, the complexity challenge has been actually addressed by electronic design automation, which is capable of transforming complex system-on-chip designs from high-level functional specifications into detailed geometric descriptions. Similarly, the uptake of emerging interconnect technologies depends not only on technology maturity, but also on the availability of tools and methodologies bridging the gap between system designers and technology developers. This chapter provides an early-phase synthesis methodology for wavelength-routed optical networks-on-chip, capturing all design points in a unified design framework, and refining them into an actual implementation.

5.1 Introduction

Photonic integration of multi- and many-core architectures has been extensively studied in literature, envisioning single-stage topologies from buses [1, 2] to crossbars [1, 3, 4], and different multistage topologies from quasi-butterflies [6, 7] to tori [8, 9]. Most proposals use different routing algorithms, flow control mechanisms, and optical wavelength organizations. However, the proposed architecture design solutions still represent a small subset of the design space. At this stage, there is only poor emphasis on design flexibility, i.e., on how to customize an architecture template for a wide range of possible design requirements. Similarly, the emphasis is not on the exploration of the design space, which is largely unknown, but rather on the comparison of (relatively few) emerging solutions against baseline electronic counterparts.

However, as the silicon photonic technology starts to consolidate, another concern comes to the forefront: bridging the gap between technology developers and system designers, who need to do design with the new devices. For this purpose, the design methodology needs to be entirely reversed with respect to the early one: system designers should not be constrained by the limited availability of architecture design points, but should rather be able to specify the high-level description of an interconnect fabric that matches the requirements of the system at hand. Then, they should rely on a synthesis methodology, and associated toolflow, operating on communication abstractions and refining an abstract solution into an actual implementation.

Not surprisingly, this is what typical electronic design automation (EDA) flows do for digital design [10]. Optical networks-on-chip (ONoCs) are still lagging far behind in this respect. The focus of this chapter is on a specific family of ONoCs, namely wavelength-routed ones (WRONoCs), which is no exception to the above trend. Unlike other optical bus, crossbar and (space-routed) network solutions, WRONoCs do not require any form of arbitration, but deliver contention-free global connectivity by construction. Although they make an extensive use of laser sources, hence suffering from scalability limitations, WRONoCs are attractive whenever applications require time predictability and/or performance guarantees. So far, only few connectivity patterns have been proposed in literature, where solutions such as λ-router [24], GWOR [14], Snake [12] or Ring [17] topologies stand out for their superior power efficiency. *However, they are isolated design points in a space that is still largely unexplored.*

Clearly, changing the ONoC concept into an actual technology with practical relevance depends not only on the maturity of the underlying technology, but also on the outcome of the design automation effort [11]. The success story of electronic NoCs is rooted exactly in this trend [35]. Interestingly, there are indirect signs of a turning point in the evolution of the ONoC concept. First, the increasing effort devoted to the exploration of multiple communication protocols for optical ring topologies [20, 21] contributes to gain visibility of a largely unknown design space. Second, the awareness of missing abstraction layers between system-level architecture design and photonic integrated circuit design (e.g., early-stage analysis of physical properties [28], placement and routing of ONoCs [19], etc.) contributes to define the intermediate steps a complete synthesis process should go through. The timely evolution and growth of this research field requires a more systematic and coherent effort though, aiming at the understanding of all design points in the context of a unified design framework [23]. This chapter intends to serve as a stepping stone into the evolution of design automation beyond its electronic roots, and lays the foundation of a synthesis methodology for future WRONoC topologies.

5.2 Analogy with EDA Flows

Given the complexity of today's systems-on-chip, modeling at register-transfer level (RTL) is not feasible as a starting point anymore. Therefore, abstractions were found to speed up simulation times and enable system designers to evaluate many design alternatives, in addition to providing a systematic path to implementation. In this direction, several abstraction layers for early-stage system design were defined: algorithmic models, untimed functional models, timed functional models, cycle accurate models, RTL models. As an example, in Figure 5.1, the algorithmic model of an MP3 audio compression system is reported. Is there an equivalent layer for optical topology synthesis? A high-level specification of a system interconnect consists of expressing at least connectivity requirements and the communication protocol. Similarly to electronic NoCs, in the future other specifications may be provided at this abstraction layer for optical NoCs, consisting for instance of the average communication bandwidths per flow, and of specific quality of service or fault-tolerance requirements. In this chapter, we make a choice to limit the scope: we focus on wavelength-selective routing, that is, on a communication protocol that delivers conflict-free all-to-all connectivity between n initiators and n targets.[1]

---

[1]The framework presented in this chapter can be easily extended to the $n \times m$ scenario, but it is omitted for lack of space.

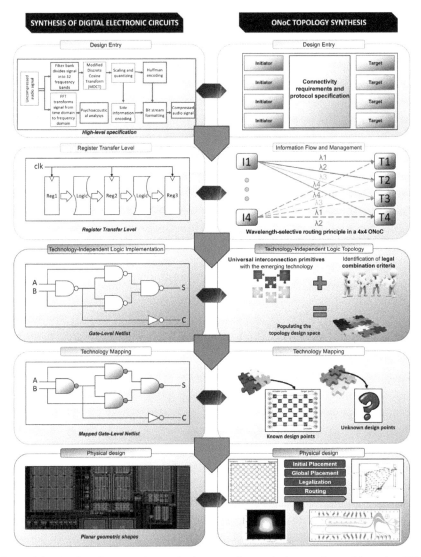

Figure 5.1　Analogy between the synthesis steps of an optical network-on-chip and an EDA flow.

As electronic system designs are refined, their first structural representation is provided at the Register Transfer Level (Figure 5.1). The analogy for WRONoC topology synthesis can be established by abstracting the behaviour of a WRONoC as a bufferless, non-blocking crossbar:

each initiator is connected with any target, with capability to transmit to each of them concurrently, since communications go over different optical channels.

Electronic RTL designs are further refined into technology-independent gate-level netlists (Figure 5.1), for instance through the technology-independent Synopsys technology library (GTECH) [16]. The underlying assumption is that the basic building blocks that compose a logic library (i.e., the logic gates) are universal, in the sense that any logic function can be mapped onto one or more combinations of such blocks. There are several possible gate-level implementations for the same logic function, which motivates the use of suitable logic synthesis algorithms driven by performance, area and/or power optimization goals. In contrast, the ONoC topology synthesis discipline is still in its infancy. The most urgent task, documented in this chapter for WRONoCs, consists of identifying abstract universal interconnection primitives with the emerging interconnect technology, in addition to the criteria for their legal combination. Addressing this challenge is a key enabler to fully utilize the expressive power of emerging devices, and essential to unlock the true value of the candidate technology. In practice, novel abstractions and synthesis techniques are needed to populate a design space which is currently still largely unknown, thus laying the foundation for its automated exploration, and for pruning strategies driven by predefined objective functions.

Technology-independent electronic gate-level netlists are then mapped to the logic cells of an actual technology library (Figure 5.1), which is closely linked to a manufacturing process in a given technology node. Again, there is no unique mapping (e.g., basic logic cells can be directly mapped to their counterparts, or can be grouped/transformed into compound cells before the actual mapping), and again specialized algorithms are needed to come up with a mapped netlist which meets the design constraints. Similarly, several solutions exist to map basic interconnection primitives to real devices, which rises a correspondent technology mapping problem for the synthesis of ONoC topologies.

As an outcome of such process, we will certainly end up again in those few WRONoC topologies that have been proposed so far in literature, that the researchers' intuition has find out because of some appealing properties (e.g., the presence of geometrical patterns in the connectivity structure, or an effective scalability algorithm, etc.). However, we will also find out a whole set of new interconnect solutions that have not been investigated by researchers yet. This is due to the fact that in the absence of a synthesis framework, researchers tend to mix descriptive information at different abstraction layers in the same hardwired design description. Unfortunately, overly coupling different design levels artificially limits the design space, and for an emerging technology there is likely even less intuition on the most promising parts of the design space. In contrast, a crisp specification of abstraction layers is the foundation for any effective synthesis methodology.

Finally, electronic designs are changed into planar geometric shapes by going through the placement and the routing steps, and final signoff loop, which are today largely automated (Figure 5.1). In the field of ONoCs, there is currently a large gap between mapped topologies and their actual physical implementations. While novel ONoC architectures and topologies are relentlessly proposed in literature, photonic integrated circuit designers are used to much smaller-scale (typically device-level) designs. Unfortunately, the gap to bridge for the physical design automation of entire optical NoCs in the context of many-core architectures with complex thermal behaviour is still huge, although there are early-stage research efforts that go in this direction [32].

Following a top-down design methodology, the most critical emerging need consists of placement and routing tools of ONoCs on the optical layer of a 3D-stacked integrated computing system. It is worth observing that state-of-the-art placers for electronics cannot be directly applied, nor efficiently adapted, to the place&route problem of optical interconnects [15], due to the different nature and constraints of the technology substrate.

5.3 Wavelength-Selective Routing

Wavelength-routed optical NoCs (WRONoCs) rely on the principle of wavelength-selective routing, which associates a wavelength channel to each source-destination pair. In particular, master $M1$ uses n wavelengths λ_1 to λ_n to reach slaves $S1$ to Sn, respectively. However, instead of allocating an additional set of wavelengths for the communications of master $M2$ to all the slaves, the initial set of n wavelengths is reused across masters. This is pictorially illustrated in the information flow and management section of Figure 5.1, which shows the wavelength-selective routing principle at work in a 4×4 ONoC. It is on burden of the topology to avoid the overlap of same-wavelength channels onto the same waveguides, which would corrupt the information. Overall, wavelength-selective routing ends up delivering contention-free all-to-all connectivity, since signal contention for resources is avoided at design time rather than solved at run time. Therefore, no arbitration of ONoC resources is needed.

The wavelength-selective routing function fulfilled by each WRONoC topology can be logically viewed as consisting of two sub-functions:

Drop function (Figure 5.2a). Each master receives from the power distribution network a wavelength-division multiplexed (WDM) optical signal consisting of multiple carriers with wavelengths λ_1 to λ_n. The topology should then resolve the individual wavelength channels from the multiplexed compound signal, so that each resolved component can be routed to a different destination. In practical terms, this task can be accomplished by using add/drop optical filters, which are tuned to a specific wavelength, and therefore split the associated optical channel from the compound signal.

Add function (Figure 5.2b). Resolved wavelength channels from the different masters and heading to the same slave should be recombined together into a WDM optical signal propagating onto the output waveguide of that slave. This way, a selective filtering stage can eject the desired wavelength

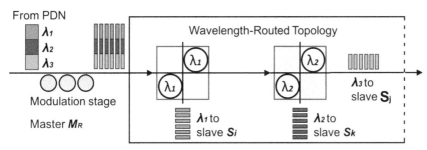

(a) Drop function of a WRONoC topology.

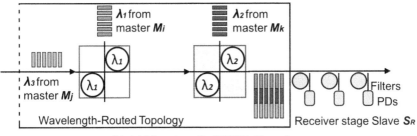

(b) Add function of a WRONoC topology.

Figure 5.2 Logical tasks performed by a WRONoC topology.

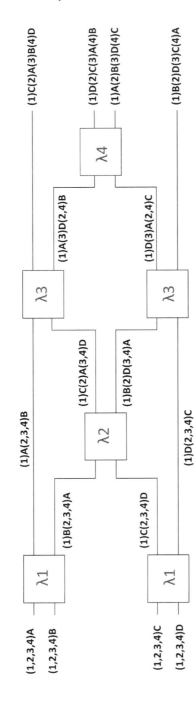

Figure 5.3 The tightly-intertwined add and drop functions at work in a λ-router WRONoC topology. Numbers refer to wavelength identifiers (IDs), while letters refer to master/slave IDs.

channel and feed it to a photodetector stage. In practice, this task can be accomplished by using different inputs of the same add/drop optical filters.

In a WRONoC topology, the add and drop functions are tightly inter-twined: as the WDM input signal from a given master propagates down the topology, its wavelength channels are progressively and selectively resolved and coupled with channels tuned on different wavelengths and originating from different masters. This process is pictorially illustrated in Figure 5.3 for the well-known λ-router WRONoC topology [24].

5.4 WRONoC Synthesis Methodology at a Glance

The key steps of a synthesis methodology for WRONoC topologies are illustrated in Figure 5.4. At first, a resolution step of WDM input signals is required, which corresponds to the drop function illustrated above. Regard-less of where the drop function of the actual components takes place in the network, all drop functions can be conceptualized together at this stage.

Next, technology mapping takes place: the basic drop operators are aggregated together for the sake of mapping them onto real optical switching devices.

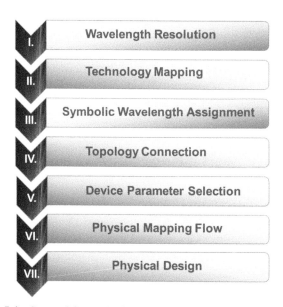

Figure 5.4 Steps of the synthesis methodology for WRONoC topologies.

The third step consists of assigning symbolic resonance wavelengths to the photonic switching elements. Legal assignments are those that perform the correct add and drop functions for all the wavelength channels in the network. For instance, optical components should not be dropped twice, and filtered components should not be recombined back with the original WDM signal.

Symbolic wavelength channels should then be refined into modulated optical carriers at an exact wavelength value to be specified. At the same time, specifying such wavelengths implies the definition of MRR radius lengths inside photonic switches, since optical channels should be on-resonance with specific MRRs, and off-resonance with all others. One constraint at this stage is to avoid routing faults, that is, the event that an optical channel is on-resonance with multiple MRR types, both the intended one and one or more unintended ones.

The synthesis methodology is finally completed by the physical mapping flow, which we fundamentally identify with the placement and routing steps in this chapter. This requires to identify a target environment for the implementation, which will be hereafter assumed to be a 3D-stacked computing system, and will dictate the floorplanning, placement and routing constraints for the physical design of ONoC topologies.

Last but not least, the back-end physical design steps take place, which end up in the mask set generation for the actual photonic integrated circuit. This will not be illustrated in this chapter, because photonic integrated circuit design is a well-established field, which relies on mainstream commercial tools [29–31,33,34], although coping with such large-scale designs as ONoCs remains a challenging task. Moreover, 3D stacking of this optical layer is a challenge in itself.

5.5 Front-End Synthesis Methodology

Next, the problem of synthesizing a generic $n \times n$ wavelength-routed ONoC topology is addressed, thus demonstrating that the topology design space can be populated much beyond the isolated design points that are available in current literature. This corresponds to the first 4 steps of the methodology reported in Figure 5.4. The reader should bear in mind that the topology schematic that is output by these synthesis steps should be regarded as a *logical topology*, to be clearly differentiated by its physical implementation.

5.5.1 Wavelength Resolution

Topology synthesis can be viewed as resulting from the combination of basic universal primitives. In Section 5.3, it became apparent that both the add and the drop functions, which are at the core of wavelength-selective routing through the optical NoC, can be ultimately decomposed into a combination of add/drop filters. More specifically, this chapter identifies the 1×2 add/drop filter (ADF) as the universal primitive to synthesize any WRONoC topology (see Figure 5.5a).

In the first synthesis step, the WDM input signal from any master should undergo the drop function by crossing $n - 1$ ADFs, each filter resolving the wavelength channel for a specific destination. Clearly, one of the wavelength channels of the WDM signal will not be on-resonance with any of the ADFs, thus the last ADF will end up resolving two optical channels. This process can be pictorially illustrated as a *wavelength resolution graph (WRG)*, reported in Figure 5.5b for a 4×4 generic WRONoC topology. At this abstraction layer, the graph can be completely generic, since it does not need to specify the resonant wavelengths of the ADFs. The graph only shows the minimum number of filtering stages the input WDM signals

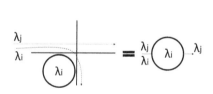

(a) An 1x2 add-and-drop filter (ADF), and its symbolic representation.

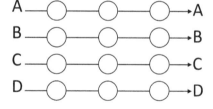

(b) Generic wavelength resolution graph.

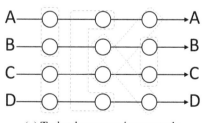

(c) Technology mapping example.

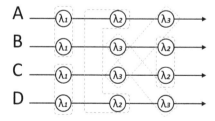

(d) Wavelength assigment example.

Figure 5.5 Front-end synthesis methodology of WRONoC topologies.

should undergo. Fixing the parameters of the WRG indirectly prunes the design space and biases the methodology toward a specific topology design point.

5.5.2 Technology Mapping

How to implement the WRG of Figure 5.5b? A common option consists of grouping the 1×2 ADFs into compact 2×2 photonic switching elements (PSEs, those used in Figure 5.2a and Figure 5.2b). Clearly, future work should explore more complex switching cells and their mix in order to give rise to efficient technology mappings, i.e., association of logic primitives (and combinations thereof) with real devices. In order to map ADFs to 2×2 PSEs, pairs of ADFs should be identified on the WRG, and replaced by a PSE. However, there exists a fundamental constraint for such mapping process: the legal mappings are the combinations without repetitions $C(n, 2)$ for picking 2 unordered outcomes from n possibilities. In practice, when assuming 4 masters (A,B,C,D) connected to 4 slaves (the same A,B,C,D), then the number of legal mappings C(4,2)=[n(n–1)....(n–k+1)]/2! is 6, namely AB, AC, AD, BC, BD, CD. Since for each master there are $n - 1$ ADFs on the associated WRG row, there are $(n - 1)x(n - 1)$ ways to group an ADF filter of A with an ADF filter of B, of which only 1 should be selected for each topology design point. One legal mapping example is illustrated in Figure 5.5c. To validate the correctness of the example, one might notice that the first filter of A is grouped with the first filter of B (option AB taken), the second of A with the second of D (option AD taken), the third of A with the second of C (option AC taken), and so on, until all the 6 legal combinations are taken.

There is a fundamental reason for posing the above constraint, associated with the add function. When the WDM signal from a generic input (say, master C) enters a 2×2 PSE, only the remaining unresolved channels propagate on the same row to complete the drop function, while the resolved wavelength channel is coupled onto another row, where it overlaps with the WDM signal from another master. Therefore, the correct operation of the add function consists of coupling a distinct wavelength channel onto each row. In contrast, when the WDM signal from a specific master is coupled onto another row twice (despite the two PSEs are correctly tuned to different wavelengths), the dropped channels end up being recombined onto the target row, which violates the philosophy of the drop function performed with the minimum number of filtering primitives.

Based on these considerations, it is possible to compute the number of required PSEs to deliver $n \times n$ contention-free full connectivity through a generic WRONoC topology:

$$C(n, 2) = n * (n - 1)/2 \qquad (5.1)$$

In fact, the λ-router topology illustrated in Figure 5.3 connects 4 communication actors with each other through 6 PSEs. Whenever the literature reports topologies with a lower number of 2×2 PSEs, it means that they are actually providing a lower connectivity. For instance, a 4×4 GWOR topology [14] is reported to require 4 PSEs instead of 6, the reason being that self-communication is not provided. Finally, the number of WRONoC topologies in the design space amounts to:

$$[(n - 1) * (n - 2) * ..... * (n - n + 2)]^n \qquad (5.2)$$

A 4×4 WRONoC can therefore be implemented in 1296 different ways, including the $\lambda-$router, the *snake* or the *GWOR* topology. For $n > 4$, the full enumeration of all possible design points becomes computationally unaffordable.

5.5.3 Symbolic Wavelength Assignment

Once mapping of ADFs to 2×2 PSEs has been performed, the actual assignment of resonant wavelengths to PSEs becomes mandatory. The basic assumption at this stage is that real wavelengths (e.g., 1550 nm) are not considered, but rather symbolic ones, such as λ_1 to λ_n in an $n \times n$ WRONoC. Actual wavelength values will be assigned in Section 5.6. Clearly, both microring resonators inside PSEs share the same resonant wavelength. In order for the wavelength assignment to be legal, one constraint should be enforced on the WRG: the resonant wavelengths of the ADFs along a single row should be different from each other. That is, $\lambda_1, \lambda_2, \lambda_3, \lambda_4$ is a valid drop order for the WDM signal from a given master, while $\lambda_1, \lambda_1, \lambda_2, \lambda_3$ is not. There is one fundamental reason for posing this constraint. When a set of unresolved wavelength channels goes through consecutive ADFs tuned to the same carrier wavelength (λ_1 in the example of Figure 5.6), the first ADF drops channel λ_1, while the set of wavelength channels entering the second ADF reaches its output unaffected. Since the minimum number $n - 1$ of ADF primitives has been instantiated for each row of the WRG, this causes two or more unresolved channels from the same master to reach one output of the

Figure 5.6 Incorrectly-populated WRG by dropping the same wavelength channel twice on the same row.

topology, as can be clearly observed in Figure 5.6. At the same time, some slaves will become unreachable.

A greedy algorithm has been proposed in [38] to perform the wavelength assignment. Without lack of generality, it will be used in the experimental section of this chapter. When applying the algorithm to the technology mapping of Figure 5.5c, the legal assignment in Figure 5.5d is derived.

5.5.4 Topology Connection

Given the synthesis stage in Figure 5.5d, it is possible to draw the connectivity pattern of the topology, as well as the features of its PSEs. In particular, the reader can easily prove himself that the topology in Figure 5.5d is exactly the λ-router topology presented in Figure 5.3. There is only a minor difference: in the baseline λ-router the last stage is tuned to λ_4, thus resulting in the use of 4 resonator types. Instead, the presented wavelength-assignment algorithm tunes the last stage to λ_2 because of its preferred choice for minimum wavelength channel identifiers. This way, the exact connectivity pattern of the λ-router can be provided while making use of 3 resonator types only.

5.6 Device Parameter Selection

The above synthesis steps still omit to specify the exact value of the carrier wavelengths on top of which optical channels are modulated. At this stage, their symbolic values need to be refined into actual wavelength values. An associated refinement issue consists of specifying the actual radius length of the MRRs building up the PSEs of the topology. This corresponds to the device parameter selection process at stage 5 of the general methodology in Figure 5.4.

Interestingly, the selection of these design parameters is key to evaluating both the feasibility and the performance of the synthesized WRONoC topology. On the one hand, parameters of wavelength channels and of the filters that selectively route them to destination should be chosen in such a way that routing faults are avoided (the problem is addressed in Section 5.6.1).

This constraint limits the level of connectivity that can be achieved, which means that large-scale topologies may turn out to be infeasible when routing fault freedom is enforced. On the other hand, even when the connectivity of the target number of nodes is feasible, the routing fault concern limits the achievable level of communication parallelism on the available wavelength channels. Overall, without proper emphasis on this topology refinement step, system designers may consider WRONoC configurations for their architecture that later turn out to be practically infeasible, in terms of levels of connectivity and/or target parallelism.

Last but not least, device parameter selection is the first design step where the parameters of the manufacturing process at hand become visible. In fact, a high level of uncertainty in that process forces the designer to make conservative design choices to meet the routing fault freedom constraint, which penalizes the performance to a significant extent.

A formal methodology is needed to select WRONoC physical parameters while maximizing communication parallelism for a specific level of network connectivity. More specifically, the following parameters should be explored:
a) The exact radius length of the MRRs inside PSEs, which typically falls in the range $5 \div 20\,\mu$m, and determines the periodic resonant wavelengths of the MRRs, as well as their inter-spacing (named free spectral range, FSR).
b) The exact value of the n wavelengths used by each initiator in an $n \times n$ WRONoC, which are typically chosen in the frequency band 1500–1600 nm. The two sets of parameters are tightly intertwined, as hereafter illustrated, and should be reliably chosen by accounting for parameter uncertainty.

5.6.1 The Routing Fault Concern

A wavelength channel must be positioned in the frequency band of interest so to be a resonant wavelength of one kind of ADFs, the one that drops (adds) that channel from (to) the input (output) WDM signal. This is pictorially illustrated in Figure 5.7a by means of an example, together with a possible inconvenient. Without lack of generality, wavelength channel λ_2 is placed on one peak of the transmission characteristic of the larger MRR with radius R_2, while a smaller MRR with radius R_1 is selected for tuning on the λ_1 channel.

The FSR of the two MRRs is not the same, since their transmission characteristics are not shifted replicas, but they are rather determined by their radius lengths. In principle, larger MRRs exhibit a lower spacing among resonances (hence more resonance peaks within the same band), while the opposite holds for small MRRs. Therefore, two resonant peaks from the two

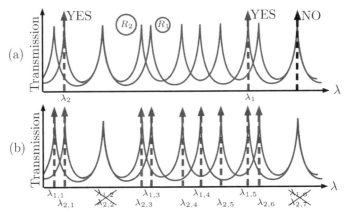

Figure 5.7   The routing fault concern: (a) constraining wavelength channel selection; (b) limiting parallelism.

MRRs might overlap, as illustrated in the figure. As a consequence, if λ_1 were chosen to coincide with the overlapped peaks, a **routing fault** would occur in the network: *when entering an ADF tuned to λ_2, channel λ_1 would be incorrectly on resonance with such filter, instead of being off-resonance, thus ending up in a misrouting.* While the problem can be easily solved in the example of Figure 5.7a by selecting another resonant wavelength of the small MRR, it can become a serious concern when increasing the network size. In fact, the proliferation of ADFs and of wavelength channels may limit the availability of non-overlapped transmission peaks, which may cause the topology to be practically infeasible.

Even for a fixed network size, the routing fault concern turns out to be a fundamental limiter for communication parallelism. In fact, the examples seen so far assume that each initiator sends one bit at a time to each target on a specific wavelength channel. Unfortunately, optical interconnect technology requires serialization of bit-parallel electronic words, which is only partially compensated by the high transmission rates of at least 10 Gbit/s. Thus, communication parallelism could be increased by allocating multiple wavelength channels to an I/O connection, provided that they are allocated to the resonant peaks of the same ADF, in order to properly and rigidly perform the add and drop functions for all of them in the same way. For instance, the former λ_1 channel for a specific 1-bit I/O connection might be augmented to $\lambda_{1,j}$ channels, with $j = 1..M$, for an M-bit extended communication. Figure 5.7b shows why the routing fault concern poses an upper bound on the communication parallelism. In the figure, the 6 (7) resonances of R_1 (R_2) are shown, however at most 4 bits of parallelism can be overall guaranteed. Denote by $\Lambda_r = \{\lambda_{r,j}\}$ the resonances of the MRR with radius length R_r ($\lambda_{r,j}$ can be used for brevity). Since $\lambda_{1,2}$ conflicts with $\lambda_{2,2}$ and $\lambda_{1,6}$ conflicts with $\lambda_{2,7}$, routing fault avoidance prevents from selecting any of these four; now at most 4 (5) peaks in R_1 (R_2) can be used at the same time. In this case, the maximum parallelism that can be sustained by *both* communication flows is 4.

The above problems are further exacerbated by the uncertainties of both the manufacturing process and of the device parameters (see Section 5.6.2), which cause even resonant peaks that are just close enough not to be usable for the routing of any wavelength channel.

5.6.2 The Role of Parameter Uncertainty

This section addresses the effect of two important parameter uncertainties that capture the fundamental dependency of architecture performance from the manufacturing process.

The first uncertainty is due to the MRRs' fabrication process. Depending on the lithography, up to 10 nm of variation is expected for MRRs using rib-waveguides that undergo a full CMOS process flow [5], i.e., given the tolerance $Rtol = 10$ nm and the nominal radius R_r, the radius length of the manufactured MRR will be between $R_r^- = R_r - R_{tol}$ and $R_r^+ = R_r + R_{tol}$; R_{tol} identifies the maximum gap between the nominal value at design time and the actual one. The transmission response of the manufactured MRR will vary according to its actual radius. Figure 5.8 shows the optical spectrums for R_1 and its maximum variations, say Λ_1^-, Λ_1, and Λ_1^+; in particular, the most noticeable effect is that the resonances are shifted from the nominal ones to the left (right) for a negative (positive) radius variation.

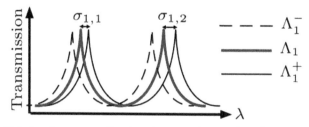

Figure 5.8 Relation between radius length uncertainty and positioning uncertainty of resonance peaks for a MRR of radius R_1.

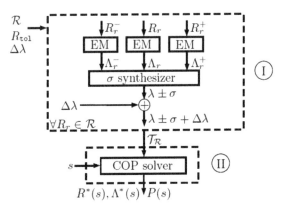

Figure 5.9 Solving architecture.

The greatest peak-to-peak distance (nm) between the nominal $\lambda_{1,j}$ and its variations is $\sigma_{1,j}$. Since a small radius variation slightly affects the FSR, in general $\sigma_{r,j} \neq \sigma_{r,h}$ for $j \neq h$ ($\sigma_{1,1} \neq \sigma_{1,2}$ in the figure). To translate the variation $Rtol$ into the many $\sigma_{r,j}$, a two-step procedure can be used: i) an Electromagnetic Model [13] (EM) computes the three transmission responses Λ_r^-, Λ_r, Λ_r^+, then ii) the responses are passed to a synthesizer that processes them and returns the σ values. This process is shown within the block I of Figure 5.9.

The other uncertainty is due to the variation of the central operating wavelength of the laser, caused by variation in temperature and driving current. Without lack of generality, in this chapter this value is set to $\Delta\lambda = 0.5$ nm.[2] Since laser uncertainty is independent from radius variation, $\Delta\lambda$ is considered in addition to σ and accordingly this quantity is processed in the block I of Figure 5.9.

Given R_1, R_2, R_{tol}, and $\Delta\lambda$, the resonances $\lambda_{1,j}$ and $\lambda_{2,h}$ may vary within the intervals:

$$I_{1,j} = [\lambda_{1,j} - \sigma_{1,j} - \Delta\lambda, \lambda_{1,j} + \sigma_{1,j} + \Delta\lambda]$$
$$I_{2,h} = [\lambda_{2,h} - \sigma_{2,h} - \Delta\lambda, \lambda_{2,h} + \sigma_{2,h} + \Delta\lambda]$$

In the most conservative hypothesis, $\lambda_{1,j}$ and $\lambda_{2,h}$ should never conflict, i.e., $I_{1,j} \cap I_{2,h} = \emptyset$.

---

[2]https://www.thorlabs.com/_sd.cfm?partNumber=SFL1550S&filename=21060-D02.pdf

Obviously, the larger the intervals, the higher the chance of overlapping, and the lower the number of available off-resonances for other radius values, to the detriment of potential parallelism.

5.6.3 Problem Formulation: A Case Study

Without lack of generality, a viable solution to tackle the device parameter selection challenge is reported in [37]. The philosophy of the approach is hereafter reported.

All topologies initially appear in the form of Figure 5.5d, from which the relevant inputs to the optimization problem are derived, i.e., the network size $n \times n$ and the number s of ADF types. Notice that the ADF types differ for the different radius length of their MRRs, while the number of times each ADF type is instantiated is irrelevant for the problem formulation that follows. The problem is to compute for each ADF type the radius length and its subset of resonances so that the communication parallelism is maximized under the most conservative assumption, i.e., the selected wavelengths (together with their uncertainty margins) never overlap. Future work might try to relax these assumptions, and to statistically project a manufacturing yield of the ONoC as a function of parameter variability.

The choice of the s radius lengths and their wavelength channels is made from a lookup table T_R, where the r-th row contains the resonances λ_{rj} of radius R_r. The radius values are given by the discrete set

$$\mathcal{R} = \left\{ Rmin + r * Rstep, \forall r \in \left[0, ..., \frac{Rmax - Rmin}{Rstep} \right] \right\}$$

where the *fabrication options* $Ropt = \{Rmin, Rstep, Rmax\}$ are the minimum radius length, the incremental step, and the maximum radius length, respectively. The resonances and their variations are computed as in block I in Figure 5.9. The example T_R in Table 5.1 is built with $Ropt = \{5, 1, 8\}$ and reports the σ values along with the resonances, with $Rtol = 0.01 \, \mu m$.

Table 5.1 T_R with $Ropt = \{5, 1, 8\} \, \mu m$, $Rtol = 0.01 \, \mu m$, and $\Delta \lambda = 0.5$ nm

r	R_r [μm]	$|\{\lambda_{r,j}\}|$	$\lambda_{i,1}\langle\sigma_{i,1} + \Delta\lambda\rangle$, $\lambda_{i,2}\langle\sigma_{i,2} + \Delta\lambda\rangle$, $\cdots$	[nm]
1	5	5	1496.4$\langle$3.5$\rangle$, 1521.3$\langle$3.6$\rangle$, 1547.1$\langle$3.6$\rangle$, 1573.8$\langle$3.6$\rangle$, 1601.4$\langle$3.6$\rangle$	
2	6	6	1500.5$\langle$3.0$\rangle$, 1521.3$\langle$3.0$\rangle$, 1542.7$\langle$3.1$\rangle$, 1564.8$\langle$3.1$\rangle$, 1587.5$\langle$3.3$\rangle$,	$\cdots$
3	7	6	1503.4$\langle$2.6$\rangle$, 1521.3$\langle$2.6$\rangle$, 1539.6$\langle$2.7$\rangle$, 1558.4$\langle$2.7$\rangle$, 1577.7$\langle$2.7$\rangle$,	$\cdots$
4	8	7	1505.6$\langle$2.2$\rangle$, 1521.3$\langle$2.2$\rangle$, 1537.3$\langle$2.2$\rangle$, 1553.7$\langle$2.2$\rangle$, 1570.4$\langle$2.0$\rangle$,	$\cdots$

Once T_R is computed, the optimal sets of radius lengths R^* and of wavelength channels Λ^* are chosen in order to maximize the parallelism. These choices are made on discrete sets; this highlights a *combinatorial* structure of the problem, meaning that the number of possible choices increases exponentially with the table size and the number of resonances. Since not any combination is feasible, due to routing fault prevention, authors of [37] are faced with a Constrained Optimization Problem (COP).

The core decisions concern which resonances should be selected from T_R. The objective function maximizes the parallelism in the selected row with the least parallelism, since the global network parallelism is bounded by the channel with lowest parallelism. In practice, the goal is to maximize the parallelism $P(s)$ that can be sustained by **all** of the communication flows at the same time. Among the other things, constraints impose to select exactly s rows (radius lengths), corresponding to the s ADF filter types that compose the topology under test, and prevent the selection of conflicting resonances not to incur routing faults.

The COP can be reformulated by way of well-known declarative technologies in Operation Research, e.g., Mixed Integer Linear Programming [25], and in Artificial Intelligence, e.g., Answer Set Programming (ASP) [27], and others. The resulting mathematical/logic programs can be solved through off-the-shelf *solvers* (e.g., Clasp [26] for ASP), which either return a provably optimal solution (R^*, Λ^*, and $P(s)$) or prove no solution exists. When non-idealities are involved, the solution consists of those radius lengths and resonances that ensure the parallelism $P(s)$ in whatever post-manufacturing condition.

The presented approach can be used also for scalability analysis of connectivity and communication parallelism. Given T_R, such an analysis can be performed by increasing the value of ADF filter types s (indirectly corresponding to larger networks), until $P(s) = 0$; for each s, the COP solver should be called, as shown in the block II in Figure 5.9, and the maximum parallelism $P(s)$ should be returned.

5.7 Physical Mapping Flow

Once device parameters have been selected, the design is ready for placement and routing (step 6 of the methodology in Figure 5.4). The common environment for the physical mapping of WRONoC topologies is a clustered 3D many-core architecture (Figure 5.10) composed of an electronic

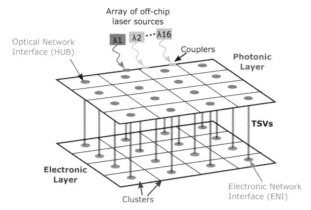

Figure 5.10   3D-Stacked clustered many-core processor.

layer and an optical layer vertically stacked on top of it. The latter might be powered by an array of off-chip continuous-wave laser sources providing multiple optical carriers multiplexed onto an input power waveguide[3]. There are vertical alignment dependencies that determine the location of the optical network interfaces (or hubs) on the optical plane. This gives rise to the real floorplanning constraints for the placement and routing of an ONoC, which make its ultimate layout profoundly different from the logical scheme. This is the root cause for the design predictability concern. For instance, a λ−router logically assumes all initiators on the lefhand side, and all targets on the righthand side (see Figure 5.3). However, in the common floorplan in Figure 5.10, hubs are both initiators and targets, and distributed on a grid, which causes a folding of the topology. In turn, a number of additional waveguide crossings arise to meet actual floorplanning constraints, which were not there in the topology logical scheme. Ultimately, the gap between the insertion loss estimated from the topology logical scheme as opposed to the post-layout one results in an increase of laser power requirements in order to meet photodetector sensitivity.

Typical place&route tools operate at the granularity of the individual PSEs to fit tight space constraints. The placement and routing problem for generic WRONoC topologies can be formulated as follows: *determine*

---

[3]Energy efficiency and proportionality arguments are currently being raised in favor of on-chip sources, which might be included in network design in the long run.

a geometrical description of all optical devices and waveguides such that:

- *all relevant objectives (e.g. insertion loss, number of crossings, number of waveguide bends, routing congestion, etc.) are minimized;*
- *all constraints (e.g. no overlap, place all optical devices inside the placement area, place all waveguides inside the routing area) are fulfilled.*

One major pillar of any optimization framework consists of selecting a suitable objective function. One key finding of current literature is that the objective function for the placement and routing of optical NoCs is tightly inter-related with the way the optical power is generated and distributed throughout the optical layer.

State-of-the-art placers typically minimize the maximum insertion loss over all optical paths. This is done mainly to make the optimization problem affordable, rather than based on technical considerations. Intuitively, an objective function targeting the minimization of the *critical path* (from an insertion loss viewpoint) is compatible with two scenarios. In the first one, off-chip multi-wavelength comb laser sources are used [22]. They can directly generate a spectral comb, and are extremely compact, although they do not easily enable power tuning of wavelength channels individually. Therefore, the output power should be tuned based on the worst-case optical path requirement. In the second scenario, the wavelength channels are generated by multiplexing independent and individually tunable DFB laser sources into the same input WDM signal for the optical NoC. While the power requirement of the wavelength channel with the highest on-chip losses is optimized, the power requirements of the other channels are indirectly derived by the place&route engine without their explicit optimization. In fact, this would require the optimization framework to target not only one critical path, but many of them (e.g., on a wavelength basis, or on an initiator basis), which would bring the place&route tool into the new and challenging ground of a multi-objective optimization framework.

Overall, current place&route optimization frameworks typically minimize only the worst-case insertion loss. The latter is mainly dominated by propagation losses and crossing losses, therefore the place&route engine should be able to infer a physical topology that strikes a good balance between path length and number of crossings on it. In fact, minimizing the propagation loss and minimizing crossing loss can be contradictory objectives, as shown in Figure 5.11. When the waveguide length is minimized,

Figure 5.11 Minimizing propagation loss or crossing loss can be contradictory objective functions.

this results in one crossing. In contrast, when crossing loss is minimized, a significantly longer waveguide length arises. In general, if a lot of detours are taken, and the waveguides are excessively long, there exist more possibilities for crossings. Thus, the algorithm should prefer short waveguides and avoid crossings only if there are short detours available.

The authors of [19] presented the first automatic placement and routing algorithm (named PROTON) for 3D ONoCs, minimizing the worst-case insertion loss. The placement problem was solved with the help of nonlinear optimization. Unfortunately, the runtime of the algorithm strongly increases when the number of optical devices and paths increases slightly. In addition, due to the high memory consumption of this approach, the algorithm is not able to place larger topologies than 16×16 WRONoCs.

A force-directed placement algorithm for 3D WRONoCs was reported in [39] (named PLATON), targeting better scalability. Despite the shorter runtimes and the more power-efficient solutions, PLATON does not improve the objection function of PROTON: they both aim at minimizing the maximum insertion loss over all paths, hence resulting in conservative values for the total laser power requirements. Also, both tools use the same maze routing framework proposed in [19]. Overall, the field is currently still in the early stage.

5.8 Experimental Results

5.8.1 Synthesis of Logical Topologies

By running the front-end synthesis methodology in Section 5.5, the complete design space for 4×4 WRONoC topologies is generated, built out of 2×2 PSEs. The design points are illustrated in Figure 5.12a. Obviously, among the generated solutions, one expects to find the topologies reported in literature and stemming from researchers' intuition, such as the λ-router and snake

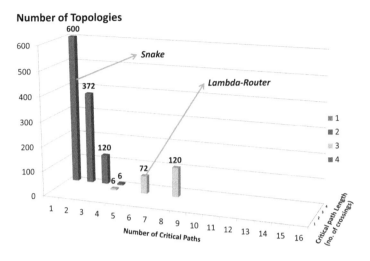

(a) Complete design space for 4×4 WRONoC topologies.

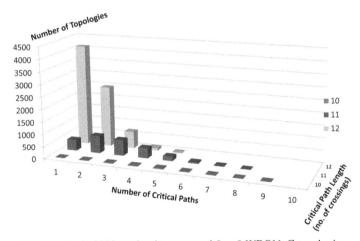

(b) Sample of 10000 randomly-generated 8×8 WRONoC topologies.

Figure 5.12 Design space of generic logical topologies based on 2×2 PSEs.

topologies. Most logical topologies have 4 crossings on the critical path, while 15% of the population has only 3 (the λ-router belongs to this latter category). Interestingly, when the critical path is the longest possible, then it is likely to be unique in the topology (46% probability). The worst case consists of 4 critical paths of 4 crossings. In contrast, when the critical path is shorter (3 crossings), the topology is likely to have more than one. In fact, when the critical path is 3-crossings deep, the probability to have 8 of them is 60%. The figure finally shows in which bar established WRONoC topologies fall.

Next, a sample of 10000 design points for 8×8 topologies is obtained by taking random decisions whenever a choice has to be made in the synthesis flow (e.g., technology mapping), thus avoiding to bias the sample. Results are reported in Figure 5.12b. Two features are noticeable. First, most of the topologies feature the longest critical path of 12 crossings. This means that without proper methods to explore and prune the design space (currently not available in literature), random walks through it are very likely to end up in a sub-optimal design point. Second, a high number of critical paths materializes only when the critical path length is shorter than the maximum one (e.g., 10 or 11 crossings).

5.8.2 The Design Predictability Gap

The previous section may lead to think that despite the large number of feasible design points (even for a small 4×4 network), most of them are basically equivalent, since they feature only few values of critical path length. This is not true for a couple of reasons. First, a realistic total static power of a topology depends not only on the critical path of the topology as a whole, but also on the critical paths for each wavelength carrier, since they may come from different sources. Second, even sticking to the worst-case assumption, the small variation of the critical path across logical topologies does not remain such as the latter undergo physical design.

To prove this, the PROTON tool is used for the place&route of generic WRONoC topologies to lay out the population of 4×4 topologies. In order to avoid trivial layouts associated with the small radix of the considered networks, they are used in the context of partitioned ONoCs. In practice, a 3D-stacked optical plane is considered, with 4 hubs (providing ONoC access to 4 clusters of processing cores in the electronic plane) and 4 memory controllers. Such floorplanning constraints are borrowed from [18]. In this context, three 4×4 networks (same topology) are used to provide 8×8 connectivity: one network for memory requests (from hubs to controllers), one for memory responses (from controllers to hubs), and one for inter-hub communications. All 1296 4×4 WRONoC logical topologies are tried in this context. Post-place&route results are reported in Figure 5.13, where

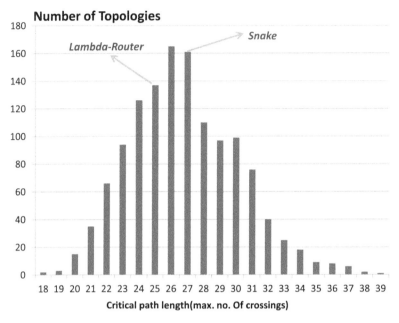

Figure 5.13 Distribution of the critical path after physical mapping.

the critical path (simplified as the maximum number of crossings across all optical paths) distribution is reported. Interestingly, now the critical path length ranges from 18 to 39 crossings, thus featuring a much larger variability than in logical schemes. This raises the issue of placement-aware logical topology synthesis, which is consolidated for electronic design [36], but is completely new for optical NoCs.

Finally, the power model from [38] is applied in order to derive the static power requirement of laser sources, which accounts not only for the critical path, since each laser source is tuned separately in [38], and only same-wavelength channels originate from the same source.

Total laser power results for each 4×4 topology, replicated three times in the complex floorplanning scenario above, is reported in Figure 5.14. Logical schemes are categorized as nx_cy, which means that a topology has x critical paths of length y crossings. Most of the topologies in each category end up in the same power range, from 50 to 130 mW. When the category is densely populated (like $n1\_c4$ or $n2\_c4$), the distribution tends to a gaussian curve, although there is a long and non-marginal tail toward high power values. For less dense categories ($n3\_c4$, $n6\_c3$, $n8\_c3$), the distribution tends to become

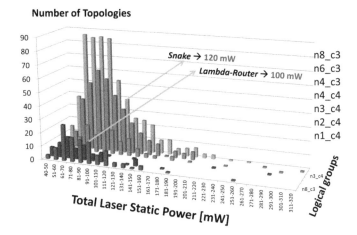

Figure 5.14   Distribution of static power requirements of topologies.

more uniform, and the tail shorter. Interestingly, both categories $n8\_c3$, $n6\_c3$ and $n1\_c4$ are able to materialize solutions with the lowest power (range 40–50 mW). Also, a few solutions exist that bring the total power to more than 300 mW, thus making the point for future smart pruning strategies of the design space toward the most power-efficient solutions.

5.8.3 Bounds on Connectivity and Parallelism

In this section, we use the framework in Section 5.6, and aim at testing how the maximum achievable parallelism $P(s)$ by a WRONoC topology scales with respect to: i) the number of ADF types s in it (each type being defined by a specific MRR radius length), ii) the fabrication options R_{opt}, iii) the manufacturing non-idealities (ring radius variability R_{tol} and laser center wavelength uncertainty $\Delta\lambda$).

In general, a higher network size requires also a higher number of filter types s to provide connectivity and wavelength routing, although the exact value is strictly topology-specific.

For the first test we set $R_{opt} = \{5, 1, 25\}\mu m$ [40], and finally $R_{tol} = 0.01\ \mu m$, $\Delta\lambda = 0.5$ nm to model worst-case variations. The ideal manufacturing process is when both R_{tol} and $\Delta\lambda$ are 0.

The chart in Figure 5.15a shows a first analytical result: the bit parallelism is larger than one ($P(s) > 1$), i.e., wavelength routing is feasible, only for a

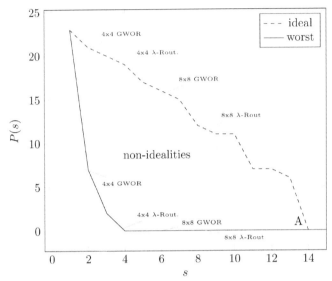

(a) Maximum parallelism $P(s)$ as a function of the number of filter types in a WRONoC topology, with fabrication options $R_{opt} = \{5, 1, 25\}\mu$m.

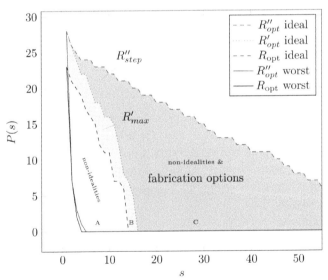

(b) Maximum parallelism achievable with fabrication options $R'_{opt} = \{5, 1, 30\}\mu$m compared with $R''_{opt} = \{5, 0.25, 30\}\mu$m.

Figure 5.15   Bounds on connectivity and parallelism as a function of device parameter uncertainty.

very limited range of the s parameter (i.e., for overly small network sizes), when worst-case conditions are considered. To make the outcome more tangible, let us associate actual topology configurations to their values of s. In fact, $s = 2$ and $s = 4$ correspond to a 4×4 GWOR and a 4×4 λ–router, respectively, since they use different numbers of MRRs types[4]. While the GWOR enables 7-bit parallelism, the larger number of MRR types limits the λ–router parallelism to zero, which means the topology is infeasible. Both 8×8 topologies (with $s = 6$ and $s = 8$, respectively) turn out to be infeasible. In the ideal case, the higher number of MRR types of the λ–router is reflected into a relative drop in achievable parallelism by 9.5% for a 4×4 WRONoC, and by 25% for an 8×8 one with respect to GWOR. In this latter case, GWOR has a potential parallelism of 16 bits: the improvement with respect to the worst-case variations is impressive, which denotes the key limiting role played by the non-idealities of the manufacturing process. There is a huge inaccessible "performance" gap (see grey area A) that improvements in fabrication and laser generation technology can reduce.

Increasing R_{max}, as well as decreasing R_{min} and R_{step}, means to increase the number of the available radius lengths and resonances in Table T_R (see Section 5.6), with a potential benefit on the achievable parallelism. Not necessarily this makes sense from a physical design viewpoint, since a lower ring radius causes a higher sensitivity to process variations, and impairs the manufacturing yield. In contrast, large rings take up more space. Finally, the step parameter depends on the lithography. In order to express the theoretical bounds, the chart in Figure 5.15b adds to the chart in Figure 5.15a the maximum parallelism $P(s)$ curves obtained with $R'_{opt} = \{5, 1, 30\}\mu m$ (ideal) and with $R''_{opt} = \{5, 0.25, 30\}\mu m$ (ideal and worst-case variation). In practice, the impact of the lithography resolution is assessed.

This chart suggests three main facts:

i) $P(s)$ with R''_{opt} (ideal) is much better than R'_{opt} (ideal), in fact area $C \gg B$, meaning that the resolution R_{step} is a better driver to increase the parallelism with respect to making available more resonant wavelengths through larger rings (i.e., the T_R table in input contains many more radii and wavelengths in the former case than in the latter one).

---

[4]The ultimate reason is because the λ–router provides self-communication, while the GWOR does not.

ii) With R''_{opt} (ideal) $P(s)$ is much higher than with R_{opt} (ideal) and the area $B + C \gg A$, meaning that the combination of multiple fabrication options yields a significant improvement of parallelism and connectivity (especially through the R_{step} parameter).

iii) In the worst variation case, both R''_{opt} and R_{opt} yield the same parallelism, meaning that WRONoC performance is highly sensitive to the uncertainty of the manufacturing process. In fact, making available multiple wavelength channel selection options is useless if the uncertainty ranges are not reduced accordingly. For instance, with R''_{opt} a 32×32 $\lambda-$router yields 14-bit parallelism in the ideal case, which technology maturity can try to materialize over time.

5.9 Conclusions

This chapter lays the foundation for the automation of the front-end and back-end synthesis steps of wavelength-routed optical NoCs. At first, the chapter illustrates a methodology that systematically populates all the points of the WRONoC topology design space, by building upon the combination and aggregation of basic filtering primitives. Then, the use of state-of-the-art placement and routing technology for the physical mapping of generic topologies on the optical layer of a 3D-stacked system is discussed. At the same time, the chapter emphasizes the role of the vertical integration of synthesis steps, since elegant logical topologies may turn out to be practically infeasible either for the large degradation of static power requirements (a side effect of crossing-dominated layouts), or for the impossibility to avoid routing faults (amplified by the uncertainties of the manufacturing process). As a result, a wide design predictability gap might arise between logical topologies and their physical implementations.

Looking forward, this chapter points out the need for pruning methods of the design space, capable of automatically inferring the most promising logical topology that meets the connectivity requirements as well as the placement constraints of the system at hand. In particular, placement-aware topology synthesis should be pursued in order to reduce the design predictability gap.

References

[1] D. Vantrease et al., "Corona: system implications of emerging nanophotonic technology," in ISCA 2008, pp. 153–164.

[2] A. N. Udipi et al., "Combining memory and a controller with photonics through 3D-stacking to enable scalable and energy-efficient systems," in ISCA 2011, pp. 425–436.

[3] N. Kirman et al., "Leveraging optical technology in future bus-based chip multiprocessors," in Int. Symp. Microarchitecture, MICRO 39, pp. 492–503, 2006.

[4] G. Kurian et al., "ATAC: A 1000-core cache-coherent processor with on-chip optical network," in PACT 2010, pp. 477–488.

[5] A. V. Krishnamoorthy, X. Zheng, G. Li, J. Yao, T. Pinguet, A. Mekis, H. Thacker, I. Shubin, Y. Luo, K. Raj, and J. E. Cunningham. "Exploiting CMOS manufacturing to reduce tuning requirements for resonant optical devices", *IEEE Photonics Journal*, 3(3):567–579, June 2011.

[6] A. Joshi et al., "Silicon-photonic clos networks for global on-chip communication," in Int. Symp. Networks-on-Chip, pp. 124–133, 2009.

[7] Y. Pan et al., "Firefly: Illuminating on-chip networks with nanophotonics," in ISCA 2009, pp. 429–440.

[8] M. J. Cianchetti, J. C. Kerekes, and D. H. Albonesi, "Phastlane: a rapid transit optical routing network," in Int. SCA2009.

[9] Z. Li et al., "Iris: A hybrid nanophotonic network design for high-performance and low-power on-chip communication," J. Emerg. Technol. Comput. Syst., vol. 7, no. 2, Article no. 8, 2011.

[10] Giovanni De Micheli, "Synthesis and optimization of digital circuits," McGraw Hill, 1994.

[11] Kahng, Andrew B. and Koushanfar, Farinaz, "Evolving EDA beyond Its e-roots: an overview," Proceedings of the IEEE/ACM International Conference on Computer-Aided Design, pp. 247–254, 2015.

[12] L. Ramini, P. Grani, S. Bartolini, D. Bertozzi: "Contrasting wavelength-routed optical NoC topologies for power-efficient 3D-stacked multicore processors using physical-layer analysis," DATE 2013, pp. 1589–1594.

[13] A. Parini, L. Ramini, G. Bellanca, and D. Bertozzi: "Abstract modelling of switching elements for optical networks-on-chip with technology platform awareness," In *Proceedings of the Fifth International Workshop on Interconnection Network Architecture: On-Chip, Multi-Chip*, INA-OCMC '11, pp. 31–34, New York, NY, USA, 2011. ACM.

[14] X. Tan et al., "On a scalable, non-blocking optical router for photonic networks-on-chip designs," SOPO 2011.

[15] A. von Beuningen et al., "PROTON+: a placement and routing tool for 3D optical networks-on-chip with a single optical layer," ACM Journal on Emerging Technologies in Computing Systems (JETC), — Vol. 12, No. 4, pp. 44:1–44:28, 2016.

[16] Kurup, Pran; Abbasi, Taher; "Logic synthesis using synopsys," Springer, 1997.

[17] S. Le Beux et al., "Optical ring network-on-chip (ORNoC): architecture and design methodology," DATE 2011, pp. 1–6.

[18] L. Ramini, D. Bertozzi and L. Carloni, "Engineering a bandwidth-scalable optical layer for a 3D multi-core processor with awareness of layout constraints," NOCS 2012, pp. 185–192.

[19] A. Boos, L. Ramini, U. Schlichtmann, D. Bertozzi "PROTON: an automatic place-and-route tool for optical Networks-on-Chip," International Conference on Computer-Aided Design (ICCAD), 2013 IEEE/ACM, pp. 138–145.

[20] S. Le Beux, Hui Li, G. Nicolescu, J. Trajkovic, I. O'Connor "Optical crossbars on chip, a comparative study based on worst-case losses," Concurrency and Computation Practice and Experience, October 2014.

[21] P. Grani and S. Bartolini, "Design options for optical ring interconnect in future client devices," (JETC): ACM Journal on Emerging Technologies in Computing Systems, Vol. 10, Issue 4, Article no. 30, 2014.

[22] M.J.R. Heck and J.E. Bowers, "Energy efficient and energy proportional optical interconnects for multi-core processors: driving the need for on-chip sources," IEEE Journal of Selectd Topics in Quantum Electronics, vol. 20, issue 4, pp. 332–343, 2014.

[23] C. Batten, A. Joshi, V. Stojanovic, and K. Asanovic: "Designing chip-level nanophotonic interconnection networks," IEEE Journal on Emerging and Selected Topics in Circuits and Systems, vol. 2, issue 2, pp. 137–153, 2012.

[24] M. Briere, B. Girodias et al., "System level assessment of an optical NoC in an MPSoC platform," DATE 2007, pp. 1–6.

[25] R.K. Martin, "Large scale linear and integer optimization: a unified approach," Springer, 1999.

[26] M. Gebser, B. Kaufmann, and T. Schaub, "Conflict-driven answer set solving: from theory to practice," *Artificial Intelligence*, 187–188: 52–89, 2012.

[27] M. Gelfond, "Answer sets." In F. van Harmelen, V. Lifschitz, and B. Porter, editors, *Handbook of Knowledge Representation*, Chapter 7, pages 285–316. Elsevier Science, 2008.

[28] Mahdi Nikdast, et al., "Crosstalk noise in WDM-based optical networks-on-chip: a formal study and comparison." IEEE Trans. on VLSI Systems, Vol. 23, issue 11, pp. 2552–2565, 2015.

[29] https://www.lumerical.com/tcad-products/interconnect/

[30] http://www.vpiphotonics.com/Tools/

[31] http://www.aspicdesign.com

[32] G. Hendry, J. Chan, L.P. Carloni, K. Bergman, "VANDAL: a tool for the design specification of nanophotonic networks", DATE 2011, pp. 1–6.

[33] http://www.phoenixbv.com

[34] https://www.lumerical.com/solutions/partners/eda/mentor_graphics/

[35] D. Bertozzi et al., "NoC synthesis flow for customized domain specific multiprocessor systems-on-chip." IEEE Transactions on Parallel and Distributed Systems, vol. 16, issue 2, pp. 113–129, 2005.

[36] S. Murali, P. Meloni, F. Angiolini, D. Atienza, S. Carta, L. Benini, G. De Micheli, L. Raffo, "Designing application-specific networks on chips with floorplan information." ICCAD 2006, pp. 355–362.

[37] A. Peano, L. Ramini, M. Gavanelli, M. Nonato, D. Bertozzi, "Design technology for fault-free and maximally-parallel wavelength-routed optical networks-on-chip," ICCAD 2016, pp. 1–8.

[38] Mahdi Tala, Marco Castellari, Marco Balboni, Davide Bertozzi, "Populating and exploring the design space of wavelength-routed optical network-on-chip topologies by leveraging the add-drop filtering primitive," IEEE/ACM Int. Symposium on Networks-on-Chip, 2016.

[39] Anja von Beuningen, Ulf Schlichtmann, "PLATON: a force-directed placement algorithm for 3D optical networks-on-chip," Proceedings of the 2016 International Symposium on Physical Design, Pages 27–34, 2016.

[40] D. Liang and J. E. Bowers. "Recent progress in lasers on silicon," Nature Photonics 4, pp. 511–517, 2010.

6

Application-Specific Mapping Optimizations for Photonic Networks-on-Chip

Edoardo Fusella[1], Alessandro Cilardo[1] and José Flich[2]

[1]Department of Electrical Engineering and Information Technologies, University of Naples Federico II, Naples, Italy
[2]Department of Computer Engineering (DISCA), Universitat Politècnica de València, Valencia, Spain

Abstract

Silicon photonics opens a promising path to energy-efficient ultra-high bandwidth on-chip communication. However, in order to fully exploit photonics, the mapping of application tasks to the NoC tiles is a crucial step: the designer needs to decide to which tile each selected task should be mapped such that the metrics of interest are optimized. In this chapter, we present a methodology which automatically maps the tasks onto a generic regular photonic NoC architecture such that the worst-case insertion loss or crosstalk noise are minimized, allowing a higher network scalability.

6.1 Introduction

The restless grow in transistor integration has pushed the semiconductor industry to a shift from the single-core era to a multi-core paradigm during the last decade. Over the next years, heterogeneous many-core systems will potentially include hundreds to thousands of general- and special-purpose cores as well as large memory elements such as shared and local caches. However, highly parallel computing systems have to face serious challenges in terms of energy and heat dissipation. Transistor power consumption does not scale with integration density and performance gain, causing a serious increase in the overall chip power dissipation. In this context, the role of the on-chip communication infrastructure is critical because it provides all

171

the required communication facilities, enabling the distributed computation among the different cores and impacting the overall system performance and energy consumption. Packet-switched networks-on-chip (NoCs) are among the most prominent paradigms for handling the on-chip communication due to their scalable and modular nature [1, 4]. Although the benefits of NoCs seem to be substantial, their potential is constrained by physical limitations in power dissipation, latency, and bandwidth. The power due to the communication architecture is nearly 30% of the whole tile power [14] and is still too high (by a factor of 10) to meet the tight energy requirements for future systems [19].

On these grounds, the optical NoC paradigm is emerging as a promising alternative solution providing low-power ultra-high bandwidth on-chip communication. Nanophotonic waveguides, the photonic counterpart of a wire, can in fact achieve bandwidths in the order of terabits per second by exploiting wavelength division multiplexing (WDM), while photonic signaling promises reduced power consumption [9]. In particular, photonics ensures bit-rate transparency and low loss in optical waveguides, meaning that the energy consumption in the optical network components, such as Microring Resonators (MRs) and Waveguides, is independent of the bit-rate and the distance between the two end-points.

6.2 Motivation: Application-Specific Mapping Optimization

For efficient design of future photonic networks-on-chip, an automatic mapping of tasks onto NoC tiles is highly desirable. The problem of mapping a set of given application tasks to the NoC tiles is illustrated in Figure 6.1. An application, previously divided into a graph of concurrent tasks, should be assigned to a set of available cores by mapping different functions to different regions of the system, so that the metrics of interest are optimized.

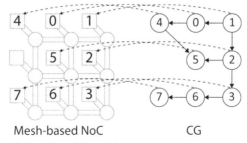

Figure 6.1 A mesh-based on-chip architecture and an example of mapping problem.

The mapping problem for electronic NoC architectures is addressed by a large body of works. Hu *et al.* [13] first showed the importance of a proper mapping strategy and introduced a branch-and-bound algorithm aiming to map tasks onto a mesh-based NoC while minimizing energy under bandwidth constraints. A mapping algorithm for NoC architectures which supports traffic splitting was presented in [16]. Differently, Srinivasan *et al.* [22] proposed optimizing the energy consumption under both bandwidth and latency constraints. Based on these previous works, a joint optimization of both mapping and routing algorithms was presented in [12]. Refer to [21] for a detailed review of mapping techniques for electronic NoC architectures.

The majority of the approaches that target the mapping problem in the electronic domain aim to optimize the performance in terms of latency or the energy consumption. This is done trying to exploit the principle of spatial locality, that is, placing the blocks that communicate more frequently closer to each other. In this way, an average reduction of the latency and energy consumption is achieved since both latency and energy grows proportionally to the hop count. Differently, the distinctive properties of photonic interconnects make the previous mapping strategies useless. In an optical path, in absence of conflicts, the propagation delay is simply the time of flight at the speed of light and hence the latency is constant as the hop count scales up. As regards the energy consumption, it is mainly consumed in the electro-optic (E-O) and O-E conversions and in the laser source. As a consequence, it is not affected by the hop count and hence by the mapping.

On the contrary, in order to exploit the potential advantages of photonics, we need to deal with specific electromagnetic effects, like insertion loss and the crosstalk noise, that potentially impact the optical NoC architecture to a large extent. Insertion loss, i.e. the power loss that a photonic element induces when it is inserted in an optical path, is one of the major limitations affecting the design of a photonic NoC [3]. In fact, the power of an optical signal must be above a certain threshold when arriving at the photodetectors in order to ensure a proper detection. As a consequence, the power injected into the chip must be higher than the photodetector sensitivity plus the worst-case power loss. However, the total power cannot exceed a certain threshold due to the non-linearities of the silicon material. In case of multi-wavelength signals, this problem is exacerbated since these considerations must apply to each individual wavelength channel. Differently, crosstalk is caused by an unfavorable coupling between optical signals [24]. In multihop photonic NoCs, two different optical signals can induce crosstalk noise to each other when reaching simultaneously a waveguide crossing or a photonic switch. In case of perfect coupling between two waveguides, optical signals propagate entirely with no reflection and with no crosstalk. However, ideal crossing is unfeasible and hence a small amount of optical power switches into the coupled waveguide.

Electromagnetic effects should be a major goal when designing a photonic NoC architecture, since a high power loss or crosstalk noise may easily result in a network leading to poor performance or even in an inoperable architecture [17, 23, 24]. As a major insight of this work, we recognize that application-specific mapping optimization is the best strategy to face these problems in application-specific multi-core systems-on-chip. The NoC architecture should be customized for a target application, when its traffic characteristics are known at design time, which is the case for most embedded applications running on multiprocessor Systems-on-Chip (MPSoCs). Differently from the mapping approaches in the electronic domain, we are interested in reducing the worst-case insertion loss or crosstalk noise instead of an average value that dramatically changes the mapping method. In addition, as explained in the following sections, power loss and crosstalk noise are marginally affected by the hop count. All these considerations diversify the mapping problem in the photonic domain from its electronic counterpart. While a large body of works addresses the mapping targeting electronic on-chip networks, to the best of our knowledge, this work is the first contribution to address the mapping problem for photonic NoC architectures and propose an efficient way to solve it. Early versions of the work presented here were reported in previous publications [7, 8, 10]. In this chapter, we collect the main results presented in these papers and we present the full methodology and a comprehensive experimental evaluation involving both insertion loss and crosstalk noise.

The rest of this chapter is organized as follows. Section 6.3 shows the target architecture and the used crosstalk model. Section 6.4 provides the problem definition to be tackled in this work and describes our implementation of the genetic algorithm to solve it. Section 6.5 discusses the results achieved by presenting the experimental setup and four real-world applications. Section 6.6 gives some final remarks.

6.3 Architecture Description

In this section, we describe the regular tile-based NoC architecture and the related optical loss and crosstalk models.

6.3.1 Architecture Overview

The optical NoC architecture is illustrated in Figure 6.2. The nodes are organized in a two-dimensional mesh topology where the connectivity is achieved using exclusive links between adjacent nodes. The architecture is composed of $m \times n$ tiles, each containing an optical router connected with an IP core and with the four neighboring tiles. Each connection is made up of two unidirectional silicon waveguides connecting two routers or a router and an IP core. Notice that, although we rely on a mesh topology, the proposed approach could be simply extended to any other topologies.

Each IP core is connected to a network interface (NI) owning the necessary logic to perform Electronic/Optical (E/O) and Optical/Electronic (O/E) conversions. The interface is made up of several components: serializer, laser source, and driver circuit in the generation stage as well as photodetector, transimpedance amplifier (TIA), limiting amplifier (LA), and deserializer in the reception stage. The generation stage handles the data transition from the electronic domain to the optical domain. The data is first serialized. Then, the driver circuit controls the optical modulator device which translates the

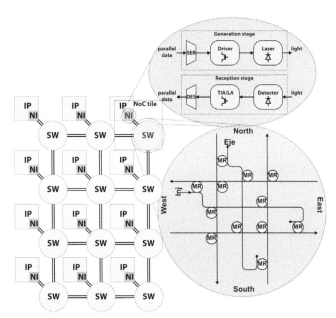

Figure 6.2 The target photonic NoC architecture. The internal architectures of a photonic router and a network interface are zoomed in on the right-hand side of the figure (SW, optical switch; IP, IP core; NI, network interface, MR, microring resonator; SER, serializer; DES, deserializer).

electrical signal into an optical signal through a laser source by modulating the light intensity according to the data bit values. Conversely, the reception stage is responsible for bringing back the optical signal into the electrical domain. The photodetector turns the light waves into an electrical current that is converted into electrical voltage by TIA. Then, LA brings the electrical voltage to the proper logic level and last, the digital signal goes through the deserializer stage.

When a message is transmitted between two tiles, we need to decide which route to take. While there are different routing algorithms proposed in the literature for electronic NoCs, dimension order routing (DOR) is the main choice in the photonic domain because:

- The photonics inability to perform inflight buffering and logic leads to preferring a static routing algorithm.
- Photonic switches for 2D direct topologies could be designed in order to take advantage of XY routing where there is at most a single turn along the path between a source and a destination. Note that each turn implies that the signal crosses a ring in the ON resonance state. As a consequence, implementing switches with straight default paths[1] leads to cross, in the worst case, only three rings in the on resonance state regardless of the network size: two in the source and destination nodes and one in the node where the turn takes place. Obviously, this contributes to reducing the power consumption, insertion loss, and crosstalk. In addition, the switches could be customized on the XY routing by providing only the turns allowed by XY routing, and removing the microrings required for the forbidden turns, ensuring a further optimization.
- An adaptive routing algorithm would require Electronic/Optical and Optical/Electronic conversions in each node that are extremely power hungry [3].

6.3.2 The Photonic Switch Element Model

As previously stated, the crosstalk noise arises when two optical signals reach simultaneously a waveguide crossing or a PSE. PSEs (Figure 6.3) are made up of a microring resonator and two waveguides. A microring resonator is a waveguide forming a closed loop, with an own resonance

---

[1]A default path is the path that the signal takes when all the rings are placed in a off resonance state.

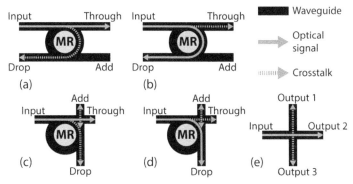

Figure 6.3 How crosstalk propagates through: (a) Parallel PSE in OFF state; (b) Parallel PSE in ON state; (c) Crossing PSE in OFF state; (d) Crossing PSE in ON state; (e) Waveguide Crossing.

Table 6.1 Loss and crosstalk parameters

Parameter	Notation	Value	Ref.
Crossing loss	L_c	-0.04 dB	[5]
Propagation Loss in Silicon	L_p	-0.274 dB/cm	[6]
Power loss per PPSE in OFF state	$L_{p,off}$	-0.005 dB	[2]
Power loss per PPSE in ON state	$L_{p,on}$	-0.5 dB	[2]
Power loss per CPSE in OFF state	$L_{c,off}$	-0.045 dB	
Power loss per CPSE in ON state	$L_{c,on}$	-0.5 dB	[15]
Crossings crosstalk coefficient	K_c	-40 dB	[5]
Crosstalk coefficient per PSE in OFF state	$K_{p,off}$	-20 dB	[2]
Crosstalk coefficient per PSE in ON state	$K_{p,on}$	-25 dB	[2]

frequency, which can be designed to perform switching functions. When an optical signal injected into the input port matches the wavelength of the resonance frequency, then it is coupled into the ring and steered to the drop port (Figure 6.3(b) and (d)). Otherwise, the signal propagates to the through port (Figure 6.3(a) and (c)).

In [23], the authors present an analytical model for characterizing the photonic switch elements. The output power at each port of both the Parallel PSE (PPSE) and the Crossing PSE (CPSE) in both the ON and OFF resonance state is evaluated as a function of the input power and the loss/crosstalk coefficients presented in Table 6.1. They assume that the signal power at the injection port of different optical switches is the same.

We estimated the insertion loss as well as the crosstalk noise using the aforementioned model with the following changes:

- The crosstalk noise on the add port as well as the light that reflects back on the input port are neglected. This is motivated by the fact that the reflection coefficient could be considered close to zero [5]. In addition, these two sources of crosstalk noise affect mainly optical signals that travel backward, exploiting the symmetric property of optical paths. This is not the case in the considered network.
- We consider only the first-order crosstalk noise and hence $K_i K_j = 0$, with $K_i, K_j \in \{K_c, K_{p,off}, K_{p,on}\}$
- We neglected the power loss that affects the crosstalk noise signal inside the optical routers where the noise arises. As a consequence, $K_i L_i = K_i$, with $K_i \in \{K_c, K_{p,off}, K_{p,on}\}$ and $L_i \in \{L_c, L_p, L_{p,off}, L_{p,on}, L_{c,off}, L_{c,on}\}$

Based on the above considerations, the equations presented in [23] are semplified as follows.

$$P_{Tppse,off} = L_{p,off} P_{in} \tag{6.1a}$$

$$P_{Dppse,off} = K_{p,off} P_{in} \tag{6.1b}$$

$$P_{Tppse,on} = K_{p,on} P_{in} \tag{6.1c}$$

$$P_{Dppse,on} = L_{p,on} P_{in} \tag{6.1d}$$

$$P_{Tcpse,off} = L_{c,off} P_{in} \tag{6.1e}$$

$$P_{Dcpse,off} = (K_{p,off} + K_c) P_{in} \tag{6.1f}$$

$$P_{Tcpse,on} = K_{p,on} P_{in} \tag{6.1g}$$

$$P_{Dcpse,on} = L_{c,on} P_{in} \tag{6.1h}$$

$$P_{out2} = L_c P_{in} \tag{6.1i}$$

$$P_{out1} = P_{out3} = K_c P_{in} \tag{6.1j}$$

Equations (6.1a) and (6.1b) and Equations (6.1c) and (6.1d) give the output powers at the through and drop ports for the PPSE respectively in the OFF and ON state, while Equations (6.1e) and (6.1f) and Equations (6.1g) and (6.1h) are the same equations for the CPSE. Finally, Equations (6.1i) and (6.1j) give the power detected at the output port of a two waveguides crossing.

6.3.3 The Router Model

The basic building block of the photonic NoC is the five-port photonic switch. In this work we rely on the *Crux* switch, first presented in [24].

Crux is a 5×5 switch whose physical layout is optimized to reduce the insertion loss and crosstalk noise. However, our approach is independent of the considered switch. In case of topologies requiring five-port photonic switches, no changes are needed. Differently, in the other cases, the router model should be adapted in order to be able to estimate the input/output power at each port of a generic n-port photonic switch.

We introduce a set $Ports = \{L, N, E, S, W\}$, which represents the set of all the ports of a switch, i.e. the local injection/ejection port as well as the four cardinal ports. In each instant of time, the switch is in a status that depends on the optical messages that are traveling from its input ports to its output ports. In that respects, $S_{in,out}$ with $in, out \in Ports$ indicates whether there is a communication between Ports in and out. Notice that, since we rely on XY routing, the switch is not allowed to U (Condition (6.2a)) and YX turns (Condition (6.2b)).

$$in \neq out \quad \forall in, out \in Ports \tag{6.2a}$$

$$in \in \{North, South\} \rightarrow out \neq East \wedge out \neq West \tag{6.2b}$$

An optical message travelling from an input port to an output port is subjected to both power loss and crosstalk noise. As in [23], the power loss $L_{in,out}$ is characterized by the following equation

$$L_{in,out} = \begin{cases} L^{sw}_{in,out} & \text{if } out = L, \\ L^{sw}_{in,out} L_p{}^d & \text{if } out \neq L. \end{cases} \tag{6.3}$$

where $L^{sw}_{in,out}$ is the insertion loss that arises when a signal crosses the switch sw from the input port in to the output port out, while L_p is the propagation loss, and $d = \sqrt{\frac{A_{die}}{(m-1)\times(n-1)}}$ is the average distance between two adjacent tiles in case of a $m \times n$ mesh. Obviously, the propagation loss due to going from a tile to the next is not considered in case the message has already reached the destination tile. Based on the above equations, the optical power at an output port is defined as

$$P^{sw}_{in,out} = P_{in} L_{in,out} \tag{6.4}$$

where P_{in} is the optical power injected in the input port.

Unlike the power loss, the crosstalk noise depends on the states of all the ports of the router in a certain instant of time. The following equation calculates the power of the crosstalk noise $N_{in,out}$ added to a signal traveling from an input to an output port.

$$N_{in,out}^{sw} = \sum_{i,j \in P} P_i S_{i,j} K_{in,out,i,j} \tag{6.5}$$

where $\{S_{i,j} : i,j \in Ports\}$ is the status of the switch, P_i is the optical power injected in the input port i and $K_{in,out,i,j}$ is the crosstalk noise coefficient that quantifies the crosstalk noise that is generated by an optical signal travelling between ports i and j affecting an optical signal travelling between ports in and out.

Based on the above model, it is possible to evaluate the insertion loss $IL(src, dst)$ affecting an optical signal when traveling between a source tile src and a destination tile dst as the product of all the losses in each hop along the path. Finally, the signal-to-noise ratio of an optical signal traveling between a source tile src and a destination tile dst is calculated as

$$SNR(src, dst) = 10 \log (P_S/P_N) \tag{6.6}$$

where P_S and P_N are, respectively, the power of the signal and of the crosstalk noise.

6.4 Methodology

6.4.1 Problem Formulation

Before presenting the design space exploration algorithms, we formulate the mapping problem. The design objective is to map a set of given application tasks, whose traffic characteristics are given in a Communication Graph (CG), to the NoC tiles in the case of a regular topology yielding the best SNR or insertion loss.

Definition 1 *A Communication Graph $CG = G(C, E)$ is a directed graph where each vertex $c_i \in C$ is an application task and $e_{i,j} \in E$ is the edge between tasks c_i and c_j characterizing the communication between them.*

Definition 2 *A Topology Graph $TG = G(T, L)$ represents how tiles are connected to each other, where $t_i \in T$ denotes a tile of the NoC and each $l_{i,j} \in L$ is a physical link connecting tiles t_i and t_j.*

Using the above graph representations, the problem addressed can be formulated as:

Given a CG and an TG satisfying

$$size(C) \leq size(T) \tag{6.7}$$

Find a mapping function $\Omega : C \rightarrow T$ which minimizes

$$\min \{ IL_{wc} = \max\{IL(\Omega(c_i), \Omega(c_j)) \quad \forall e_{i,j} \in E\}\} \tag{6.8}$$

in case of power loss optimization or maximizes

$$\max \{SNR_{wc} = \min\{SNR(\Omega(c_i), \Omega(c_j)) \quad \forall e_{i,j} \in E\}\} \tag{6.9}$$

in case of crosstalk noise optimization, where IL_{wc} and SNR_{wc} are respectively the worst-case insertion loss and the worst-case SNR

Such that:

$$\forall c_i \in C, \quad \Omega(c_i) \in T \tag{6.10}$$
$$\forall c_i \neq c_j \in C, \quad \Omega(c_i) \neq \Omega(c_j) \tag{6.11}$$

The condition (6.10) means that each task should be mapped to one tile, while condition (6.11) guarantees that each tile will host at most a single task.

Notice that, unlike the other mapping problems addressed in the literature, the crosstalk-aware mapping problem requires at each step a holistic view of the network status since a communication between a source and a destination is affected by a crosstalk noise that not only depends on the mapping of these two nodes, but also on the mapping of all the nodes of the system whose communications generate additional noise.

6.4.2 Genetic Algorithm

The problem of application mapping is NP-hard [20] so practical sizes of mapping problems can only be solved using constructive or transformative heuristics. Evolutionary techniques such as genetic algorithms are one of the most popular choices [21]. A genetic algorithm (GA) is a stochastic search algorithm, exploiting the principle of natural selection, where a fixed-sized population of candidate solutions (called phenotypes) evolves over a number of generations toward better solutions. Each phenotype has a set of properties (its genotype) which can be mutated and altered and an associated fitness measure.

GA requires a representation of the population that supports an efficient application of genetic operators. Each phenotype is represented as a string of chromosomes. Each string, called *mapping vector*, is an integer array where each entry, that is in one-to-one correspondence with a tile of the NoC, contains an identifier of the task mapped to that tile. For instance the p_1 mapping vector $\{6, 3, 4, 5, 1, 0, 7, 8, 3\}$ (the first step of Figure 6.5) means that task 6 is mapped on tile 0, task 3 on tile 1, task 4 on tile 2 and so on. Obviously, the phenotype must meet some legality criteria and a violation of the criteria results in an infeasible solution. Therefore, the cardinality of the mapping vector, and hence the number of chromosomes, should be equal to the number of nodes in the CG and lower than or equal to the number of nodes in the TG. This ensures that each task is mapped in a tile and there are enough tiles for all the tasks. In addition, every element must occur exactly once since a single task could not be mapped in two different tiles. These two criteria are equivalent to the conditions (6.10) and (6.11) in Section 6.4.1.

The fitness function makes use of the optical models in Section 6.3 in order to calculate for each genotype the worst-case power loss or SNR between all the communications that are specified in the CG.

Figure 6.4 shows the optimization process. An initial population of size P_{size} is generated randomly. Then, new populations are generated by modifying the genotypes using two operators: crossover and mutation. Crossover consists in taking multiple parent solutions and generating a child solution from them, while mutation is the process of taking a single parent solution and producing a child solution by altering one or more gene values in its genotype. The Selection operator for choosing the genotypes to modify is the roulette wheel selection. In roulette wheel selection, each individual i has a probability p_i of being selected that is directly proportionate to its fitness f_i:

$$p_i = \frac{f_i}{\sum_{j=1}^{P_{size}} f_j} \tag{6.12}$$

In this way, solutions with a lower fitness have a higher probability of being eliminated. However, although solutions with a higher fitness will be less likely to be eliminated, there is still a chance that they may be selected. This should allow the algorithm to avoid local minima by preventing some phenotypes from remaining in the population of candidate solutions for too long.

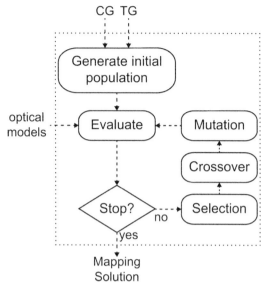

Figure 6.4 Flowchart of the genetic algorithm.

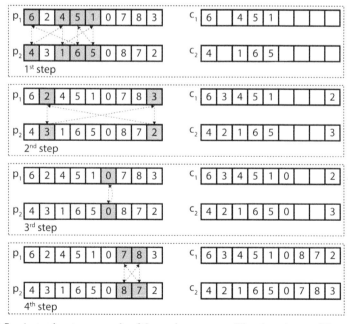

Figure 6.5 A step by step example of the cycle crossover. The phenotypes of the parents and childs are respectively in the left and right. p_1 and p_2 are the two parents and c_1 and c_2 are the two childs. The dashed arrows represent cycles.

The crossover operator produces two new offspring genotypes by inheriting partial characteristics from two parent genotypes. In order to meet the legality criteria, we need to design a crossover operator able to guarantee that every element occurs exactly once. In that respect, it is not possible to use the single- or two-points crossover, traditionally used in GA-based approaches, since it is possible that different tasks might be allocated into the same tile. We rely instead on the cycle crossover [18], that is more appropriate for the mapping problem. The crossover consists of several steps and each step has two phases: (i) finding a cycle and (ii) mixing cycles. Figure 6.5 explains how to mutate parent genotypes by means of an example. In order to identify a cycle, we proceed as follow:

1. Select a random position in parent p_1, e.g., position 1. In this position in parent p_1, we have the element 6 corresponding in parent p_2 (at the same position) to the element 4 (the arrow $6 \rightarrow 4$). Mark position 1 as taken.

2. Consider the corresponding element in parent p_2 and select its occurrence in parent p_1. In our example, select the 4 in parent p_1 at position 3 (the arrow $4 \rightarrow 4$). The corresponding element in parent p_2 is 1 (the arrow $4 \rightarrow 1$). Mark position 3 as taken

3. Continue this procedure until you get an element of parent p_2 that is the first element you considered in parent p_1. When this happens, we have found a cycle. In our example, the last element of the first cycle in parent p_2 is 6 that is the first element we considered in parent p_1. So we found the cycle: $6 \rightarrow 4$, $4 \rightarrow 1$, $1 \rightarrow 5$, $5 \rightarrow 6$. Note that each cycle involves the same tasks in both parents. This is a key feature of cycle crossover, which guarantees the feasibility of the child solutions.

In order to allow a selection of the chromosomes for each child, in the second phase, we mix the parent cycles in the following way. In the odd steps, the chromosomes of parent p_1 belonging to the cycle are copied to child c_1 and the chromosomes of parent p_2 are copied to child c_2. Differently, in the even steps, the chromosomes of parent p_1 are copied to child c_2 and the chromosomes of parent p_2 are copied to child c_1.

The mutation operator takes a genotype as input and generates a new genotype by modifying the chromosome in order to provide a new and feasible solution, thereby increasing the exploration of search space. The mutation is implemented by selecting a genotype, that is output of the crossover operator, and changing some of its portions in the following way: two tiles are randomly chosen and the tasks mapped on it are swapped. Mutation is essential since it allows avoiding local minima by preventing the population of phenotypes from becoming too similar to each other, thus slowing or even stopping evolution.

Finally, we define a stop criterion based on distance convergence. Basically, we stop the evolution when there is no longer any appreciable improvement in the consecutive phenotypes of the new populations that are being created. The main advantage is that it is not necessary to determine the exact number of iterations required to reach a satisfactory solution.

6.5 Results

For our experimental evaluation, we relied on *PhoNoCMap* [11], a Java-based toolset for the design space exploration of optical NoCs mapping solutions. We consider eight real streaming video and image processing applications [21], whose CGs are shown in Figure 6.6, namely: *263dec mp3dec*, which is a H.263 video decoder and MP3 audio decoder (decomposed in 14 tasks); *263enc mp3enc*, which is a H.263 video encoder and MP3 audio encoder (12 tasks); *DVOPD*, which is a dual video object plane decoder (32 tasks); *MPEG-4*, which is a MPEG4 decoder (12 tasks); *MWD*, which is a multi-window display (12 tasks); *PIP*, which is a picture-in-picture application (8 tasks); *VOPD*, which is a video object plane decoder (16 tasks); and *Wavelet*, which is a wavelet transform application (22 tasks).

We assumed mesh topologies with a size adequate to the number of tasks, ranging from 3×3 in case of application *PIP* to 6×6 in case of application *DVOPD*. We considered a $400 \ mm^2$ CMP die area to evaluate the distance between adjacent tiles. In order to prove that the mapping choice heavily affects the worst-case crosstalk noise and hence the minimum SNR, we generate randomly 100000 mapping solutions for each of the eight applications and, using our tool, we evaluated the worst-case SNR and power loss related to each mapping solution. Figure 6.7 shows the probability distribution of SNR and power loss values corresponding to the random generated mapping solutions. It can be easily recognized that the power loss and SNR of the best and worst solution may differ significantly. This experiment points out the high variability of power loss and crosstalk noise according to the different mapping solutions.

Then, we used our genetic algorithm to find the best mapping solution for each application. In addition, a random search was implemented and used as

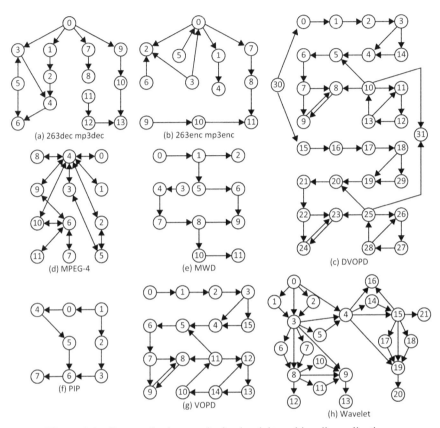

Figure 6.6   Communication graphs for the eight multimedia applications.

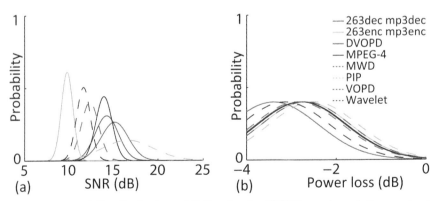

Figure 6.7   Probability distributions of SNRs related to 100000 mapping solutions randomly generated for eight multimedia applications.

Table 6.2 Algorithms comparisons

Application	Improvement		Non-opt		RS		GA	
	SNR	Loss	SNR	Loss	SNR	Loss	SNR	Loss
263dec mp3dec	4.46×	2.30×	8.66	−3.50	20.21	−2.04	38.67	−1.52
263enc mp3enc	3.74×	1.74×	10.33	−3.38	38.29	−2.04	38.63	−1.94
DVOPD	3.11×	1.88×	5.20	−4.05	12.65	−2.79	16.19	−2.15
MPEG-4	1.85×	1.66×	10.33	−3.38	19.06	−2.35	19.16	−2.04
MWD	3.74×	2.12×	10.33	−3.38	20.24	−1.81	38.63	−1.59
PIP	3.62×	1.92×	10.64	−3.22	38.58	−1.90	38.58	−1.68
VOPD	4.37×	1.78×	8.66	−3.50	18.66	−2.27	37.83	−1.96
Wavelet	5.70×	1.76×	6.67	−3.78	14.58	−2.46	37.95	−2.15

a baseline approach. The random search (RS) generates a random population of a configurable size and then picks the best individual. To guarantee a fair comparison, the running times of the different algorithms are the same. The worst-case power loss and SNR values, in case of a non-optimized optical NoC, are added in order to show the benefits of using application-specific mapping optimizations. The results are summarized in Table 6.2.

In general, compared to the non-optimized architectures, the mapping solutions found with our approach exhibit an average gain of $3.82\times$ and $1.89\times$ when optimizing respectively the crosstalk noise and the power loss. Differently, the random search reaches poor mapping solutions in most cases. Only for small-size networks (such as application PIP mapped on a 3×3 topology), it is able to find an appropriate solution. Compared to RS, the genetic algorithm performs better (up to 50–60%) when optimizing the crosstalk noise, while, in case of a power loss optimization, this gain is reduced to an average 15%. Notice that both the crosstalk noise and the power loss scale up with the network size: the worst-case values are reached in case of the DVOPD application that is mapped on the bigger topology. Also, applications that are more constrained due to their CGs, such as the MPEG-4 (26 edges), are subjected to a higher power loss and crosstalk noise compared to other applications that are less constrained and mapped on a topology of the same size, such as the 263enc mp3enc (12 edges) and the MWD (12 edges) applications.

6.6 Conclusion

For efficient design of future photonic networks-on-chip, an automatic mapping of tasks onto NoC tiles is highly desirable. Crosstalk noise and power loss should be a primary goal when designing a photonic NoC architecture. To this aim, we proposed an application-specific mapping optimization methodology for regular tile-based photonic architectures. By using a random search, we have shown that the worst-case crosstalk noise and signal-to-noise ratio are highly dependent on the mapping choices. Then, we proposed an efficient genetic algorithm which automatically maps the tasks to the tiles maximizing the worst-case SNR or minimizing the worst-case power loss. We plan to extend this research along several directions. Taking into consideration possible dependency constraints between communication tasks is essential to understanding the actual communication timing. As a consequence, a joint communication scheduling and mapping will be investigated as part of our future work.

References

[1] Luca Benini and Giovanni De Micheli. Networks on chips: A new soc paradigm. *Computer*, 35(1):70–78, 2002.

[2] Johnnie Chan, Gilbert Hendry, Keren Bergman, and Luca P Carloni. Physical-layer modeling and system-level design of chip-scale photonic interconnection networks. *IEEE Transactions on Computer-Aided Design of Integrated Circuits and Systems*, 30(10):1507–1520, 2011.

[3] Johnnie Chan, Gilbert Hendry, Aleksandr Biberman, and Keren Bergman. Architectural exploration of chip-scale photonic interconnection network designs using physical-layer analysis. *Journal of Lightwave Technology*, 28(9):1305–1315, 2010.

[4] William J Dally and Brian Towles. Route packets, not wires: On-chip interconnection networks. In *Proceedings of the 38th Design Automation Conference (DAC'01)*, pages 684–689. IEEE, 2001.

[5] Weiqiang Ding, Donghua Tang, Yan Liu, Lixue Chen, and Xiudong Sun. Compact and low crosstalk waveguide crossing using impedance matched metamaterial. *Applied Physics Letters*, 96(11):111–114, 2010.

[6] Po Dong, Wei Qian, Shirong Liao, Hong Liang, Cheng-Chih Kung, Ning-Ning Feng, Roshanak Shafiiha, Joan Fong, Dazeng Feng, Ashok V Krishnamoorthy, et al. Low loss silicon waveguides for application of optical interconnects. In *Proc. IEEE Photon. Soc. Summer Topical Meeting Ser*, pages 191–192, 2010.

[7] Edoardo Fusella and Alessandro Cilardo. Crosstalk-aware mapping for tile-based optical network-on-chip. In *High Performance Computing and Communications, 2015 IEEE 7th Intl Symp on Cyberspace Safety and Security, 2015 IEEE 12th Intl Conf on Embedded Software and Syst (HPCC, CSS, ICESS), 2015 IEEE Intl Conf on*. IEEE, 2015.

[8] Edoardo Fusella and Alessandro Cilardo. Crosstalk-aware automated mapping for optical networks-on-chip. *ACM Transactions on Embedded Computing Systems (TECS)*, 16(1):16, 2016.

[9] Edoardo Fusella and Alessandro Cilardo. Lighting up on-chip communications with photonics: Design tradeoffs for optical noc architectures. *IEEE Circuits and Systems Magazine*, 16(3):4–14, thirdquarter 2016.

[10] Edoardo Fusella and Alessandro Cilardo. Minimizing power loss in optical networks-on-chip through application-specific mapping. *Microprocessors and Microsystems*, 43:4–13, 2016.

[11] Edoardo Fusella and Alessandro Cilardo. PhoNoCMap: an application mapping tool for photonic networks-on-chip. In *Proceedings of the Conference on Design, Automation and Test in Europe (DATE16)*, pages 289–292. IEEE, 2016.

[12] Andreas Hansson, Kees Goossens, and Andrei Rădulescu. A unified approach to mapping and routing on a network-on-chip for both best-effort and guaranteed service traffic. *VLSI design*, 2007.

[13] Jingcao Hu and Radu Marculescu. Energy-aware mapping for tile-based NoC architectures under performance constraints. In *Proceedings of the 2003 Asia and South Pacific Design Automation Conference*, pages 233–239. ACM, 2003.

[14] Nevin Kırman, Meyrem Kırman, Rajeev K Dokania, Jose F Martinez, Alyssa B Apsel, Matthew A Watkins, and David H Albonesi. On-chip optical technology in future bus-based multicore designs. *IEEE Micro*, 27(1):56–66, 2007.

[15] Benjamin G Lee, Aleksandr Biberman, Po Dong, Michal Lipson, and Keren Bergman. All-optical comb switch for multiwavelength message routing in silicon photonic networks. *IEEE Photonics Technology Letters*, 20(10):767–769, 2008.

[16] Srinivasan Murali and Giovanni De Micheli. Bandwidth-constrained mapping of cores onto noc architectures. In *Proceedings of the Conference on Design, Automation and Test in Europe-Volume 2*, page 20896. IEEE Computer Society, 2004.

[17] Mahdi Nikdast, Jiang Xu, Xiaowen Wu, Wei Zhang, Yaoyao Ye, Xuan Wang, Zhehui Wang, and Zhe Wang. Systematic analysis of crosstalk

noise in folded-torus-based optical networks-on-chip. *IEEE Transactions on Computer-Aided Design of Integrated Circuits and Systems*, 33(3):437–450, 2014.

[18] IM Oliver, DJd Smith, and John RC Holland. Study of permutation crossover operators on the traveling salesman problem. In *Genetic algorithms and their applications: proceedings of the second International Conference on Genetic Algorithms: July 28–31, 1987 at the Massachusetts Institute of Technology, Cambridge, MA*. Hillsdale, NJ: L. Erlhaum Associates, 1987.

[19] John D Owens, William J Dally, Ron Ho, DN Jayasimha, Stephen W Keckler, and Li-Shiuan Peh. Research challenges for on-chip interconnection networks. *IEEE Micro*, 27(5):96, 2007.

[20] Ruxandra Pop and Shashi Kumar. A survey of techniques for mapping and scheduling applications to network on chip systems. *School of Engineering, Jonkoping University, Research Report*, 4:4, 2004.

[21] Pradip Kumar Sahu and Santanu Chattopadhyay. A survey on application mapping strategies for network-on-chip design. *Journal of Systems Architecture*, 59(1):60–76, 2013.

[22] Krishnan Srinivasan and Karam S Chatha. A technique for low energy mapping and routing in network-on-chip architectures. In *Proceedings of the 2005 International Symposium on Low Power Electronics and Design*, pages 387–392. ACM, 2005.

[23] Yiyuan Xie, Mahdi Nikdast, Jiang Xu, Xiaowen Wu, Wei Zhang, Yaoyao Ye, Xuan Wang, Zhehui Wang, and Weichen Liu. Formal worst-case analysis of crosstalk noise in mesh-based optical networks-on-chip. *IEEE Transactions on Very Large Scale Integration (VLSI) Systems*, 21(10):1823–1836, 2013.

[24] Yiyuan Xie, Mahdi Nikdast, Jiang Xu, Wei Zhang, Qi Li, Xiaowen Wu, Yaoyao Ye, Xuan Wang, and Weichen Liu. Crosstalk noise and bit error rate analysis for optical network-on-chip. In *Proceedings of the 47th Design Automation Conference*, pages 657–660. ACM, 2010.

7

Integrated Photonics for Chip-Multiprocessor Architectures

Sandro Bartolini and Paolo Grani

Dipartimento di Ingegneria dell'Informazione e Scienze Matematiche, Università degli Studi di Siena, Italy

7.1 Introduction

In current Chip Multi Processor (CMP) systems, the potential advantages given by an on-chip optical communication can be completely exploited only if requirements and behaviors of the various parts of such complex systems are analyzed together in an integrated and synergistic fashion. The general outcome from overall results in literature suggest that only in this way a global optimum design point can be reached. In fact, photonic technology has very different features and constraints from the well-known electronic counterpart. These features include, but are not limited, to:

- End-to-end communication and consumption practically insensitive to on-chip distance, with circuit switched nature compared to the typical store-and-forward electronic transmission paradigm;
- Extremely fast intrinsic signal propagation, but need for conversions to/from electronic domain as well as possible arbitration of shared optical resources, which can add latency;
- Integration problems especially in active components like lasers and photodetectors, along with 3D stacking requirements;
- Multiple optical channels possible through Wavelength Division Multiplexing (WDM) inside the same physical waveguide, but actual topologies exploiting WDM can incur in significant insertion loss.

Due to the interference between different design layers, an integrated holistic approach appears to be crucial to obtain the most out from every considered optical architecture [13, 16]. Indeed, especially in a CMP system,

191

the traffic generated by the running cores can be very challenging and performance can be very sensitive to the design choices performed at the various levels, from the application coding, operating system/runtime support and memory hierarchy organization [4] (e.g., specifically tuned software optimizations to better exploit the optical communication characteristics [20]), down to network choices [49, 50], interconnection management [17, 19] as well as raw technology options (e.g., Network interface issues [51], WDM and optical resource sharing [18]). In this chapter we want to exemplify this scenario with two cases we have recently worked on and in which these vertical design challenges have clearly emerged as crucial.

We summarize here the organization of chapter topics in order to support the reader also in a non-sequential approach to the presented material. In Section 7.2 we recall some basic facts about tiled design paradigm in CMPs and the Network on Chip (NoC) role within them. Section 7.3 highlights the key features of cache coherence in CMPs and, specifically, directory-based approaches along with the consequent challenges in the resulting NoC traffic. Section 7.4 describes the experimental methodology adopted for obtaining the results and analyses shown. Then, Section 7.5 describes the results of our design space exploration of *ring-based* optical networks for CMPs and highlights promising performance and consumption tradeoffs. In Section 7.6 we describe a novel design for optical-switch-based networks which dramatically reduces path-setup time and thus enables the support of cache-coherence traffic. Finally, Section 7.7 concludes the chapter.

The reader already accustomed to general motivation and introduction topics on CMPs, cache coherence and limits of electronic NoCs, can thus skip Sections 7.2 and 7.3.

7.2 Tiled Architectures and Networks on Chip

CMP systems are evolving towards parallel architectures to keep the pace of Moore's law and the tiled paradigm is expected to enable a sustainable design scalability [53]. State-of-the-art commercial processors already comprise numerous cores into a single die both in general-purpose (e.g., up to 16 in recent AMD Opteron 6000 models [1]) and in high-end embedded (e.g., up to 8 in Samsung Exynos Octa [54]) domains. Intel Polaris [61] prototype reached 1 TFlop some years ago thanks to an 80-core tiled architecture built around an ad-hoc interconnection network. Ideally, tiled CMP architectures comprise a number of identical tiles, each one having computational capabilities (e.g., a core), some private cache structures (e.g., L1 instruction and data) as well as distributed, possibly shared, last-level cache resources.

Tiled approach is crucial for scalability as tiles are tied up with an on-chip interconnection network (NoC) that allows parallel threads to exchange information and to synchronize [16, 36, 41]. Nowadays, in CMP architectures the latency for accessing on-chip memory hierarchy and off-chip main memory resources is the key to application performance. In the meanwhile, the aggregate requests from the numerous on-chip cores push towards increased chip I/O bandwidth support [21]. With the increase of the number of processor cores, the tiled design approach will be unavoidable and will imply a central role of NoCs in future CMP processors from both performance and power consumption points of view.

In particular, the traffic to be supported is determined by the heterogeneous memory access patterns from the cores, cache communication requirements, coherence protocol management, inter-core synchronization, the memory model, the runtime and OS support for parallelism and the global orchestration of power/performance. Latency and energy issues due to the on-chip communication are becoming increasingly critical in the design of modern architectures as overall instruction execution is becoming dramatically dominated by the cost of moving data and signals (e.g., for coherence), instead of by the computation [12].

7.2.1 Limits of Electronic Wires and Benefits of Silicon Photonics from the Architectural Viewpoint

Current and future CMPs running parallel applications require high-bandwidth and low-latency interconnections for enabling efficient and effective synchronization and operand/results/state transfer towards neighboring cores, as well as possibly managing resource and memory management traffic (e.g. for coherence/consistency). However, due to the emerging wire delay and energy scaling issues [31], and to the increasing number of on-chip cores, traditional electronic NoC designs are expected to fall short in fulfilling these requirements while maintaining an acceptable power consumption [41].

CMP systems are very demanding from the memory bandwidth point of view because of the aggregate requirements of the constantly increasing number of cores per chip [46, 52]. On top of that, latency for accessing the memory hierarchy is typically the key to application performance and in some cases even more important than the available bandwidth [11].

A related problem faced by current multi-core architectures, and that will be exacerbated by the ongoing technological trends, is the significant speed gap between cores and memory (*memory wall*). The conventional approach to alleviate off-chip memory bandwidth pressure consists of adopting larger on-chip caches [26], but technology scaling forces designers to cope with increasing wire delay effects, which limit access latencies of large caches, and affect power-consumption. Non-Uniform Cache Architecture (NUCA) cache organization is a design option for large, scalable, and high performance caches aiming at lowering the average access time compared to conventional ones. NUCA caches have been successful deployed in CMP systems [6, 27, 38, 58], and the optimal NUCA architecture (in terms of banks and data management policies) depends on the workload features [3, 15], the latency induced by wire technology, and the behaviors of the underlying interconnection network [39]. Furthermore, NUCA organization topology potentially fits the tiled multi-core design paradigm, where each tile, or group of tiles, can host one or a few NUCA cache banks. Regarding power consumption, the large die area occupied by Last Level Caches (LLC) and the scaling trends of deep submicron CMOS technologies, increase the role of static power due to leakage currents [32]. In NUCA and tiled architectures, power is spent in both cache banks and in the interconnection network between tiles. Existing solutions for mitigating static power consumption in caches typically change the average access latency to the banks and/or in traversing the NoC switches. Consequently, they affect the organization of NUCA banks and NoC topology to achieve optimal performance-power compromise. Therefore, NoC latency, bandwidth and consumption play a crucial role in the optimal tuning of NUCA caches in CMPs.

Summarizing, current and foreseeable trends in the development of CMPs expose a physical parallelism both on the computational side (multi-cores) and on the on-chip LLC (NUCA organization) through spatial resource replication on-chip. NoCs allow to interconnect such resources in a scalable way but intrinsically introduce locality effects, and thus non-uniformity, in the communication/collaboration effectiveness among all resources on-chip. However, designers are still aiming to expose to the programmer a uniform abstraction of both cores and LLC space due to the widely and convenient adoption of the shared-memory programming paradigm. This setting, constrained by the increasing wire-delay problems associated with technology scaling, poses tremendous latency pressure to the effective communication and synchronization between on-chip cores and to the coherent LLC management.

Integrated photonics is emerging as a promising technology that can theoretically introduce a positive discontinuity in the depicted scenario. In fact, its extremely fast propagation delay [1], which can be up to 10× faster than electronics [56], is extremely interesting for low-latency on-chip communications. Furthermore, photonics features an intrinsic end-to-end communication pattern, which can positively affect the power consumption of chip-wide transmissions. In fact, compared to the store-and-forward multi-hop pattern for on-chip electronic NoCs, where each hop consumes power, optical communication power can be considered insensible to on-chip distance.

Photonics have been used for board-to-board communication for a long time, aimed at inter-/intra-rack connectivity. However, since silicon-photonic integration became a feasible solution for CMP interconnects, a myriad of photonic networks have been proposed in scientific literature also as a solution to the lack of scalability of electrical NoCs. Various nanophotonic-based network topologies have been studied, from simple photonic rings [4, 5, 33, 43, 44, 64, 67] that operate like a crossbar, to complex articulated topologies [19, 55, 56, 62] that require or combine different transmission technologies, and to logical all-to-all interconnect designs [21, 22, 40]. More detailed and extensive discussion on these, and other, related works can be found in [18]. Previous works on photonic NoCs [4, 47, 64] have mainly explored the effect of photonics on CMP performance and dynamic power consumption, without accounting also for the combined effect of on-chip cache organization and cache-coherence requirements, static power consumption and heterogeneous-technology NoC architectures.

In the results presented here, we will propose examples of optical NoCs solutions and approaches stemming from a quite vertical perspective across design layers: from parallel application requirements within a shared memory cache-coherent CMP down-to physical level design choices in the optical communication channels.

7.3 Coherence Protocols

In this section we recall some background on coherence protocols for CMPs in the perspective of on-chip interconnections, both electronic and photonic.

Traditionally, coherence protocols have been divided in two broad categories: bandwidth intensive *snooping-based* coherence protocols, and *directory-based* protocols which achieve important traffic and energy savings. Although broadcast based protocols work well in small-scale systems, their

---

[1]The speed of light in silicon: 16 ps/mm group velocity.

performance is highly compromised beyond a small number of cores due to their prohibitive requirements of network bandwidth. Therefore, directory-based protocols are preferable in current and future CMPs. Plenty of research has been done, and is being done, to make this class of protocols more scalable and energy efficient [61]. Since on-chip optical interconnects became a possibility, many proposals of coherence protocols taking advantage from their expected performance characteristics have been proposed. The abundant bandwidth has encouraged the design of broadcast based protocols like ATAC [33] or C^3 [67], while the low latency can be used to build simpler protocols [63].

In all the analyses and results presented in this chapter we relied on MOESI coherence protocol, which encompasses all of the states of the MESI protocol [45], plus a fifth *Owned* state representing data that is both modified and shared. This state avoids the need to write a dirty cache line back to the next memory level when another processor tries to read it. In fact the *Owner* node, that holds the valid copy of a block, can *forward* the data directly to the requesting core through the NoC. The specific solutions proposed in this chapter are quite general and, for example, can fit also the MESI protocol.

In a CMP, coherence protocol evolution over a NoC can require numerous message transmissions according to the state of the involved modules (i.e., L1 and L2 caches, and directories). Even basic operations of the cores (Load, Store) can induce quite articulated sequences of NoC messages and protocol events, involving also conceptually multicast messages (e.g., inval-idations) [4, 10, 35]. Most of these messages are *control* messages, so don't carry data. From another perspective, the time to handle such protocol opera-tions and message transmissions can significantly contribute to the execution time of processor operations and are perceived by the processor as part of the memory hierarchy latency. Thus, speeding up the protocol evolution through a faster delivery of coherence control messages is crucial for the overall CMP performance.

Furthermore, control messages, being typically far smaller than data ones (e.g., 8 bytes compared to 64 bytes of data messages) can *benefit more from decreasing end-to-end transmission latency than from increasing bandwidth* (transmission parallelism and frequency). In fact the total time for a message to reach the destination in a digital network can be modeled as:

$$T_{tot} = T_{hfl} + \left(\left\lceil \frac{M_s}{par} \right\rceil - 1 \right) \cdot t_{mod} \tag{7.1}$$

where par is the channel bit parallelism, T_{hfl} is the head-flit latency (i.e., the time for the first par bits to reach the destination), M_s is the message size in bits and t_{mod} is the data modulation period (e.g., flit transmission frequency in a NoC). If message size is a small multiple of channel parallelism as it is the case for control messages, T_{tot} can be dominated by T_{hfl}, especially for long-distance on-chip transmissions. While data messages can benefit more from improvement of the second term.

7.4 Experimental Methodology

In our analysis we model tiled CMP architectures with up to 64 cores. Each core has private L1 caches (Instruction+Data) and a slice of a shared, distributed L2 cache. Directory information is distributed as well. Tiles are tied together by on-chip networks: electronic ones, hybrid electronic+optical ring-based networks, all-optical ring-based and optical-switch-based instances have been modeled, analyzed and compared. Our main focus was on novel future CMP architectures endowed with integrated optical networks, which could be applied in near future to a high-performance general purpose processor for a workstation or a rich *client* device.

7.4.1 Simulator

Performance evaluation made in this chapter are obtained with GEM5 simulator [9], in which we modeled both the ring-based *Passive Optical Network* network setups (Section 7.5), the *Optical Dynamic Reconfigurable* circuit-switched architectures (Section 7.6) and the various *Electronic-NoC* baseline configurations. The simulator is a modular platform for computer system architecture research, encompassing system-level architecture as well as processor microarchitecture. We adopted a full-system simulation approach running an unmodified Linux 2.4/2.6 operating system. Table 7.1 summarizes the main architectural details and parameters that have been considered in this work. Specifically, the table shows the common parameters, as well as the specific architectural features and numeric values of the two macro-cases analyzed in Sections 7.5 and 7.6. The *Passive Optical Networks* case in Section 7.5 employ a 2D-Mesh electronic NoC and/or a multi-waveguide ring-based optical network. Then, the *Optical Dynamic Reconfigurable Networks* case in Section 7.6 adopts an all-optical network split into path-setup ring-based network, a folded Torus optical switch-based network for message transmission and a *same-topology* electronic network for baseline reference.

Table 7.1 Parameters of the simulated architectures. The top section highlights the common values, while the following sections show the architectural differences related to the indicated chapter sub-sections

Cores	8/16/32/64 Cores (64 bit), 4 GHz
L1 caches	16 kB (I) + 16 kB (D), 2-way, 1 cycle hit time
L2 cache	16 MB, 8-way, shared and distributed 8x2MB, 16x1MB, 32x512kB, 64x256kB banks, 3/12 cycles tag/tag+data
Directory	MOESI protocol, 8/16/32/64 slices, 3 cycles
Main memory	4 GB, 300 cycles
Parameters related to Section 7.5: "Passive Optical Networks for CMPs"	
ENoC	2D-Mesh, 4 GHz, 5 cycles/hop, 32 nm, 1 V, 128 bit/flit
Photonic ring-based	3D, 1-9 parallel waveguides, 30 mm length, 8/16/32 I/O ports, 10 GHz, 64/70 (16 and 32) wavelengths, 460 ps full round
Parameters related to Section 7.6: "Optical Dynamic Reconfigurable Networks for CMPs"	
Core Clusters	4 (4 cores/cluster [4x4], 8 cores/cluster [4x8], 16 cores/cluster [4x16]) for the default configuration
Directory	MOESI protocol (Two-level for ONoC)
Baseline ENoC	electronic folded Torus (2x4/4x4/4x8/8x8), 5 GHz, 64 bits, 2 cycles
Path-setup ONoC	electronic folded Torus (2x4/4x4/4x8), 5 GHz, 8 bits, 2 cycles
Path-setup Arbiter	3D-stacked ring(s), 30 mm, 10 GHz, 1/2x32 DWDM, 460 ps full round latency, degenerate wavelength-router
Data ONoC/Arbiter	optical folded Torus (2x4/4x4/4x8/8x8), 10 GHz, 32 DWDM, 2 cycles, default path-multiplicity = 2

7.4.2 Benchmarks

Results are evaluated using the PARSEC 2.1 benchmark suite, a collection of heterogeneous multi-threaded applications spanning different emerging application domains [2, 8]. The benchmark set comprises applications for artificial vision, media processing, 3D and physical animation, similarity search and that feature parallel algorithms with various degrees of sharing and different parallelization schemes [7, 8]. So, these applications are representative of the current and near future datacenter workloads as well as high-end *client* devices, such as smart phones and tablets [14]. All benchmarks were instantiated with a degree of parallelism equal to the number of available cores (i.e., $n = 8, 16, 32, 64$) and this has resulted in at least n threads concurrently in execution on the CMP. Many benchmarks instantiate $n + 1$ threads, while others spawn more threads: e.g., *ferret* generates $3 + 4 \cdot n$ threads. Benchmarks were modified to enforce that each spawned thread is pinned to a fixed core of the processor (i.e., *core affinity*). This approach allows avoiding some non-determinism in the parallel benchmark execution because the Linux booted in our simulator assigns each thread to the same core across successive executions. So network performance differences are not affected by possibly different scheduling and thread-to-core assignment decisions. We used the medium input-set size to maintain a reasonable simulation time while using a good amount of executed instructions. Parsec benchmarks are composed of a) an initialization portion, in which the data structures are prepared and the required threads are spawned, b) the parallel region (called *Region Of Interest*, ROI) and c) the final part in which benchmark resources are released. For performance metric, in line with similar works, we considered the execution time of the entire parallel region (ROI) of each benchmark as representative of the end-user perceived parallel performance.

The total dynamic and static energy dissipated by all optical (lasers, modulators, photo-detectors and microring resonators) and electronic modules have been taken into account for comparing the consumption of the discussed architectures. Specifically, energy consumption of modules in the baseline ENoCs are derived from ORION 2.0 [30] and DSENT [60] for 32 nm technology. Then, static and dynamic consumption of the analyzed ONoCs is calculated based on state-of-the-art optical devices and parameters. In case of the ONoCs, additional electronic modules of the optical network interface were considered and, for the *Distributed arbiter* case, also the electronic arbiter(s) and optical switch driving circuitry were modeled and accounted.

Further details on performance/consumption models and parameters adopted in two cases discussed in this chapter can be found in [18, 19].

7.5 Passive Optical Networks for CMPs

Simple photonic topologies are the most likely to be first integrated in near future CMP systems, both in general-purpose and high-end client devices, and this section aims at dissecting the design tradeoffs around the adoption of a few photonic rings 3D-stacked [37] to a standard electronic NoC, and also as its substitute if sufficient optical bandwidth is deployed.

The main focus here is to reach significant energy reduction and then, possibly, performance improvements in tiled CMPs. As part of the crucial design options, we consider the different behavior of three arbitration techniques [44], and associated overhead, to access the shared photonic resources, in conjunction with other design parameters like bandwidth, number of cores as well as various traffic selection strategies in case of hybrid electronic/photonic NoC solutions. The considered solutions span from a highly parallel but mutually exclusive and power-hungry one, where only one message at a time flows quickly but needs full arbitration, to low-parallelism conflict-free and more energy efficient ones, where many transmissions can occur in parallel but each one experiences longer transmission time due to its lower bandwidth (see Equation (7.1)).

7.5.1 Analyzed Architecture

Figure 7.1 shows our baseline tiled architecture in case of a 16-core tiled setup where a 2D electronic Mesh (ENoC) connects all the tiles. Figure 7.1 shows also the logical ring-based photonic path (3D-stacked), which augments the baseline to obtain a photonic-enhanced architecture or completely substitutes the electronic Mesh. The depicted architecture can be trivially scaled up and down to 8 and 32 cores, respectively. We have analyzed different *ring-based* network configurations. Given current trends [34], Dense Wavelength Division Multiplexing (DWDM) is expected to scale up to at least 64 wavelengths per waveguide [64]. For this reason we have chosen as reference design value around 64 lambdas (64–70 range) but, for completeness, we have also analyzed the more conservative 16 and 32 WDM degrees. Spatial division multiplexing and/or mode-division multiplexing could provide bandwidth scaling also in case DWDM progression would result harder than expected. In the following, we define as *1-ring* (1R) configuration an one optical ring employing 64–70 wavelengths. Then, an *N-ring* (*N*R) configuration comprises N distinct photonic rings, each like the *1-ring* one, used together to increase parallelism in data transmission. In fact, the same set of wavelengths

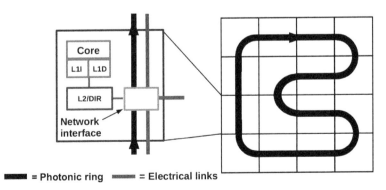

Figure 7.1 Logical view of the considered architecture for a 16-core setup. Each tile has access to the optical ring.

can be reused in different waveguides. We assume to use concentric rings in order to limit the number of crossings between waveguides as to limit the overall insertion loss of the optical paths. However also adopting this layout, a limited number of crossings is unavoidable. For instance, depending on the physical topology and routing, the laser light supply to internal waveguides may need to cross the surrounding ones, as it is the case in our configuration.

7.5.2 Arbitration Strategies and Physical Layout Implications

The considered medium-access strategies rely on shared optical resources (i.e., wavelengths in a specific waveguide) and employ a *token ring* [59] arbitration mechanism. Token ring guarantees starvation free mutual exclusion on an ordered shared *medium* relying on a special frame called *token*. In our case the token is a one-bit light packet circulating on the ring[2]. A node that wants to transmit, waits for the token, destructively reads it from the ring (light detection) and releases it again (token regeneration) after *medium* usage. In general, an effective arbitration technique should allocate the shared *medium* to the nodes fairly and should allow high channel utilization (high throughput) with low latency and low overhead. The quick round trip time of the photonic token (i.e., less than two clock cycles @4 GHz in idle conditions), plays a crucial role in this. We evaluated three well-known access strategies to the photonic ring resources [62, 64, 44]: *Multiple Writers Multiple Readers* (MWMR), *Single Writer Single Reader* (SWSR) and *Multiple Writers Single Reader* (MWSR, one independent bus per destination).

In MWSR, given only one waveguide the available colors are split and associated to the receivers (e.g., eight lambdas for each receiver in the 8-core, 1-ring setup, DWDM degree (64)) and receiver-specific tokens allow arbitration of senders. Thus, transmissions to all receivers can happen simultaneously without interference. For control-flow, also in MWMR case, we assume that receivers can issue a *Nack* signal towards the sender through the complementary part of the ring using a specific wavelength.

Figure 7.2 exemplifies the physical organization and optical resources needed for receiving (on the waveguide) and transmitting (before the coupler) a message in case of the MWSR scheme.

SWSR strategy enables the wavelength-routed [42] connectivity within a ring-shaped waveguide. In principle, it has one wavelength statically assigned to a sender-receiver pair (e.g., 56 lambdas for full connectivity in an 8-core

---

[2]Token is regenerated periodically by the last issuer.

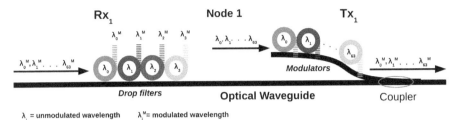

λ_i = unmodulated wavelength λ_i^M= modulated wavelength

Figure 7.2 (MWSR), possible physical organization of the optical receiving (Rx) and transmitting (Tx) endpoints in a CMP node.

CMP). Thus it does not require arbitration (no token) and allows concurrent communications between all cores (non-blocking network). The main issue of this scheme is the limited optical parallelism in each communication and scalability. Both MWSR and SWSR schemes statically allocate channels (wavelengths) to specific destinations or source-destination pairs, respectively. The drawback of static wavelength allocation is that, on average, the connectivity is over-provisioned in time because coherence traffic is typically unbalanced and bursty, as shown for SPLASH-2 [65] benchmarks in [43] and as we verified for the PARSEC 2.1 [8] too. Under this kind of traffic, the use of wider photonic paths on-demand (e.g., MWMR or MWSR), can be more efficient even at the cost of reduced transmission concurrency and some arbitration overhead.

Single waveguide MWMR aims at improving optical resource utilization by letting all the nodes use all available ring wavelengths in mutual exclusion. MWMR requires a single arbitration token and, before sending the message, it needs also a receiver-selection phase to get the destination node ready for receiving (i.e., tune the microring drop filters). Then, all the available wavelengths in the waveguide are used for message transmission with maximum parallelism and thus minimum latency.

Figure 7.3 sketches the considered physical organization of the optical resources needed for receiving (left) and transmitting (right) a message in each node using the MWMR access scheme within a 1-ring setup. Each node needs 64 microring modulators and, diversely from MWSR, 64 microring drop filters and photo-detectors to transmit and receive on all available colors. In each node, a splitter is employed to spill a fraction of the light toward the microring drop filters. This way, thanks to splitter tuning, all cores can potentially receive a similar amount of light and the worst case insertion loss is maintained far smaller than applying all microrings directly on the main

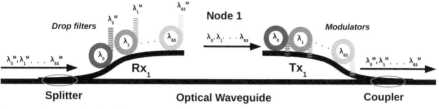

Figure 7.3   (MWMR), possible physical organization of the optical receiving (Rx) and transmitting (Tx) endpoints in a CMP node.

waveguide, as it can be feasible for MWSR scheme. Despite this, MWMR insertion loss is still quite bigger than in simpler schemes and thus induces higher laser power requirements. Furthermore, MWMR requires a far more microrings as every node has to modulate and receive light in all wavelengths and microrings consume static power for their thermal tuning.

The relative performance and energy consumption of such techniques are not obvious to predict. In fact going from MWMR to MWSR and SWSR, message concurrency increases but optical parallelism per transmission decreases, which in turn implies that each communication suffers from longer serialization (transmission) time.

7.5.3 Results

In the following we will discuss the execution time and energy consumption results of MWSR and MWMR access schemes for 8-/16-core architectures and for different overall available optical bandwidth. We will highlight the cross-interference between all these facets. Further details on 32-core architecture and, furthermore, on the most suitable traffic fraction that can be fruitfully routed on the photonic path in each considered configuration can be found in [21].

Figure 7.4 shows average benchmark results for 8- and 16-core architectures (left and right sections of the graphs, respectively) when the available rings for message transmission increase from one to two, four and eight/nine with an overall wavelength parallelism growing up to 512/576. For MWMR scheme we show 9-ring (9R) results instead of 8-ring ones (8R) because our data coherence messages are 72-byte long (576 bits) and in MWMR all wavelengths are used together for message transmission. Thus with 9-ring configuration MWMR needs only one optical aggregated flit, among

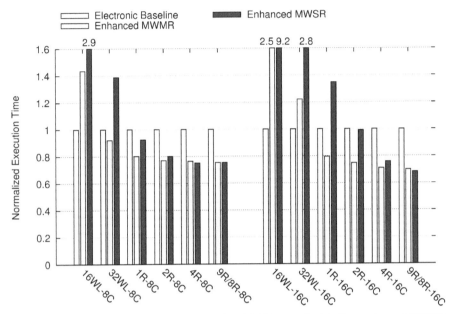

Figure 7.4   (8 and 16 cores, indicated as 8C and 16C respectively) Comparison of MWSR and MWMR access schemes normalized to the electronic baseline for 16, 32 wavelengths (16WL, 32WL), 1-, 2-, 4- and 9/8-ring setups (1R, 2R, 4R and 9/8R histograms respectively).

the nine rings, to transfer a full message. Conversely MWSR splits the available aggregated DWDM capability among the ring endpoints, therefore the parallelism of each transmission is far below data message size even with the 8-ring setup. Then, the first two bar-groups of Figure 7.4 show results for 1-ring (1R) with more conservative 16- and 32-wavelength parallelism (WDM), instead of 64/70 (DWDM). In these cases MWMR starts to have problems in managing the traffic, especially for the 16-core setup, delivering more than 20% and 2.5× slowdown for 32 and 16 DWDM, respectively. For the 8-core, 32-wavelength case MWMR still grants 8% speedup while the narrower configuration slows down by more than 40%. On the other hand MWSR is unusable for these narrower configurations despite the higher potential concurrency of transmissions. The available message bandwidth per endpoint is too low (1/2/4 bits) and each transmission suffers long serialization, and consequently token-conflict, times. With higher optical bandwidth available, MWMR scheme improves performance 20% with 1-ring configuration and up to about 25% at 2-ring in the 8-core case and about 30% at 4-ring for 16-core case. MWSR improves about 20% (2-ring) and 25%

(4-ring) for the 8-core setup and speeds up 25% (4-ring) and 30% (8-ring) in the 16-core case.

From the energy point of view, Figure 7.5 shows that the fewer and simpler structures required for MWSR scheme allow to maintain the ever dominant static power to promising values even when using high optical parallelism (4-ring). Neglecting the low DWDM cases which are unusable performance-wise, the figure shows that MWMR dramatically worsen energy consumption for 4-ring and 9-ring configurations. Then it doesn't deliver a huge improvement even in the 2-ring case and improves 70% and 60% only for 1-ring in 8- and 16-core configurations.

Overall, MWSR appears to be the best tradeoff at 2-ring (8-core) and 4-ring (16-core) to provide about 80% energy reduction while delivering almost all the potential performance speedup. In fact, aiming to achieve energy improvements, the combined performance-energy design sweet spot for MWMR scheme is the 1-ring case where similar speedups as MWSR can be achieved but with far less energy advantage. This result is remarkable

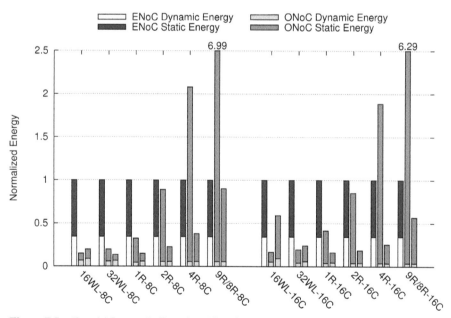

Figure 7.5   (8 and 16 cores, indicated as 8C and 16C respectively) Comparison of the access management strategies normalized to the electronic baseline for 16, 32 wavelengths (16WL, 32WL), 1-, 2-, 4- and 9/8-ring (1R, 2R, 4R and 9R/8R histograms respectively).

as MWSR adopts token arbitration to each destination and its consumption comprises also the additional ring for multiple token management.

7.6 Optical Dynamic Reconfigurable Networks for CMPs

This section proposes a novel arbitrated all-optical path-setup scheme for tiled CMPs adopting a dynamic circuit-switched optical network. It aims at significantly reducing *path-setup* latency and overall energy consumption as to effectively enable the support of cache coherence traffic (e.g., 8/72 byte messages). In fact, existing circuit-switched optical networks configure the optical switches in a sequential fashion, through the same network [25, 66] or assisted by an overlayed ENoC [56, 69], and their time overhead can be amortized only with thousands of bytes per transmission.

Our proposed reservation scheme is able to configure multiple photonic switches simultaneously through an orchestrated centralized or distributed arbitration. We make use of a specifically tailored optical *ring-based* path-setup network to connect requesting cores to the arbiter(s), and among arbiters, where multiple (distributed). This allows achieving the decoupling between the path-setup network's and optical data network's topologies.

7.6.1 Background on Circuit Switched Optical NoCs

Our proposal is built around a dynamic circuit-switched optical network [69, 56, 25, 24] which lays on a separate 3D-stacked layer and serves a tiled CMP architecture. This kind of networks employs optical switches [57, 23, 24, 48, 68], which are able to either let light go through from one direction towards another (e.g., east to west) or divert it towards a different exit (e.g., from east to south). Using these optical switching elements, optical resources (e.g., links, switches and thus paths and sub-paths) can be reused to serve different communications (paths) at different times.

A number of network topologies can be constructed, such as Mesh, Torus, Fat-Tree. In this section we will mainly focus on folded Torus organizations even if the proposed solutions are very general and can be applied with negligible modifications to other cases. Figure 7.6 shows a simple exemplification on how to obtain a complete optical dynamic topology starting from the simplest building block: a Micro-Ring Resonator (MRR). Specifically, interconnected MRRs can be arranged to form an optical router and a composition of the latters, interconnected via waveguides, can easily generate various, scalable, optical dynamic topologies. Some optical switches are

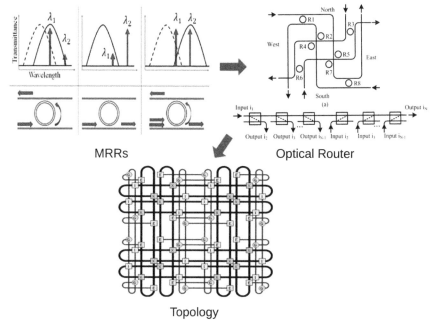

Figure 7.6 Dynamic circuit switched optical network composition scheme.

actually based on 1×2 optical Photonic Switching Elements (PSE), which implement a basic switch functionality where the light from an input port can be routed onto one of the two exit ports [24, 28, 29, 57].

One of states of the optical switch is typically stable and does not require to be maintained. The other state requires some action (i.e., ring tuning), and therefore energy and time, to be triggered and hold.

7.6.2 Limitations of Existing Solutions

PSE configuration, and thus, optical switch configuration, requires an electronic control circuit to drive the microrings and such circuit needs to be reached by configuration signals in order to properly setup an end-to-end path. This determines the need of a path-setup network, which is typically overlaid to the optical one, and has the same topology as the data transmission network for efficiency and simplicity reasons [56, 69]. Consequently, using such kind of network, the path-setup procedure is sequential (one hop at a time) and when the end-to-end path is reserved, an acknowledgment (*Ack*) signal is sent back usually with a small optical packet [56, 69] to the requestor so that it can start the transmission. After usage, the path is released through a *path-tear-down* packet sent into the electronic path-setup network [69] or also through a special optical packet traversing the path [56].

Figure 7.7 shows the flow of sub-path reservations, and local optical switch configurations, on such a path-setup network. Optical switches are visited and configured one after the other along the required path from source to destination according to the adopted routing (e.g., XY). Some time-overlap can be exploited advancing the reservation message while the optical switch is configured but overall latency is linear with the number of hops to be done.

Before the path is all reserved, the actual data message transmission cannot start and, for this reason, the potentially long path-setup latency can significantly compromise performance. Furthermore, when the network is loaded, paths are not always immediately available and need even more than one reservation attempt. Proposals exist for employing the same optical links of the circuit switched network also for path-setup [25]. In any case, sequentiality of the path-setup procedure remains the main overhead and in the latter case, despite of taking advantage from fast optical link traversal during path-setup steps, the scheme incurs also in multiple EO and OE conversion times, for triggering the switch reservation logic. Approximatively, path-setup latency in such networks can be modeled by the following equation:

$$Path\_setup\_latency = (H - 1) \times Tp + Tq \qquad (7.2)$$

where H is the number of hops, Tp is the perceived processing latency in each step and Tq is the total additional latency due to potential contention along the links to be reserved in the successive stages. When conflicts occur, Tq can be even the dominating component, especially considering the tear-down of partially reserved path and subsequent retry.

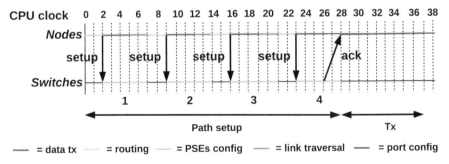

Figure 7.7   Sequential path-setup procedure for the Optical Baseline architecture.

To overcome this limitation due to such serial approach, we propose a simple optical path-setup network, used in conjunction with an arbiter, that enables simultaneous configuration of the required optical switches. With this solution we aim at both reducing the path-setup latency and making it not dependent on the number of hops to be traversed. The latency of the path-setup procedure for the worst case (32-core setup), which is shown in Figure 7.8, can then be expressed as:

$$Path\_setup\_latency = Tca + Ta + Tas + Tpse' + Tq' \qquad (7.3)$$

where *Tca* is the traversal time of a path-setup request from a core to the arbiter, *Ta* is the arbiter processing time, *Tas* is the traversal time of arbiter commands towards optical switches, and *Tpse'* is the configuration time of the Photonic Switching Elements (PSE) in the optical switches. Then, *Tq'* is the additional path-setup latency due to possible conflicts in reserving the required resources. It is typically quite smaller than *Tq*, because the arbitrated path-setup procedure exposes lower probability, and time penalty, in case

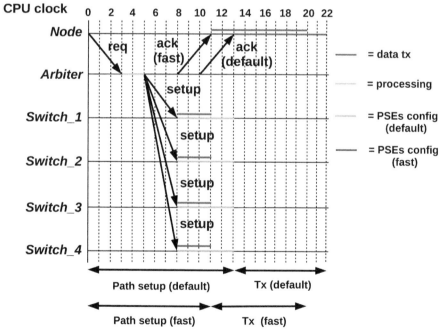

Figure 7.8 Simultaneous path-setup procedure for an arbiter-based configuration (intra-cluster).

of conflicts. Importantly, the equation does not depend on the number of hops of the requested path. Furthermore, our proposal allows decoupling the topologies of the path-setup and the main reconfigurable network.

7.6.3 Analyzed Architecture

We propose here an arbitrated solution where the main idea is to split the architecture into logical clusters each one managed by a separated arbiter (*Distributed Arbiter*). Each arbiter is able to reserve paths autonomously within its cluster and can collaborate with other ones to reserve inter-cluster paths.

Figure 7.9 gives a logical overview of a 4-cluster, 64-core overall (16 cores per cluster) CMP architecture served by the Distributed Arbiter for simultaneous path-setup. The right side of the figure shows the details of the intra-cluster architecture. In the figure, each cluster is associated to 16 contiguous cores and all 64 cores are interconnected through a folded Torus optical data transmission network, thus the topology is logically but not physically clustered.

A photonic ring provides dedicated channels for each core in a cluster to send requests to their arbiter and to receive *Acks* from it, along with the configuration commands for the optical switches (PSE-config) of the

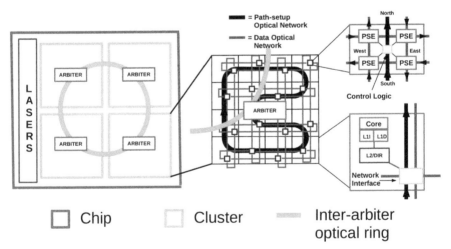

Figure 7.9 Overview of the Distributed Arbiter architecture for a 16-core setup. Each tile has access to the optical ring.

node. Therefore, it implements a contention-free point-to-point, not all-to-all, connectivity on top of WDM resources inside the ring-based waveguides (degenerate wavelength-routed [42]).

The arbiter receives the destination identifier, the transmission type (control/data message) and implicitly the sender ID according to incoming wavelength. Then, for intra-cluster domain, it checks if the required path from source to the destination is free and, in the positive case, it issues the configuration commands to all the needed switches at once. If at least one link in the path is not available, the reservation procedure aborts and the arbiter keeps the request into a requestor-specific buffer for later round-robin retry.

Inter-cluster paths need to be reserved through a distributed protocol involving two or more arbiters. Arbiters communicate among them through a fast *ring-based* network (green circle in Figure 7.9). As a technical detail, we associate inter-cluster links randomly to one of the two connected clusters, thus to one of the two associated arbiters, only paying attention to distribute such links in a balanced way.

7.6.3.1 Logical operating scheme

The Distributed Arbiter protocol operation for an inter-cluster path-setup can be seen as a hierarchical extension of the intra-cluster path-setup procedure (shown in Figure 7.8). As Figure 7.10 shows, source nodes always interact with the arbiter of their cluster (*local arbiter*) also for inter-cluster paths. In this case, the local arbiter determines the local sub-path and the list of *remote* arbiters required to reserve the whole path. This is done using the source and destination IDs, which are associated with the path-setup request, and a lookup table.

At this point, if the local sub-path is free, the *initiating* arbiter pre-reserves it, sends a remote request to the other arbiters and waits for their answer. In the meanwhile, it can serve other requests which don't need the pre-reserved links. The other arbiters threat the remote request in a similar way as a local one. They determine the respective required sub-path and they notify the initiating arbiter the outcome of their reservations through an *Ack/Nack* message. In the positive case they also pre-reserve the links.

Within a few cycles, depending on how busy the remote arbiters are, the initiating arbiter will receive all the *Acks/Nacks* responses. If they are all positive, the global path can be finally reserved and so it commits the local path reservation and sends the confirmation to the remote arbiters for reserving their parts. If one or more messages carry a *Nack*, the global

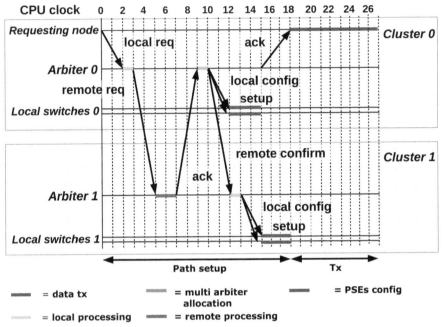

Figure 7.10   Successful inter-cluster path-setup procedure for the Distributed Arbiter architecture.

reservation cannot proceed and the arbiter sends back to the remote ones a clearing message to make the pre-reserved links available again. Similarly, it clears the local ones as well, and waits for a later retry. We would like to remark that in the considered setup with *directory-based* cache coherent load, the average traffic is not very intense and it is quite *bursty* (see Section 7.3). Therefore, the number of missed reservations and *Nacks* is very limited.

In case of global allocation possible, all the arbiters, local and remotes, then send the configuration messages to the required optical switches in their clusters. Finally, Figure 7.10 highlights that the initiating arbiter sends back an *Ack* to the requestor to enable the transmission. Paths are freed by the arbiters through timeouts based on the size of the message to be transmitted.

In order to limit inter-cluster traffic to the distributed home-nodes of the shared last-level cache, it was crucial to adopt a two-level modification of the adopted MOESI coherency protocol. This way, cores first try to reach a *cluster-local* home node for directory information and only in case of a local miss, they revert to the standard *global*, and unique, addressing of the home node within the cache. The high ratio of hits within the local cluster allows the majority of path reservations to be served locally and independently by single arbiters.

7.6.4 Results

Figure 7.11 compares the execution time results between the *Electronic Baseline* and our proposed *Distributed Arbiter* optical architecture for the considered 16-, 32- and 64-core cases. For the 16- and 32-core configurations we present the results achieved by the *Optical Baseline*. Then, we compare also against a simple *Centralized Arbiter* [19] which manages the requests from all the cores and is suitable for a smaller scale systems. For the biggest 64-core setup these two last configurations are performing so badly that we opted not to show it for the sake of graph clarity. The figure shows that the Distributed Arbiter is capable to achieve an around 10% speedup over the Electronic Baseline for the 16-core and 32-core cases. The Distributed Arbiter reaches the average break-even for the 64-core setup, for which some benchmarks show also good speedup (40% for *streamcluster*) while

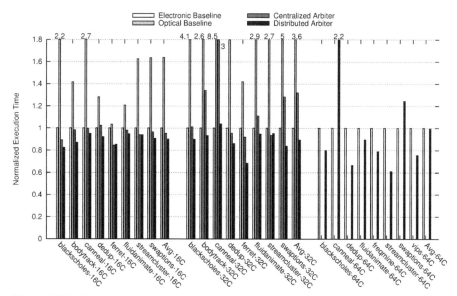

Figure 7.11 Execution time of the Distributed Arbiter scheme (4 arbiters) normalized to the Electronic Baseline for the 16- (left), 32- (middle) and 64-core (right) setups. For the 16- and 32-core configurations results are shown also for Optical Baseline adopting a sequential path-setup technique, and for a Centralized Arbiter scheme.

a few others expose slowdown. *canneal* drops performance more than $2\times$ and *swaptions* slows down by about 25%. These benchmarks expose high degree of data sharing among all the cores and, specifically for *canneal*, also quite fine-grained sharing. For these reasons they are less compatible with the arbitrated solutions than other benchmarks.

Finally, the most important observation is that the Distributed Arbiter solution is able to outperfom by far the Optical Baseline, in all the cases. The latency introduced by the sequential path-setup procedure (electronic path-setup, as shown in Figure 7.7) of this latter solution, is irrecoverably detrimental when applied to short packets. Indeed, the path-setup latency is way longer than the data transmission itself, producing useless results. These results confirm the limitations introduced by classical electro/optical recon-figurable solutions when dealing with cache coherence traffic, as discussed in Section 7.6.2.

Figure 7.12 shows the energy consumption of the Centralized (first columns) and Distributed (second columns) Arbiter setups normalized to the Electronic Baseline. As a general comment, the Distributed Arbiter is able to achieve energy reductions of around 45% for 16-core setup, of around 40% for 32-core configuration, and a slight improvement for the 64-core one, making it very appealing in comparison to the Electronic Baseline. Furthermore, as already discussed, the Centralized solution does not scale well to the biggest 64-core setup, losing, on average, a 30% in comparison to the Electronic Baseline. Furthermore, Figure 7.12 high-lights that, as it is typical in optical networks for CMPs using current technology, the biggest fraction of the energy spent, is by far the static component. Therefore, a performance speedup/slowdown typically reflects positively/negatively also in the energy consumption. For this reason, the Optical Baseline is absolutely not competitive from the energy point of view because of the tremendous induced performance slowdown. Even in terms of instantaneous power, despite lacking some optical structures for the Distributed optical arbitration, the Optical Baseline needs a full electronic NoC, in parallel to the optical one, for the path setup. Hence, its instanta-neous average power cannot be dramatically different from the one of the Arbitrated solutions and, most importantly, such slight difference doesn't change the dominant effect in increased static energy due to its significant slowdown.

These results prove the goodness of the proposed Distributed Arbiter solution that is delivering both performance and significant energy improve-ments over a competitive Electronic Baseline.

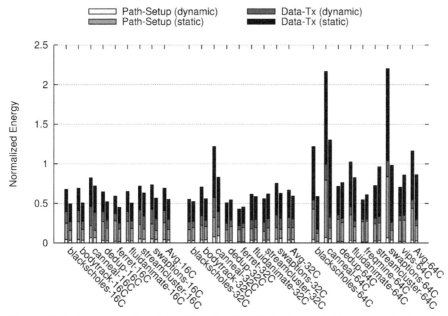

Figure 7.12 Energy results of the Centralized and Distributed Arbiter normalized to the Electronic Baseline for the 16- (left), 32- (middle) and 64-core (right) setups.

7.7 Conclusions

Silicon photonics has low-latency and low-energy raw features. However integrated photonic interconnection is very different from the familiar electronic on-chip networks counterparts and adds a number of new design space directions, interacting with the existing ones. For these reasons, at the moment, only through a comprehensive vertical analysis of such options, and of their implications in the CMP scenario (computer architecture and benchmark facets), it is possible to aim at exploiting its full potential and, most of all, avoid suboptimal designs. Most probably, as time passes and integrated photonics technologies will stabilize, more stable design criteria will emerge and more independent layers will be identified, as it is happening now in the electronic domain.

This chapter highlighted a couple of examples where the vertical interaction between highly abstract issues (e.g., shared memory and parallel applications) and very detailed physical facets of the integrated optical interconnections are exposed. Results show where performance/consumption sweetspots can be met and highlight also some counter-intuitive outcomes deriving by the specific interference between layers. We investigated CMP setups from 8 up to 64 cores in order to cover current and near future design requirements and two distinct kinds of optical networks: ring-based topology and dynamic circuit-switched.

We first addressed the various design choices around the usage of a simple photonic network (ring topology, passive) mainly aimed at improving energy consumption and, where possible, execution time in high-end tiled CMP systems. The main outcome was that the bursty cache-coherent traffic can be effectively served by assigning optical resources dynamically through a token arbitration protocol to promote transmission parallelism. Specifically, a Multiple Writers Single Reader (MWSR) scheme showed the best equilibrium between power consumption and performance speedup through its reduced insertion loss, higher message transmission concurrency and a shorter arbitration time compared to Multiple Writers Multiple Readers (MWMR) technique.

Then, we presented a novel optical path-setup scheme for circuit-switched optical networks (reconfigurable), based on a distributed arbitration which, un-intuitively, relies on the partial serialization of path-setup requests within arbiters. This way we achieve a fast path-setup procedure that, for the first time, enables this kind of networks to sustain cache coherence traffic and decouples the topologies of path-setup and data networks. The scheme is based on the simultaneous configuration of the optical switches and a clustered, hierarchical, architecture along with a specific cache coherence protocol tuning. The analyzed performance and consumption results up to 64-cores CMP configurations confirm a promising applicability of the proposed approach.

References

[1] AMD. Amd opteron 6000 series platform, 2013.

[2] N. Barrow-Williams, C. Fensch, and S. Moore. A communication characterisation of splash-2 and parsec. In *Proceedings of the 2009 IEEE International Symposium on Workload Characterization (IISWC)*, IISWC '09, Washington, DC, USA, 2009. IEEE Computer Society.

[3] S. Bartolini, P. Foglia, M. Solinas, and C. A. Prete. Feedback-driven restructuring of multi-threaded applications for nuca cache performance in cmps. In *SBAC-PAD*, 2010.

[4] S. Bartolini and P. Grani. A simple on-chip optical interconnection for improving performance of coherency traffic in cmps. In *15th Euromicro Digital System Design Conf.*, Cesme, Turkey, Sept. 2012.

[5] S. Bartolini and P. Grani. Co-tuning of a hybrid electronic-optical network for reducing energy consumption in embedded cmps. In *Proceedings of the First International Workshop on Many-core Embedded Systems*, MES '13, New York, NY, USA, 2013. ACM.

[6] B. M. Beckmann and D. A. Wood. Managing wire delay in large chip-multiprocessor caches. In *Proceedings of the 37th Annual IEEE/ACM International Symposium on Microarchitecture*, MICRO 37, pages 319–330, Washington, DC, USA, 2004. IEEE Computer Society.

[7] C. Bienia, S. Kumar, and K. Li. Parsec vs. splash-2: A quantitative comparison of two multithreaded benchmark suites on chip-multiprocessors. In *Workload Characterization, 2008. IISWC 2008. IEEE International Symposium on*, pages 47–56, Sept 2008.

[8] C. Bienia, S. Kumar, J. P. Singh, and K. Li. The parsec benchmark suite: characterization and architectural implications. In *Proceedings of the 17th international conference on Parallel architectures and compilation techniques*, PACT '08, New York, NY, USA, 2008. ACM.

[9] N. Binkert, R. Dreslinski, L. Hsu, K. Lim, A. Saidi, and S. Reinhardt. The m5 simulator: Modeling networked systems. *Micro, IEEE*, 26(4), 2006.

[10] T. Bui, B. Greskamp, I.-j. Liao, and M. Tucknott. Measuring and characterizing cache coherence traffic, 2004.

[11] P. Conway and B. Hughes. The amd opteron northbridge architecture. *Micro, IEEE*, 27(2), March 2007.

[12] W. Dally. Keynote: "power, programmability, and granularity: The challenges of exascale computing". In *International Parallel & Distributed Processing Symposium*, Anchorage (Alaska) USA, 2011.

[13] P. Dong, W. Qian, H. Liang, R. Shafiiha, N.-N. Feng, D. Feng, X. Zheng, A. V. Krishnamoorthy, and M. Asghari. Low power and compact reconfigurable multiplexing devices based on silicon microring resonators. *Opt. Express*, 18(10), May 2010.

[14] H. Falaki, R. Mahajan, S. Kandula, D. Lymberopoulos, R. Govindan, and D. Estrin. Diversity in smartphone usage. In *Proceedings of the 8th international conference on Mobile systems, applications, and services*, MobiSys '10, New York, NY, USA, 2010. ACM.

[15] P. Foglia, F. Panicucci, C. Prete, and M. Solinas. Analysis of performance dependencies in nuca-based cmp systems. In *Computer Architecture and High Performance Computing, 2009. SBAC-PAD '09. 21st International Symposium on*, pages 49–56, Oct 2009.

[16] H. Franke, M. Kirkwood, and R. Russell. Fuss, futexes and furwocks: Fast userlevel locking in linux. In *Ottawa Linux Summit*, 2002.

[17] A. Garca-Guirado, R. Fernndez-Pascual, J. M. Garca, and S. Bartolini. Managing resources dynamically in hybrid photonic-electronic networks-on-chip. *Concurrency and Computation: Practice and Experience*, 26(15):2530–2550, 2014.

[18] P. Grani and S. Bartolini. Design options for optical ring interconnect in future client devices. *J. Emerg. Technol. Comput. Syst.*, 10(4), June 2014.

[19] P. Grani and S. Bartolini. Simultaneous optical path setup for reconfigurable photonic networks in tiled cmps. In *The 2014 International Conference on High Performance Computing and Communications*, Paris, France, 2014.

[20] P. Grani, S. Bartolini, E. Furdiani, L. Ramini, and D. Bertozzi. Integrated cross-layer solutions for enabling silicon photonics into future chip multiprocessors. In *The 19th Annual International Mixed-Signal, Sensors and Systems Test Workshop*, Porto Alegre, Brazil, 2014. Optical Society of America.

[21] P. Grani, R. Hendry, S. Bartolini, and K. Bergman. Boosting multi-socket cache-coherency with low-latency silicon photonic interconnects. In *Computing, Networking and Communications (ICNC), 2015 International Conference on*, pages 830–836, Feb 2015.

[22] P. Grani, R. Proietti, S. Cheung, and S. Yoo. Flat-topology high-throughput compute node with awgr-based optical-interconnects. *Lightwave Technology, Journal of*, PP(99):1–1, December 2015.

[23] W. M. Green, M. Yang, S. Assefa, J. van Campenhout, B. G. Lee, C. Jahnes, F. E. Doany, C. Schow, J. A. Kash, and Y. Vlasov. Silicon electro-optic 4x4 non-blocking switch array for on-chip photonic networks. In *Optical Fiber Communication Conference/National Fiber Optic Engineers Conference 2011*. Optical Society of America, 2011.

[24] H. Gu, J. Xu, and Z. Wang. A novel optical mesh network-on-chip for gigascale systems-on-chip. In *Circuits and Systems, 2008. APCCAS 2008. IEEE Asia Pacific Conference on*, Nov 2008.

[25] H. Gu, J. Xu, and W. Zhang. A low-power fat tree-based optical network-on-chip for multiprocessor system-on-chip. In *Design, Automation Test in Europe Conference Exhibition, 2009. DATE '09.*, 2009.

[26] N. Hardavellas, M. Ferdman, B. Falsafi, and A. Ailamaki. Toward dark silicon in servers. *Micro, IEEE*, 31(4):6–15, July 2011.

[27] J. Huh, C. Kim, H. Shafi, L. Zhang, D. Burger, and S. Keckler. A nuca substrate for flexible cmp cache sharing. *Parallel and Distributed Systems, IEEE Transactions on*, 18(8):1028–1040, Aug 2007.

[28] R. Ji, L. Yang, L. Zhang, Y. Tian, J. Ding, H. Chen, Y. Lu, P. Zhou, and W. Zhu. Five-port optical router for photonic networks-on-chip. *Opt. Express*, 19(21), Oct 2011.

[29] R. Ji, L. Yang, L. Zhang, Y. Tian, J. Ding, H. Chen, Y. Lu, P. Zhou, and W. Zhu. Microring-resonator-based four-port optical router for photonic networks-on-chip. *Opt. Express*, 19(20), Sep 2011.

[30] A. Kahng, B. Li, L.-S. Peh, and K. Samadi. Orion 2.0: A power-area simulator for interconnection networks. *Very Large Scale Integration (VLSI) Systems, IEEE Transactions on*, 20(1), Jan 2012.

[31] C. Kim, D. Burger, and S. W. Keckler. An adaptive, non-uniform cache structure for wire-delay dominated on-chip caches. *SIGOPS Oper. Syst. Rev.*, 36(5), Oct. 2002.

[32] N. Kim, T. Austin, D. Baauw, T. Mudge, K. Flautner, J. Hu, M. Irwin, M. Kandemir, and V. Narayanan. Leakage current: Moore's law meets static power. *Computer*, 36(12):68–75, Dec 2003.

[33] G. Kurian, J. E. Miller, J. Psota, J. Eastep, J. Liu, J. Michel, L. C. Kimerling, and A. Agarwal. Atac: a 1000-core cache-coherent processor with on-chip optical network. In *Proceedings of the 19th international conference on Parallel architectures and compilation techniques*, PACT '10, New York, NY, USA, 2010. ACM.

[34] S. Le Beux, J. Trajkovic, I. O'Connor, G. Nicolescu, G. Bois, and P. Paulin. Optical ring network-on-chip (ornoc): Architecture and design methodology. In *Design, Automation Test in Europe Conference Exhibition (DATE), 2011*, march 2011.

[35] D. Lenoski, J. Laudon, K. Gharachorloo, A. Gupta, and J. Hennessy. The directory-based cache coherence protocol for the dash multiprocessor. *SIGARCH Comput. Archit. News*, 18(2SI), May 1990.

[36] Z. Li, J. Wu, L. Shang, A. R. Mickelson, M. Vachharajani, D. Filipovic, W. Park, and Y. Sun. A high-performance low-power nanophotonic on-chip network. In *Proceedings of the 14th ACM/IEEE international symposium on Low power electronics and design*, ISLPED '09, New York, NY, USA, 2009. ACM.

[37] G. H. Loh. 3d-stacked memory architectures for multi-core processors. In *Proc. of the 35th Annual International Symposium on Computer Architecture*, Washington, DC, USA, 2008. IEEE Computer Society.

[38] J. Merino, V. Puente, and J. Gregorio. Esp-nuca: A low-cost adaptive non-uniform cache architecture. In *High Performance Computer Architecture (HPCA), 2010 IEEE 16th International Symposium on*, pages 1–10, Jan 2010.

[39] N. Muralimanohar, R. Balasubramonian, and N. Jouppi. Optimizing nuca organizations and wiring alternatives for large caches with cacti 6.0. In *Microarchitecture, 2007. MICRO 2007. 40th Annual IEEE/ACM International Symposium on*, pages 3–14, Dec 2007.

[40] C. Nitta, M. Farrens, and V. Akella. Dcaf - a directly connected arbitration-free photonic crossbar for energy-efficient high performance computing. In *Proceedings of the 2012 IEEE 26th International Parallel and Distributed Processing Symposium*, IPDPS '12, pages 1144–1155, Washington, DC, USA, 2012. IEEE Computer Society.

[41] I. O'Connor and F. Gaffiot. On-chip optical interconnect for low-power. In E. Macii, editor, *Ultra-Low Power Electronics and Design*, Kluwer, Dordrecht, 2004.

[42] I. O'Connor, D. Van Thourhout, and A. Scandurra. Wavelength division multiplexed photonic layer on cmos. In *Proceedings of the 2012 Interconnection Network Architecture: On-Chip, Multi-Chip Workshop*, INA-OCMC '12, New York, NY, USA, 2012. ACM.

[43] Y. Pan, J. Kim, and G. Memik. Flexishare: Channel sharing for an energy-efficient nanophotonic crossbar. In *High Performance Computer Architecture (HPCA), 2010 IEEE 16th International Symposium on*, jan. 2010.

[44] Y. Pan, P. Kumar, J. Kim, G. Memik, Y. Zhang, and A. Choudhary. Firefly: illuminating future network-on-chip with nanophotonics. *sigARCH Comp. Arch. News*, 37(3), 2009.

[45] M. S. Papamarcos and J. H. Patel. A low-overhead coherence solution for multiprocessors with private cache memories. In *Proceedings of the 11th Annual International Symposium on Computer Architecture*, ISCA '84, pages 348–354, New York, NY, USA, 1984. ACM.

[46] M. Pavlovic, Y. Etsion, and A. Ramirez. On the memory system requirements of future scientific applications: Four case-studies. In *Workload Characterization (IISWC), 2011 IEEE International Symposium on*, Nov 2011.

[47] M. Petracca, B. G. Lee, K. Bergman, and L. P. Carloni. Design exploration of optical interconnection networks for chip multiprocessors. In *Proceedings of the 2008 16th IEEE Symposium on High Performance Interconnects*, Washington, DC, USA, 2008. IEEE Computer Society.

[48] A. W. Poon, F. Xu, and X. Luo. Cascaded active silicon microresonator array cross-connect circuits for wdm networks-on-chip. *Proc. SPIE*, 6898, 2008.

[49] L. Ramini, H. T. Fankem, A. Ghiribaldi, P. Grani, , M. Ortin-Obon, A. Boos, and S. Bartolini. Towards compelling cases for the viability of silicon-nanophotonic technology in future manycore systems. In *The 8th International Symposium on Networks-on-Chip*, Ferrara, Italy, 2014.

[50] L. Ramini, P. Grani, S. Bartolini, and D. Bertozzi. Contrasting wavelength-routed optical noc topologies for power-efficient 3d-stacked multicore processors using physical-layer analysis. In *Proceedings of the Conference on Design, Automation and Test in Europe*, DATE '13, pages 1589–1594, San Jose, CA, USA, 2013. EDA Consortium.

[51] L. Ramini, P. Grani, H. T. Fankem, A. Ghiribaldi, S. Bartolini, and D. Bertozzi. Assessing the energy break-even point between an optical noc architecture and an aggressive electronic baseline. In *Proceedings of the Conference on Design, Automation & Test in Europe*, DATE '14, pages 308:1–308:6, 3001 Leuven, Belgium, Belgium, 2014. European Design and Automation Association.

[52] B. M. Rogers, A. Krishna, G. B. Bell, K. Vu, X. Jiang, and Y. Solihin. Scaling the bandwidth wall: Challenges in and avenues for cmp scaling. *SIGARCH Comput. Archit. News*, 37(3), June 2009.

[53] A. Ros, M. E. Acacio, and J. M. Garca. Scalable directory organization for tiled cmp architectures. In *CDES'08*, 2008.

[54] Samsung. Samsung exynos 5 octa, 2013.

[55] M. Schulz, B. S. White, S. A. McKee, H.-H. S. Lee, and J. Jeitner. Owl: Next generation system monitoring. In *Proceedings of the 2Nd Conference on Computing Frontiers*, CF '05, pages 116–124, New York, NY, USA, 2005. ACM.

[56] A. Shacham, K. Bergman, and L. P. Carloni. Photonic networks-on-chip for future generations of chip multiprocessors. *Computers, IEEE Transactions on*, 57(9), 2008.

[57] N. Sherwood-Droz, H. Wang, L. Chen, B. G. Lee, A. Biberman, K. Bergman, and M. Lipson. Optical 4x4 hitless slicon router for optical networks-on-chip (noc). *Opt. Express*, 16(20), Sep 2008.

[58] B. Sinharoy, R. Kalla, W. J. Starke, H. Q. Le, R. Cargnoni, J. A. Van Norstrand, B. J. Ronchetti, J. Stuecheli, J. Leenstra, G. L. Guthrie, D. Q. Nguyen, B. Blaner, C. F. Marino, E. Retter, and P. Williams. Ibm power7 multicore server processor. *IBM Journal of Research and Development*, 55(3):1:1–1:29, May 2011.

[59] C. Smythe. Iso 8802/5 token ring local-area networks. *Electronics Communication Engineering Journal*, 11(4), aug 1999.

[60] C. Sun, C.-H. O. Chen, G. Kurian, L. Wei, J. Miller, A. Agarwal, L.-S. Peh, and V. Stojanovic. Dsent - a tool connecting emerging photonics with electronics for opto-electronic networks-on-chip modeling. In *Proceedings of the 2012 IEEE/ACM Sixth International Symposium on Networks-on-Chip*, Washington, DC, USA, 2012. IEEE Computer Society.

[61] S. Vangal, J. Howard, G. Ruhl, S. Dighe, H. Wilson, J. Tschanz, D. Finan, A. Singh, T. Jacob, S. Jain, V. Erraguntla, C. Roberts, Y. Hoskote, N. Borkar, and S. Borkar. An 80-tile sub-100-w teraflops processor in 65-nm cmos. *Solid-State Circuits, IEEE Journal of*, 43(1), 2008.

[62] D. Vantrease, N. Binkert, R. Schreiber, and M. H. Lipasti. Light speed arbitration and flow control for nanophotonic interconnects. In *Proceedings of the 42nd Annual IEEE/ACM International Symposium on Microarchitecture*, MICRO 42, New York, NY, USA, 2009. ACM.

[63] D. Vantrease, M. Lipasti, and N. Binkert. Atomic coherence: Leveraging nanophotonics to build race-free cache coherence protocols. In *High Performance Computer Architecture (HPCA), 2011 IEEE 17th International Symposium on*, pages 132–143, Feb 2011.

[64] D. Vantrease, R. Schreiber, M. Monchiero, M. McLaren, N. P. Jouppi, M. Fiorentino, A. Davis, N. Binkert, R. G. Beausoleil, and J. H. Ahn. Corona: System implications of emerging nanophotonic technology. *sigARCH Comp. Arch. News*, 36(3), 2008.

[65] S. C. Woo, M. Ohara, E. Torrie, J. P. Singh, and A. Gupta. The splash-2 programs: characterization and methodological considerations. *SIGARCH Comput. Archit. News*, 23(2), May 1995.

[66] X. Wu, Y. Ye, W. Zhang, W. Liu, M. Nikdast, X. Wang, and J. Xu. Union: A unified inter/intra-chip optical network for chip multiprocessors. In *Nanoscale Architectures (NANOARCH), 2010 IEEE/ACM International Symposium on*, 2010.

[67] Y. Xu, Y. Du, Y. Zhang, and J. Yang. A composite and scalable cache coherence protocol for large scale cmps. In *Proceedings of the international conference on Supercomputing*, ICS '11, New York, NY, USA, 2011. ACM.

[68] Y. Ye, X. Wu, J. Xu, W. Zhang, M. Nikdast, and X. Wang. Holistic comparison of optical routers for chip multiprocessors. In *Anti-Counterfeiting, Security and Identification (ASID), 2012 International Conference on*, Aug 2012.

[69] Y. Ye, J. Xu, X. Wu, W. Zhang, W. Liu, and M. Nikdast. A torus-based hierarchical optical-electronic network-on-chip for multiprocessor system-on-chip. *J. Emerg. Technol. Comput. Syst.*, 2012.

PART III

Challenges in Performance Analysis and Design Solutions

8

Thermal Management of Silicon Photonic NoCs in Many-core Systems

**Tiansheng Zhang[1], Jonathan Klamkin[2], Ajay Joshi[1]
and Ayse K. Coskun[1]**

[1]Dept. of ECE, Boston University, Boston, MA, U.S.A.
[2]Dept. of ECE, University of California Santa Barbara, Santa Barbara, CA, U.S.A.

Abstract

Silicon-photonic network-on-chip (PNoC) is a promising candidate to provide reliable and high-bandwidth on-chip communication for future many-core systems. Compared to electrical network-on-chip, PNoC has lower data-dependent energy, higher bandwidth density and lower global communication latency. The optical devices that form silicon-photonic links in the PNoC, however, are highly sensitive to thermal and manufacturing process variations, which can lead to a shift in the optical frequency of these devices. This could lead to data loss and transmission errors, especially in many-core systems since they are exposed to higher thermal and process variations due to their large chip area. Thus, thermal management of on-chip optical devices is essential for reliable PNoC communication in many-core systems.

This chapter first provides background on PNoC, including a description of the basic components, their operating mechanism, and their sensitivity to thermal and process variations. Then, we discuss existing design-time and runtime techniques for PNoC thermal management. The chapter especially focuses on techniques with a design automation and/or software optimization component. Such techniques provide low-cost opportunities to support efficient integration of the PNoC technology with the traditional CMOS technology. Finally, we summarize this chapter and present the remaining challenges in this field.

227

8.1 Introduction

Computer systems are moving from the multi-core era to the many-core era to support the ever-increasing thread-level parallelism exhibited by today's applications (e.g., cyber-physical and big data). In tandem, to support the large on-chip core counts, a high-bandwidth network-on-chip (NoC) is required. However, it is becoming increasingly difficult to meet these high-bandwidth requirements using the traditional electrical link technology. Silicon-photonic NoC (PNoC) has been projected as a promising replacement to electrical NoC due to its significantly higher bandwidth density, lower global communication latency, and lower data-dependent power [1–7].

A PNoC is composed of silicon-photonic links, which use optical signals (after combined with some electrical circuits or links) for on-chip communication. In a generic silicon-photonic link, a laser source emits an optical wave that is coupled into a waveguide. Then, the optical wave is modulated at the transmitter side in a ring modulator using an electrical driver circuit. In this way, data is converted from the electrical domain to the optical domain. At the receiver, the modulated optical wave is filtered using a ring filter. The filtered wave is then incident on a photodetector, whose output (in the form of an electrical current) is fed to an electrical receiver. To ensure reliable optical wave modulation and filtration, the ring modulator and ring filter are designed to operate at the same optical frequency, which also must match the frequency of the optical wave emitted by the laser source. A mismatch in optical frequency among these devices can introduce data transmission errors or can potentially break the link entirely [2].

Though every ring resonator and laser source in the PNoC are designed to operate at a specific optical frequency, the actual optical frequency does not always remain at the designed value, mainly because of two factors: process variations and thermal variations. The optical frequency of a ring resonator is highly sensitive to its physical dimensions such as height, width, and radius. Within-die process variations during manufacturing cause variations in these physical dimensions and, thus, lead to a mismatch between the designed and the actual optical frequencies of ring resonators. Also, the variations in optical devices' temperatures significantly affect their optical frequencies due to the thermo-optic effect [8]. In a many-core system, the impact of process and thermal variations on optical frequency mismatch is further amplified due to the large chip area and more diverse on-chip thermal conditions.

It is possible to (at least partially) counter the impact of both types of variations by managing the thermal conditions of optical devices. Thus, there is a critical need for thermal management techniques that are able to help align on-chip optical devices' frequencies within each silicon-photonic link to guarantee reliable operation of a many-core system.

The work that has been carried out in the area of PNoC thermal management can be categorized as design-time techniques and runtime techniques. During chip design stage, extra hardware such as micro-heaters [9] can be added to control optical devices' temperatures. Designers could also leverage athermal devices [10, 11] to make the PNoC less sensitive to thermal variations. Alternatively, chip floorplanning techniques [12, 13] that explicitly consider optical device locations can help lower the impact of process and thermal variations. As for runtime thermal management techniques, workload allocation techniques [14–16] and NoC routing algorithms [17] can be engineered to manage the thermal conditions of optical devices.

In this chapter, we elaborate on the challenges and the cutting-edge work in thermal management of PNoCs in many-core systems. The organization of this chapter is as follows:

- We first discuss the thermal sensitivity of on-chip optical devices and the potential impacts of thermal variations on the PNoC functionality.
- We then present a detailed overview of design-time thermal management approaches for PNoCs in many-core systems. We mainly review device-level and chip floorplanning techniques.
- We carry out a detailed survey of runtime thermal management approaches for PNoCs in many-core systems. In addition, we also introduce the simulation frameworks that are used for developing and evaluating PNoC thermal management policies.

8.2 Thermal Sensitivity of Optical Devices in PNoCs

A PNoC is composed of silicon-photonic links, each of which contains the following devices: (1) a laser source that emits optical waves, (2) a coupler that couples optical waves from the laser source to a waveguide, (3) a waveguide that carries optical waves, (4) a driver that receives electrical signals from the circuit side, (5) a ring modulator inside whom the driver modulates optical waves at the transmitter side based on the electrical signals of the driver, (6) a ring filter that filters optical waves at the receiver side, (7) a photodetector that converts optical signals into electrical signals, and (8) an amplifier that amplifies electrical signals, as shown in Figure 8.1. To transmit data without errors or loss, the ring modulator and filter must resonate at the same frequency as the optical frequency of the corresponding laser source.

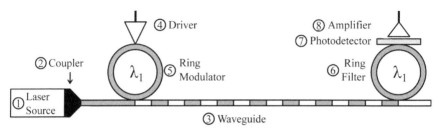

Figure 8.1 A silicon-photonic link.

Optical devices such as ring resonators, however, are highly sensitive to temperature variations due to the thermo-optic effect (i.e., thermal modulation of the refractive index of a material). This is because ring resonators are manufactured using silicon, which possesses a relatively large thermo-optic coefficient (TOC). The larger a TOC is, the more significantly the refractive index changes under a given temperature change, and the larger optical frequency shift happens as a result. Thus, a temperature mismatch between a ring modulator and a ring filter leads to their optical frequency mismatch. During data transmission, it is necessary to ensure the ring modulator and the ring filter match in their optical frequency for link integrity.

The thermal sensitivity of ring resonators depends on the design choices such as center resonant wavelength and ring radius. For example, ring resonators designed around a center wavelength of 1550 nm (λ_0) have a thermal sensitivity of 78 pm/K ($\Delta\lambda_R$) [18], which corresponds to 9.7 GHz/K (Δf_R) shift in optical frequency based on the following equations.

$$F_0 = \frac{c}{\lambda_0} = 193 \; THz \tag{8.1}$$

$$\frac{\Delta\lambda_R}{\lambda_0} = \frac{\Delta f_R}{F_0} \tag{8.2}$$

For every degree of temperature difference between a ring modulator and its corresponding ring filter, there is a 9.7 GHz mismatch in resonant frequency. Depending on the frequency spacing between optical waves multiplexed on one waveguide, the mismatch in resonant frequency may cause either data loss or data read by a different ring filter from intended.

The frequency spacing between optical waves is determined by the free spectral range (FSR) of the selected ring resonator design and the number of optical waves multiplexed in a waveguide as shown in Equations (8.3) and (8.4).

$$FSR = \frac{c}{2\pi r n_g} \tag{8.3}$$

$$F_{spacing} = \frac{FSR}{n_\lambda} \tag{8.4}$$

In these equations, n_g is the group index, c is the speed of light, n_λ is the number of optical waves per waveguide, and $F_{spacing}$ is the frequency spacing between two adjacent optical waves in a waveguide. The impact of frequency mismatch on data transmission is shown in Figure 8.2, where FWHM refers to "full width at half maximum". In Case 1, where the frequency mismatch is small, a ring filter receives only a small portion of the optical signal power, resulting in a weaker electrical signal at the output of the photodetector, which results in data loss. While in Case 2, where the mismatch is larger, the ring filter filters out the optical wave corresponding to its neighboring optical channel.

Other than the ring resonators, the laser sources are also sensitive to temperature variations [19]. There are two major ways of integrating laser sources with the chip package: off-chip integration and on-chip integration [20]. For off-chip laser sources, to maintain the frequencies of emitted optical waves, the laser source temperatures are typically controlled, or a frequency locking circuit is employed. The operation of these off-chip laser sources is agnostic to on-chip temperatures. On the other hand, on-chip laser sources' temperatures are affected by chip thermal conditions due to their close proximity to the computational components [21]. Thus, for a PNoC with on-chip laser sources, one must control the optical frequencies of both ring resonators and laser sources for reliable silicon-photonic link communication.

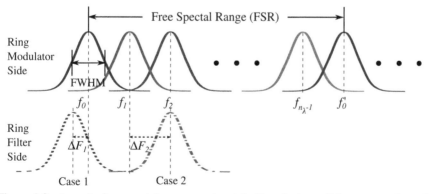

Figure 8.2 Impact of resonant frequency mismatch. Case 1: A small frequency mismatch reduces the filtered optical power; Case 2: A large frequency mismatch may result in a ring filtering the data of its neighboring ring.

8.3 Design Methods for Thermal Management in PNoCs

During the design stage of many-core systems with PNoC, there are two major groups of approaches to decrease the computation components' thermal impacts on optical devices: (1) at the device level, there are techniques to actively control the temperatures of optical devices as well as techniques that render optical devices less thermally sensitive, (2) at the chip scale, some recent methods have focused on reducing the thermal coupling between on-chip logic components and optical devices and balancing the thermal conditions of optical devices through chip floorplanning.

8.3.1 Device-level Techniques

When designing and selecting optical devices for a PNoC, there are two common ways to protect the optical devices from on-chip temperature variations: (1) actively control the temperatures of ring resonators; (2) choose optical devices for ring resonators that are less thermally sensitive.

Active PNoC thermal management is mainly carried out by integrating micro-heaters with ring resonators. During operation, each micro-heater can heat up its ring resonator, to a fixed temperature that is equal to the maximum temperature allowed for logic components. This forces all ring resonators to have the same amount of frequency shift and prevents them from being affected by on-chip thermal conditions. It is also possible to tune a ring resonator's temperature to achieve a particular frequency (e.g., using a feedback loop that measures the frequency and adjusts temperature accordingly). A more advanced design [9] integrates a ring modulator with both a micro-heater and a temperature sensor, as shown in Figure 8.3. This additional temperature sensor allows the OS to acquire the runtime temperatures of ring

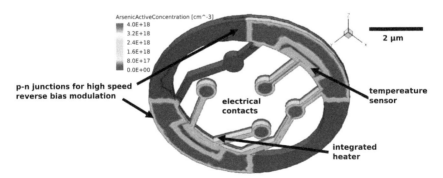

Figure 8.3   A schematic of a ring resonator with integrated heater and temperature sensor [9].

resonators, which enables sophisticated thermal management at OS level. The measurement results demonstrated a fast response ($\sim$1 μs) temperature control loop with a temperature range of 85 $^\circ C$, which is sufficient for runtime thermal management.

Other than actively tuning the temperature, it is also possible to design and manufacture athermal ring resonators. As introduced in Section 8.2, most ring resonators are fabricated using silicon, which possesses relatively large positive TOC. One approach to make ring resonators less sensitive is to clad silicon with negative TOC materials. For example, Djordjevic et al. [10] use amorphous titanium dioxide (a-TiO$_2$), which is a material with high negative TOC that can compensate for silicon's positive TOC. Ring resonators utilizing such a structure have been fabricated and the measurement results demonstrate a thermal sensitivity of 0.2 $GHz/^\circ C$.

One other way of athermalizing ring resonators is to couple them with other devices such as Mach-Zehnder interferometers (MZI). One novel design [11] couples a ring resonator to a MZI, which compensates temperature by tailoring the optical confinement in waveguides. The schematic of the proposed device is shown in Figure 8.4(a). The waveguide widths and lengths of the two arms of a MZI are chosen for a balanced transmission while having a strong negative temperature sensitivity overall. Since the ring resonator has a strong positive thermal sensitivity, the net device resonates at the center wavelength regardless of temperature changes. The fabricated devices show temperature stability over a temperature range of over 80 $^\circ C$.

Generally speaking, active thermal-control techniques are easier to implement, are able to provide flexibility in runtime thermal management, and

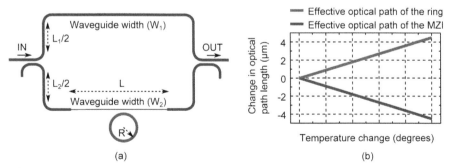

Figure 8.4 (a) Schematic of the device showing the various waveguide lengths and widths. The MZI is highlighted in blue and the ring in red. (b) Change in optical path length with temperature for the ring and MZI. The devices are designed to have opposite and equal phase shifts increase in temperature [11].

are more mature compared to ring resonator athermalization techniques. However, it is energy-inefficient when the temperature gradient among ring resonators is high. On the other hand, passive techniques that make ring resonators athermal do not require extra energy for thermal management, however, they are usually not compatible with traditional CMOS manufacturing process. In addition, their process variations cannot be modulated via thermal tuning (i.e., such devices may end up suffering more significantly from process variations as a result). Passive techniques may also add additional design constraints to optical devices, which limit the design flexibility.

8.3.2 Chip-level Techniques

Integrating micro-heaters is a more realistic option from the implementation perspective compared to using athermal optical devices. However, such an approach induces high thermal tuning power (e.g., 2.6 mW/nm for every ring resonator [18]). Thus, significant research efforts have been devoted to decrease the thermal tuning power of ring resonators. During the design stage, there are novel chip designs and place-and-route (P&R) strategies that can be used for preemptive PNoC thermal management.

One approach [22] integrates redundant ring resonators in order to provide higher tolerable temperature differences among ring modulators and filters within a given thermal tuning budget. A ring group is defined as a collection of co-located ring resonators used to implement a communication interface. Every ring modulator has one associated ring filter in a silicon-photonic link, and the ring modulator and ring filter of a silicon-photonic link are in different ring groups. If there is a temperature difference between these two ring groups, optical signals are either lost or filtered by neighboring ring filters. Adding extra ring filters creates a sliding ring window, which leverages the fact that the resonant frequencies of all ring resonators in a ring group shift the same amount during temperature changes and allowing the optical signals to be all filtered out by the neighboring ring filters. The simulation results demonstrate that under 10 W thermal tuning power budget, adding two extra ring filters per photonic channel can increase the tolerable temperature gradient from 2.1 ^oC to 5.6 ^oC for a 64-bit crossbar network.

Another chip design thermally decouples the processor die from the silicon-photonic die by inserting an insulator layer in between to make active thermal tuning more power-efficient [23], as shown in Figure 8.5. This design assumes that each ring resonator is only integrated with a micro heater (without a temperature sensor) and all the ring resonators work at a fixed

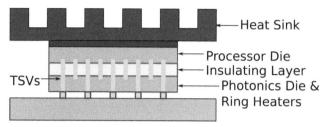

Figure 8.5 Cross section of a many-core system with thermal insulator layer between silicon-photonic die and processor die [23].

temperature (90 ^oC) using localized thermal tuning. Normally, the thermal tuning power is mostly wasted as it dissipates through the chip stack and also heats up the processor die. To prevent such power waste, this design adds an insulator layer between the processor die and the silicon-photonic die, which keeps the temperatures for optical devices more stable and minimizes the spatial and temporal thermal coupling between logic components and optical devices. The insulation layer is implemented by etching a 5 μm thick air or vacuum cavity inside a 150 μm porous Si layer [24]. The experimental results demonstrate that the design with an insulator layer can reduce the thermal tuning power by a factor of 3.8–5.4 on average based on real-life workloads compared to the design without an insulator layer.

Other than chip stack design, P&R of PNoCs also affects the thermal coupling of on-chip logic components and optical devices. A good P&R scheme can help alleviate the thermal issues and result in lower power consumption for thermal management. For example, a CAD flow [12] has been proposed to provide a global routing solution for optical-electrical interconnects with the following thermal and power considerations:

- Thermal reliability and functionality;
- Minimal optical driving power;
- Signal integrity and data conversion quality;
- Timing considerations and wavelength-division multiplexing (WDM) channel utilization;
- Legalization based on optical-electrical interconnect design rules.

Figure 8.6 shows an example solution of the waveguide routing issue. A WDM trunk is an optical waveguide and a WDM channel refers to the carrier of a modulated optical wave on a WDM trunk. The target of this CAD flow is to route the optical waves from off-chip laser sources to connect the driver pins in a low-power fashion while avoiding over-heated blocks. The routing

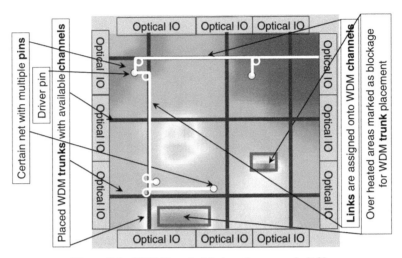

Figure 8.6   WDM based global routing scenario [12].

problem is formulated and solved using integer linear programming (ILP). The objective function is as follows:

$$Minimize\{P_{total}\} \qquad (8.5)$$

$$where \quad P_{total} = P_{loss} + P_{dynamic} \qquad (8.6)$$

$$P_{loss} = P_{cross} + P_{trunk\_thm} + P_{ring\_thm} + P_{path} \qquad (8.7)$$

In this objective function, P_{total} is the total power required to drive a PNoC. P_{loss} represents the optical power needed to compensate the on-chip optical loss so as to guarantee the detection on the photodetector side. P_{loss} can be divided into four items: waveguide cross power (P_{cross}), thermal related WDM trunk power (P_{trunk_thm}), thermal related ring resonator power (P_{ring_thm}) and the power to compensate the propagation loss in waveguides (P_{path}). $P_{dynamic}$ is the signal switching power on WDM channel carriers and contains two parts: the base power consumption for each WDM trunk and the switching power on all WDM channels. The test cases for this CAD flow are derived from ISPD07-08 global routing contest benchmarks [25, 26]. The results demonstrate that compared to a greedy routing approach that assigns silicon-photonic links to WDM trunks in a sequential manner, the proposed ILP formulation achieves 23% to 50% total power reduction.

The above technique attempts to solve the thermal challenges by routing optical waveguides away from over-heated regions. However, the thermal map of a chip is not fixed and it depends on the runtime workload allocation. Thus, having active runtime control over ring resonators' temperatures is necessary to guarantee PNoC functionality. At the same time, design-time techniques can preemptively prevent thermal imbalance among ring resonators, and thus, lower the thermal tuning power. For example, an ILP-based cross-layer thermally-aware optimizer [13] was designed for PNoC P&R. This optimizer minimizes PNoC power by explicitly considering laser source power consumption (including all optical losses), electrical-optical-electrical conversion power and thermal tuning power. As a result, it outputs the placement of ring groups, core clusters and routing of waveguides simultaneously. The optimization flow is shown in Figure 8.7.

This optimizer takes as input a parameter file (*.param* in Figure 8.7) containing core parameters, possible chip aspect ratios and PNoC-related parameters. A simplified core impact matrix is extracted for the simulated floorplans using HotSpot-3D [27] to represent the thermal impact of each core on each potential ring group location to balance the thermal conditions of ring groups. An example of core impact matrix generation is shown in Figure 8.8. The optimization results show that the optimal PNoC power is very sensitive to the system power profiles, optical data rate, and system

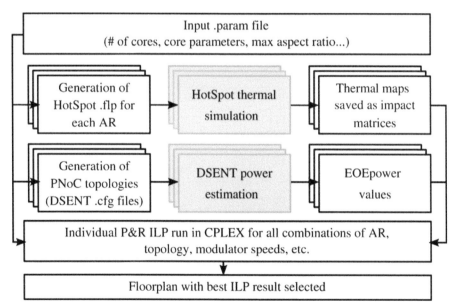

Figure 8.7 The P&R optimization flow [13]. AR: Aspect Ratio.

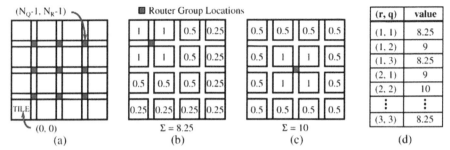

Figure 8.8 Core impact matrix generation: (a) illustrative floorplan with 16 tiles (64 cores) and nine potential ring group positions; (b) sample core impact calculation for ring group (1,3); (c) sample core impact calculation for ring group (2,2); (d) a 1x9 core impact array generated for the floorplan.

architecture (number of cores and chip aspect ratios). Compared to thermally-agnostic P&R solutions, this ILP-based optimizer saves up to 15% PNoC power.

In addition to methods for P&R of waveguides and ring resonators, there are also methods focusing on thermally-aware placement of on-chip laser sources. Li et al. [28] propose a methodology (see Figure 8.9) enabling thermally-aware design for PNoCs with CMOS-compatible vertical-cavity surface-emitting lasers (VCSELs). This requires steady-state thermal simulations and signal-noise-ratio (SNR) analysis that consider the temperatures of VCSELs and ring resonators. It provides a platform that allows for design

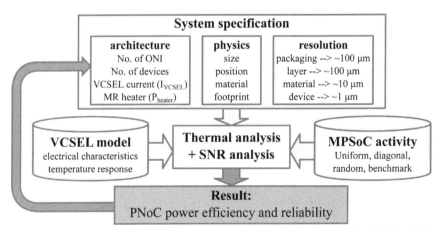

Figure 8.9 Thermally-aware design methodology for PNoCs with on-chip VCSELs [28].

space exploration on placement of optical devices in order to minimize PNoC power while maintaining PNoC reliability.

This methodology takes the system-level information as input for steady-state thermal simulations and SNR analysis, and outputs PNoC power efficiency and reliability. The system-level information includes packaging details (heat sink and fan specifications, etc.) and target architecture details (die size, optical network interface (ONI) locations, and materials and thickness of each die within the chip stack, etc.). The thermal simulator has a built-in VCSEL model and can simulate power profiles based on user's need. Based on the output thermal map, SNR analysis can be conducted to estimate PNoC reliability under a given chip activity. This methodology could be used to evaluate power and reliability trade-offs in PNoC design.

8.4 Runtime Methods for Thermal Management in PNoCs

During the runtime of many-core systems with PNoC, the on-chip thermal conditions change substantially depending on the workload allocation and chip utilization. Integrating micro-heaters with ring resonators can provide a steady thermal environment for these optical devices and guarantee PNoC functionality, but at the same time it also consumes a significant amount of thermal tuning power. Researchers have proposed runtime thermal management techniques such as workload allocation and migration policies specifically for PNoCs to reduce the power consumption spent on matching the optical frequencies of on-chip optical devices.

RingAware [14] is a workload allocation policy to balance the ring resonator temperatures by maintaining similar power profiles around each ring group. For a given layout, this policy categorizes ring groups' neighboring cores based on the distance of each core from its closest ring group, as shown in Figure 8.10. $RD\#$ represents the category for each core, where $\#$ is the

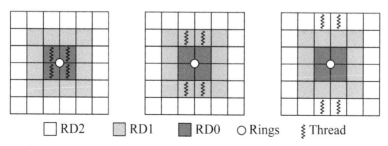

☐ RD2   ▨ RD1   ▮ RD0   ○ Rings   ⌇ Thread

Figure 8.10 Core classification based on the distance to ring groups [14].

cores' relative distance to the ring group. Since for each ring group, its $RD0$ cores have the highest impact on this ring group's temperature, *RingAware* maintains similar power dissipation across the $RD0$ regions for all ring groups to balance their temperatures. When there are S threads to allocate, if S is larger than the total number of *non-RD0* cores, $RingAware$ utilizes the $RD0$ cores during workload allocation. Otherwise, the $RD0$ cores are left idle. The $RD0$ regions of all ring groups need to have the same active core count to maintain similar power profile. *RingAware* starts with allocating threads to the $RD0$ regions, then, it partitions the many-core chip into four equal quadrants and the rest of the threads are evenly assigned to each quadrant. If there are any residual threads, they are allocated to the quadrants in a round-robin fashion. For each quadrant, *RingAware* allocates threads to *non-RD0* cores alternately from the outer boundaries to the inner part of the chip, starting from the corner core.

RingAware allocation effectively reduces the ring group temperature difference, which leads to a low resonant frequency gradient when the system does not have process variations. When compared to two other workload allocation algorithms: (1) $Clustered$, which allocates threads to cores starting from one corner of a chip to the diagonal corner; (2) $Chessboard$, which allocates threads to alternate cores starting from two opposite sides of a chip, $RingAware$ shows its advantages in balancing ring groups' temperatures. The experimental results in Figure 8.11 shows that for single application cases, $RingAware$ achieves lower maximum temperature compared to $Clustered$, and reduces the ring temperature gradient lower than 2.5 ^oC for most of the cases. For multi-program workloads, $RingAware$ also outperforms $Clustered$ and $Chessboard$.

Although *RingAware* is effective at reducing the temperature difference among ring groups, it only considers the workload allocation statically and does not consider application phases. During system runtime, the various phases of applications might significantly impact the on-chip thermal conditions and result in imbalance in ring group temperatures. One thermally-aware runtime workload migration approach named *Therma* [15] balances the temperatures of ring groups by moving threads across cores based on the core/ring group temperature history.

Therma assigns a thermal index to each core to indicate its proneness to hot spots and an opto-thermal index to each ring group to represent its resonant frequency shift, and it also sets up target temperatures for cores and ring groups, respectively. The thread migration happens based on a preset interval. For each interval, to make migration decisions, *Therma* calculates

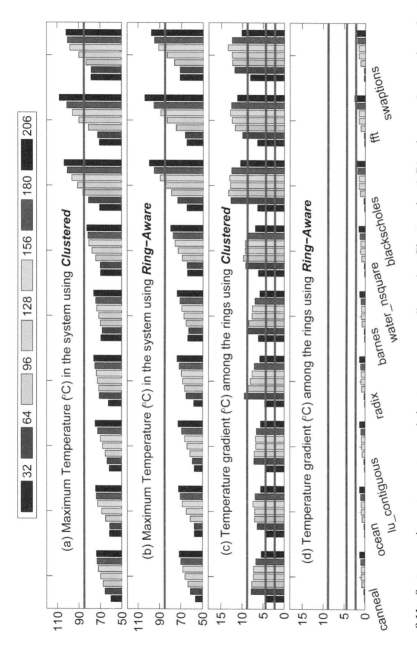

Figure 8.11 System maximum temperature and ring temperature gradient using *Clustered* and *RingAware* when running selected benchmarks from SPLASH2 [29] and PARSEC [30] benchmark suite [14].

a score for each core, which indicates how urgent a core needs to be cooled down/heated up. The calculation of a core's score depends on its score in the previous interval and a weight factor of its cluster. The cluster weight factor is computed using the thermal and opto-thermal indices of cores and ring groups within this cluster as well as their temperature history. After updating the scores for all cores in the system, cores with highest and lowest scores swap their threads to balance their thermal conditions. The results show that compared to the heat-and-run workload migration policy [31], *Therma* is able to reduce the PNoC power consumption by 6.1% in single thread mode (every core is running only one thread) and 10.3% in multi thread mode (one core can run more than one thread).

Aurora [17] is an approach leveraging both localized tuning and workload allocation techniques. This is a cross-layer approach at the device, architecture and OS levels. At the device level, *Aurora* controls small temperature variations by applying a bias current through the ring resonators [32]. For larger temperature changes, *Aurora* reroutes messages away from hot regions, and uses dynamic voltage and frequency scaling to reduce the temperature of hot areas at the same time. Two techniques are tested based on shortest-distance (SD) algorithm: shortest-path first (SPF) and temperature first (TF). SPF selects the routing path that traverses the least hot regions among the shortest ones while TF selects the shortest path that traverses the least hot regions. Figure 8.12 shows the path selection of these two algorithms under various thermal scenarios. Additionally, at the OS level, *Aurora* provides a job allocation policy that prioritizes jobs to the outer cores of the chip. The experimental results show that although TF has the longest latency among the applied three routing algorithms, it reduces bit error rate by 49% compared to SD due to its thermal consideration.

There are two common drawbacks of the above policies: (1) they only consider thermal variations' impact on the resonant frequency while ignoring the fact that with-in-die process variations also result in significant resonant frequency shift; (2) they only discuss the frequency matching of ring modulators and ring filters, but do not consider the matching of the optical frequency of laser sources with resonant frequency of ring modulators/filters. These two drawbacks have been addressed in a newly proposed on-chip optical device runtime tuning approach [16], which is composed of *FreqAlign* workload allocation and migration policy and an adaptive frequency tuning (*AFT*) policy. The mechanism of *FreqAlign* and *AFT* is shown in Figure 8.13. *FreqAlign* aims at reducing the resonant frequency difference among ring groups instead of only balancing the temperatures during

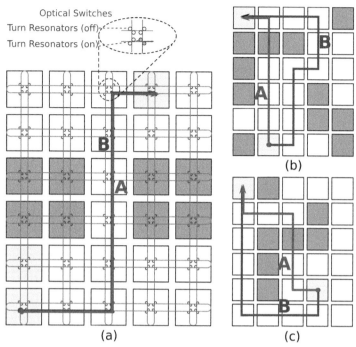

Figure 8.12 Paths selected by the routing algorithm in *Aurora* under various thermal scenarios (Path A: SPF, Path B: TF) [17].

workload allocation. It defines a weight array to represent each ring group's resonant frequency shift compared to its designed value. If there are process variations within the ring groups after manufacturing the chip, values in this weight array are set based on the chip measurement. *FreqAlign* allocates threads one by one to the available cores in a system. When allocating every thread, *FreqAlign* estimates the resultant frequency gradient among ring groups and selects the allocation that leads to the minimum frequency gradient. Such estimation is done by utilizing the thread's power consumption and a core impact matrix, which contains every core's thermal impact on every ring group per power unit. Such core impact matrix is specific to every system and can be extracted using thermal simulators such as HotSpot [33]. After the workload allocation, *AFT* sets the lowest resonant frequency among all ring groups as the target optical frequency and tunes all the other ring groups and on-chip laser sources to this target frequency. The results are twofold. From the resonant frequency matching perspective, *FreqAlign* is able to reduce the

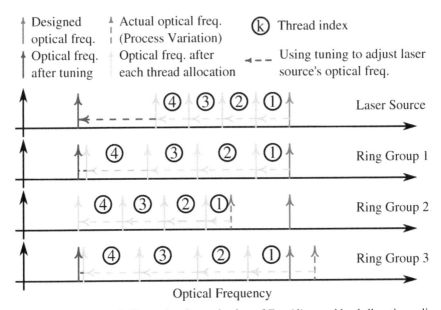

Figure 8.13   An example illustrating the mechanism of *FreqAlign* workload allocation policy and adaptive frequency tuning policy [16]. The actual resonant frequencies of ring groups are different from the designed ones due to process variations. Every thread allocated by *FreqAlign* increases the temperatures of ring groups and causes a downward shift in their frequencies. When all threads are allocated, thermal tuning is used to bring all ring groups to the lowest common resonant frequency. Above, ring groups 1 and 3, as well as the laser source, are tuned to match the resonant frequency of ring group 2.

resonant frequency gradient among all ring groups by around 50% to 60% for different PNoC logical topology and physical layout combinations. From the tuning power perspective, *FreqAlign+AFT* reduces thermal tuning power by 19.28 W on average, and is capable of saving up to 34.57 W in a 256-core system without any performance degradation, compared to tuning all optical devices to the temperature threshold of a system.

8.5 Conclusion

PNoC is an excellent candidate for on-chip communication in future many-core systems. Silicon-photonic links can provide higher bandwidth density, lower global communication latency, and lower data-dependent power compared to traditional electrical links. This chapter has reviewed both design-time and runtime approaches for PNoC thermal management. These device-level, chip-level, or OS-level approaches improve PNoC energy efficiency and, at the same time, guarantee PNoC functionality.

There are still many open challenges in the area of PNoC architecture design and optimization. Due to the packaging difficulties and long control loops, off-chip laser sources are less preferred in real many-core systems. On-chip laser sources are more promising for the future PNoC, however, laser efficiency decreases due to high on-chip temperatures. Thus, placement and integration methods as well as thermal characteristics of on-chip laser sources must be considered for future designs and optimizations of PNoC. Additionally, most of the current work only discuss homogeneous many-core systems, which are composed of a single type of cores. For heterogeneous many-core systems with more than two types of cores and/or hardware accelerators, the complexity of P&R increases and the system power profiles change dramatically. As a result, both P&R and workload allocation problems will be more complicated to solve, which calls for more research effort in PNoC thermal management.

Acknowledgement

The authors thank Prof. Jose Abellan, prof. Andrew B. Kahng, Anjun Gu, John Recchio and Vaishnav Srinivas for their invaluable contributions in the collaborative project that produced the research papers [13, 16] that have formed the basis of this chapter. This work was partially supported by the NSF grants CNS-1149703 and CCF-1149549.

References

[1] M.J. Cianchetti, J.C. Kerekes, and D.H. Albonesi. Phastlane: a rapid transit optical routing network. In *Proceedings of International Symposium on Computer Architecture*, pages 441–450, 2009.

[2] A. Joshi, C. Batten, Y. Kwon, S. Beamer, I. Shamim, K. Asanovic, and V. Stojanovic. Silicon-photonic Clos networks for global on-chip communication. In *Proceedings of International Symposium on Networks-on-Chip*, pages 124–133, 2009.

[3] N. Kirman, M. Kirman, R.K. Dokania, J.F. Martinez, A.B. Apsel, M.A. Watkins, and D.H. Albonesi. Leveraging optical technology in future bus-based chip multiprocessors. In *Proceedings of International Symposium on Microarchitecture*, pages 492–503, 2006.

[4] Y. Pan, P. Kumar, J. Kim, G. Memik, Y. Zhang, and A. Choudhary. Firefly: illuminating future network-on-chip with nanophotonics. In *Proceedings of International Symposium on Computer Architecture*, pages 429–440, 2009.

[5] L. Ramini, D. Bertozzi, and L.P. Carloni. Engineering a bandwidth-scalable optical layer for a 3D multi-core processor with awareness of layout constraints. In *Proceedings of International Symposium on Networks-on-Chip*, pages 185–192, 2012.

[6] A. Shacham, K. Bergman, and L.P. Carloni. On the design of a photonic network-on-chip. In *Proceedings of International Symposium on Networks-on-Chip*, pages 53–64, 2007.

[7] D. Vantrease, R. Schreiber, M. Monchiero, M. McLaren, N.P. Jouppi, M. Fiorentino, A. Davis, N. Binkert, R.G. Beausoleil, and J.H. Ahn. Corona: System implications of emerging nanophotonic technology. In *Proceedings of International Symposium on Computer Architecture*, pages 153–164, 2008.

[8] I. Kiyat, A. Aydinli, and N. Dagli. Low-power thermooptical tuning of SOI resonator switch. *IEEE photonics technology letters*, 18(2):364–366, 2006.

[9] C.T. DeRose, M.R. Watts, D.C. Trotter, D.L. Luck, G.N. Nielson, and R.W. Young. Silicon microring modulator with integrated heater and temperature sensor for thermal control. In *Proceedings of Conference on Lasers and Electro-Optics and Quantum Electronics and Laser Science*, pages 1–2, 2010.

[10] S.S. Djordjevic, K. Shang, B. Guan, S.T. Cheung, L. Liao, J. Basak, H. Liu, and S.J.B. Yoo. CMOS-compatible, athermal silicon ring modulators clad with titanium dioxide. *Optics Express*, 21(12):13958–13968, 2013.

[11] B. Guha, B.B. Kyotoku, and M. Lipson. CMOS-compatible athermal silicon microring resonators. *Optics Express*, 18(4):3487–3493, 2010.

[12] D. Ding, B. Yu, and D.Z. Pan. Glow: A global router for low-power thermal-reliable interconnect synthesis using photonic wavelength multiplexing. In *Proceedings of Asia and South Pacific Design Automation Conference*, pages 621–626, 2012.

[13] A.K. Coskun, A. Gu, W. Jin, A. Joshi, A.B. Kahng, J. Klamkin, Y. Ma, J. Recchio, V. Srinivas, and T. Zhang. Cross-layer floorplan optimization for silicon photonic NoCs in many-core systems. In *Proceedings of Design, Automation & Test in Europe Conference and Exhibition*, pages 1309–1314, 2016.

[14] T. Zhang, J.L. Abellan, A. Joshi, and A.K. Coskun. Thermal management of manycore systems with silicon-photonic networks. In *Proceedings of Design, Automation and Test in Europe Conference and Exhibition*, pages 1–6, 2014.

[15] M.V. Beigi and G. Memik. Therma: Thermal-aware run-time thread migration for nanophotonic interconnects. In *Proceedings of International Symposium on Low Power Electronics and Design*, pages 230–235, 2016.

[16] J.L. Abellan, A.K. Coskun, A. Gu, W. Jin, A. Joshi, A.B. Kahng, J. Klamkin, C. Morales, J. Recchio, V. Srinivas, and T. Zhang. Adaptive tuning of photonic devices in a photonic NoC through dynamic workload allocation. IEEE Transactions on Computer Aided Design of Integrated Circuits and Systems, 36(5):801–814, May 2017.

[17] Z. Li, A. Qouneh, M. Joshi, W. Zhang, X. Fu, and T. Li. Aurora: A cross-layer solution for thermally resilient photonic network-on-chip. *IEEE Transactions on Very Large Scale Integration Systems*, 23(1):170–183, 2014.

[18] J.S. Orcutt, B. Moss, C. Sun, J. Leu, M. Georgas, J. Shainline, E. Zgraggen, H. Li, J. Sun, M. Weaver, S. Urošević, M. Popović, R.J. Ram, and V. Stojanović. Open foundry platform for high-performance electronic-photonic integration. *Optics Express*, 20(11):12222–12232, 2012.

[19] T. Kimoto, T. Shinagawa, T. Mukaihara, H. Nasu, S. Tamura, T. Numura, and A. Kasukawa. Highly reliable 40-mW 25-GHz× 20-ch thermally tunable DFB laser module, integrated with wavelength monitor. *Furutaka Review*, 24:1–5, 2003.

[20] M.J.R. Heck and J.E. Bowers. Energy efficient and energy proportional optical interconnects for multi-core processors: Driving the need for on-chip sources. *IEEE JSTQE*, 20(4):1–12, July 2014.

[21] C. Chen, T. Zhang, P. Contu, J. Klamkin, A.K. Coskun, and A. Joshi: Sharing and Placement of On-chip Laser Sources in Silicon-photonic NoCs. In *Proc. International Symposium on Networks-on-Chip*, pp. 88–95, 2014.

[22] C. Nitta, M. Farrens, and V. Akella. Addressing system-level trimming issues in on-chip nanophotonic networks. In *Proceedings of International Symposium on High Performance Computer Architecture*, pages 122–131, 2011.

[23] Y. Demir and N. Hardavellas. Parka: Thermally insulated nanophotonic interconnects. In *Proceedings of International Symposium on Networks-on-Chip*, pages 1–8, 2015.

[24] Y. Zhang, H. Oh, and M. Bakir. Within-tier cooling and thermal isolation technologies for heterogeneous 3D ICs. In *Proceedings of IEEE International 3D Systems Integration Conference*, pages 1–6, 2013.

[25] ISPD 2007 global routing contest. http://archive.sigda.org/ispd2007/contest.html. Accessed: 2016-08-24.

[26] ISPD 2008 global routing contest. http://archive.sigda.org/ispd2008/contests/ispd08rc.html. Accessed: 2016-08-24.

[27] J. Meng, K. Kawakami, and A.K. Coskun. Optimizing energy efficiency of 3-D multicore systems with stacked DRAM under power and thermal constraints. In *Proceedings of Deisgn Automation Conference*, pages 648–655, 2012.

[28] H. Li, A. Fourmigue, S. Le Beux, X. Letartre, I. O'Connor, and G. Nicolescu. Thermal aware design method for VCSEL-based on-chip optical interconnect. In *Proceedings of Design, Automation & Test in Europe Conference and Exhibition*, pages 1120–1125, 2015.

[29] S.C. Woo, M. Ohara, E. Torrie, J.P. Singh, and A. Gupta. The SPLASH-2 programs: characterization and methodological considerations. In *Proceedings of International Symposium on Computer Architecture*, pages 24–36, 1995.

[30] C. Bienia, S. Kumar, J.P. Singh, and K. Li. The PARSEC benchmark suite: Characterization and architectural implications. In *Proceedings of International Conference on Parallel Architectures and Compilation Techniques*, pages 72–81, 2008.

[31] M. Gomaa, M.D. Powell, and T.N. Vijaykumar. Heat-and-run: leveraging SMT and CMP to manage power density through the operating system. In *ACM SIGARCH Computer Architecture News*, volume 32, pages 260–270. ACM, 2004.

[32] S. Manipatruni, R.K. Dokania, B. Schmidt, N. Sherwood-Droz, C.B. Poitras, A.B. Apsel, and M. Lipson. Wide temperature range operation of micrometer-scale silicon electro-optic modulators. *Optics Letters*, 33(19):2185–2187, 2008.

[33] K. Skadron, M.R. Stan, W. Huang, S. Velusamy, K. Sankaranarayanan, and D. Tarjan. Temperature-aware microarchitecture. In *Proceedings of International Symposium on Computer Architecture*, pages 2–13, 2003.

9

Thermal-Aware Design Method for On-Chip Laser-based Optical Interconnect

Hui Li[1], Alain Fourmigue[2], Sébastien Le Beux[1], Xavier Letartre[1], Ian O'Connor[1] and Gabriela Nicolescu[2]

[1]Lyon Institute of Nanotechnology, INL-UMR5270, Ecole Centrale de Lyon, Ecully, F-69134, France
[2]Computer and Software Engineering Dept., Polytechnique Montréal, Montréal (QC), Canada

Abstract

Optical Network-on-Chip (ONoC) is an emerging technology considered as one of the key solutions for future generation on-chip interconnects. On-chip laser sources, such as Vertical-Cavity Surface-Emitting Lasers (VCSELs), provide improved energy efficiency compared to off-chip sources. However, silicon photonic devices are highly sensitive to temperature variation, which leads to a lower efficiency of Vertical-Cavity Surface-Emitting Lasers (VCSELs) and a shift of resonant wavelength of Microring Resonators (MRs). In turn, this results in a lower Signal to Noise Ratio (SNR). In this work, we propose a methodology enabling thermal-aware design for on-chip optical interconnects. Thermal analysis allows the design of ONoC interfaces with low gradient temperature, by describing the studied architecture from device level to system level. Analytical models allow the influence of thermal effects to be evaluated, in particular on the SNR, based on the obtained thermal map.

9.1 Introduction

Technology scaling down to the ultra deep submicron domain provides for billions of transistors on chip, enabling the integration of hundreds of cores. Many-core designs, integrating interconnect that can support low latency and high data bandwidth, are increasingly required in modern embedded systems to address increasingly stringent power and performance constraints of embedded applications. Designing such systems with traditional electrical interconnect presents a significant challenge: both interconnect noise and propagation delay of global interconnects increase due to capacitive and inductive coupling [1]. The increase in propagation delay requires global interconnect to be clocked at a low rate, which limits the achievable bandwidth and system performance.

In this context, Optical Network-on-Chip (ONoC) is an emerging technology considered as one of the key solutions for future generations of on-chip interconnects. It relies on optical waveguides to carry optical signals, so as to replace electrical interconnect, and provide the low latency and high bandwidth characteristic of optical interconnect. ONoCs can be generally sorted into two types of approach, which are based on: i) wavelength-routing [2–4] or ii) optical switching (circuit switching [5–8] or packet switching [9, 10]).

Among the proposed ONoCs, the wavelength-routing based interconnect solutions are of considerable interest to the major players in the field, since they do not require any arbitration [1, 3, 4] to propagate the optical signals. They rely on passive Microring Resonators (MRs) that filter the optical signals based on their resonant wavelengths.

As for the laser sources, these are generally divided into two kinds: off-chip lasers and on-chip lasers. For energy efficiency, on-chip lasers are more advantageous since no optical coupling into the chip, typically a lossy operation, is required [11]. Moreover, due to the fact that no power waveguides are required in optical networks and on-chip lasers can be placed at any desired position, this increases the design flexibility of layout, while reducing optical loss at the same time.

Among the available laser sources candidates, CMOS-compatible Vertical-Cavity Surface-Emitting Lasers (VCSEL) are of high interest despite the use of less mature technologies (they usually require the inclusion of III–V semiconductors). Indeed CMOS-compatible VCSELs allow direct modulation of optical signals and can thus be *dedicated* to a communication channel. Their size is of the same order of magnitude as the size of a MR used to modulate continuous waves emitted by *shared* lasers. VCSELs are thus sufficiently compact to be implemented in a large number and at any position, which leads to the following key advantages. Firstly, integration is easier since layout constraints are relaxed. Secondly, power saving is expected since waveguide lengths are reduced and waveguide crossings are avoided. Thirdly, higher scalability is obtained since the laser sources are fully distributed.

However, silicon photonic devices are highly sensitive to temperature variation, which leads to drift in the MR resonant wavelengths. As a result, the power of the desired signal received at the destination *decreases* while that of the crosstalk from other undesired signals *increases*, consequently resulting in a lower Signal to Noise Ratio (SNR) at the destination. Moreover, the power efficiency of integrated laser sources decreases with higher temperature, which also has a negative impact on SNR and lowers the interconnect power efficiency. Consequently, the *average* temperature and *gradient* temperature both influence the SNR and power efficiency.

In order to compensate the impact of average temperature and gradient temperature, run-time solutions can be used to dynamically adjust the ONoC to measured variations in temperature. However, this can defeat the purpose of improving power efficiency since the extent of tuning can be very high (and even beyond the tuning capabilities of the devices), and also requires power to control and to carry out. We believe that it is necessary to address the temperature issues at the design stage to minimize the impact of temperature on SNR and power efficiency and also to minimize the need for run-time solutions. In this chapter, we propose a thermal-aware methodology to design ONoCs relying on fully distributed CMOS-compatible on-chip lasers (e.g., VCSELs). The methodology relies on i) steady-state thermal simulations and ii) SNR analyses taking into consideration the temperature of the on-chip lasers and the MRs. Design space exploration on MR heating power and laser current modulation allow ONoC gradient temperature to be reduced while holding the on-chip laser temperature to an acceptable range. SNR analyses allow the estimation of the ONoC reliability and power efficiency for a given level of chip activity.

The rest of this chapter is structured as follows. Section 9.2 presents the related work. Then, the considered architectural models and laser sources are described in Section 9.3. Section 9.4 details the design methodology while Section 9.5 presents the case study and Section 9.6 gives results. Section 9.7 concludes the chapter.

9.2 Related Work

Optical NoCs are often based on 3D architectures, with the optical layer placed above the electrical layer (e.g., SUOR [12] and Firefly [13]). However, the temperature of the optical layer is dependent on that of the electrical layer, which dissipates electrical power in the form of heat such that the temperature of a given point on the chip can vary with time, and be different from the temperature of other points on the chip. The typical impact of a 1°C temperature variation on the optical layer is that MR resonant wavelength will shift by ~0.11 nm [14].

Reducing the thermal sensitivity in optical communication has been already been addressed at both device and system levels.

At the ***device level***, basic solutions rely on voltage tuning [15] and local heating [16] to tune MR resonant wavelengths. Voltage tuning is fast but more suitable for small tuning ranges. In [17], the electro-optic response of a microdisk modulator is shown to be controllable over a 7.5°C temperature range with little additional energy dissipation. Local heating tunes the resonant wavelength of MR with the thermo-optic effect, by using a MR heater. This method is slower compared to voltage tuning but useful for a larger tuning range. In addition, other solutions, such as feedback control schemes [18] and athermal devices [19] have been explored to limit the thermal impact on MRs or to control the resonant wavelength of MRs.

At the ***system level***, some works have targeted reducing the influence of the MR resonant wavelength drift. These contributions can be categorized into two kinds: design-time and run-time.

- *Design-time*: In [20], system level analyses for a single-wavelength based ONoC allows the evaluation of the influence of temperature variation on the optical signal power received by the photodetectors. Based on the same principle, i.e., the minimum signal power required at receiver, further work was carried out for WDM-based ONoCs in [21].
- *Run-time*: in order to counter-balance the temperature variation, communication channels can be remapped through ONoC reconfiguration [22], and DVFS and workload migration techniques can be applied [14].

These techniques depend on redundant resources to re-map the wavelength channel or to detect run-time temperature variations. In [23], the authors proposed a thermal-aware job allocation policy to minimize the gradient temperatures among MRs. This work relies on device characteristics such as the Free Spectral Range (FSR), the number of waveguides and wavelengths.

Before exploring run-time solutions, in our work, we focus on the design-time methodology and evaluate the influence of the temperature variation on the SNR, taking into account the actual efficiency of the on-chip laser (e.g., VCSEL) for a given chip activity. In [24], the authors explore the placement of shared on-chip lasers on a layer located on top of the optical interconnect. In our work, we focus on CMOS-compatible VCSELs (allowing the allocation of dedicated communication channels) distributed on a layer dedicated to optical interconnects.

9.3 3D Architecture

This section presents the system architecture including the ONoC. The challenges addressed in this chapter are then introduced.

9.3.1 Architecture Overview

Figure 9.1(a) illustrates the MPSoC architecture we consider. It is composed of i) an electrical layer implementing processors (in tiles) and memories and ii) an optical layer dedicated to the implementation of a ring-based ONoC. The activity of the processors leads to local and global communications that are implemented with electrical interconnect and ONoC respectively. The communication hierarchy is defined at design time and depends mainly on the number of processors and the ONoC complexity and bandwidth. The CMOS-compatible silicon photonic fabrication process allows the integration of on-chip lasers (e.g., VCSELs), waveguides, MRs and photodetectors. These devices are gathered into a so-called Optical Network Interface (ONI), which is responsible for emitting the light, modulating optical signals with the data to be transmitted, transporting the modulated signals and receiving them on the destination side (Figure 9.1(c)). VCSELs and photodetectors are respectively connected to CMOS drivers and CMOS receivers through TSVs (Figure 9.1(c)).

We choose the ORNoC architecture [3] for the implementation of optical communications. ORNoC is a ring-based network allowing point-to-point communications between source and destination, with passive MRs. As reported in [25], ORNoC demonstrates reduced worst-case and average insertion losses compared with related optical crossbars including Matrix [26], λ-router [1] and Snake [4] (e.g., on average, 42.5% reduction for worst-case and 38% for average in 4×4 scale), which is a significant advantage to reduce the laser power consumption.

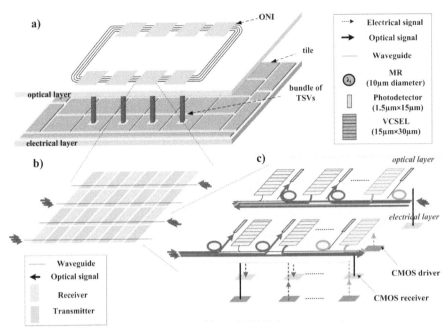

Figure 9.1   Considered 3D architecture: a) MPSoC with a stacked ONoC, b) ONI layout and c) implementation of an ONI and its operations.

9.3.2 ONoC Interface and Thermal Sensitivity

The ONIs are responsible for emitting, transmitting and receiving the optical signals on the optical layer, as illustrated in Figure 9.1(c). For this purpose, the signals are injected into waveguides and modulated directly with on-chip lasers (e.g., VCSELs). The signals then propagate on the waveguide, crossing the intermediate ONIs. Once they reach their destination, the signals are dropped from the waveguide by passive MRs toward large band photodetectors. For proper filtering operation, it is mandatory that resonant wavelengths of destination MRs be well aligned with the wavelength of the emitted signals. However, the resonant wavelengths of silicon based optical devices are sensitive to temperature (typically a shift of 0.1nm/°C occurs), which leads to a reduced SNR due to reduced coupling of signals into MRs (i.e. lower signal power and higher crosstalk noise at the photodetector).

Device-level calibration processes [14] help to improve the SNR by aligning the resonant wavelength. However, such techniques suffer from significant power consumption overhead: voltage tuning and heat tuning of MRs (that allow blue shift and red shift of the resonant wavelength respectively) lead to 130 mW/nm and 190 mW/nm respectively in consumed power per wavelength shift, as reported in [27]. For large-scale ONoCs (e.g. Corona [27] including approximately 1.1×10^6 MRs), the power dedicated to the calibration process represents over 50% of the total network power consumption. Since the calibration comes with performance overheads due to algorithm execution and heating latency, they are generally coupled to MRs clustering techniques. Indeed, clustering the MRs helps to reduce the algorithm complexity by assuming identical local temperatures among MRs that are sufficiently close. However, this technique requires careful design of the ONIs to ensure a homogeneous temperature under different processing activities.

Maintaining a low gradient temperature within an ONI including on-chip lasers is a challenging task since they dissipate relatively high power. Heating MRs to reduce the gradient temperature is thus mandatory, which can be obtained by implementing a local heater on top of each MR [28]. This allows the resonance wavelengths of MRs to be thermally tuned. In addition, alternatively placing on-chip lasers (i.e., VCSELs) and MRs contributes to reducing the MR heating power through a better initial distribution of the heat generated by on-chip lasers (i.e.,VCSELs), and also minimizes the signal crosstalk. This assumption leads to a chessboard-like layout illustrated in Figure 1(b): 4 waveguides propagating signals in clockwise and counter-clockwise rotation are placed alternately and, for each waveguide, 4 receivers and 4 transmitters are placed alternately.

9.3.3 CMOS-Compatible On-Chip Lasers

While on-chip laser sources require the use of less mature technologies compared to their off-chip counterpart, they have the potential to provide the following three key advantages:

- *Easier and more efficient integration* by relaxing layout constraints: in case of on-chip lasers, it is not necessary to distribute the light from an external source to the modulators (e.g. through the so called *power waveguide* in Corona [29]). Relaxing such constraints contributes to reducing the number of waveguide crossings or even to avoiding them altogether in the ring topology.
- *Higher scalability* by keeping the architecture fully distributed, which is not achievable by considering centralized off-chip lasers.

- *Lower power* by reducing the worst-case communication distance. This corresponds to the distance from the source IP to the destination IP for on-chip laser based architectures, while for off-chip laser based architectures, this distance also includes the distance from the off-chip laser to the source IP. Shorter distance consequently reduces the optical losses and hence the minimum required laser output power. Moreover, the power consumption can be further improved by locally turning off the laser when no communication is required.

VCSEL-based lasers [17, 30] offer a direct emission of data through current modulation. While the fabrication processes for CMOS-compatible VCSELs are less mature than those for microdisk lasers [31], they offer significant advantages in terms of scalability (a higher laser output power is achievable) and spectral density due to their small 3 dB bandwidth (typically 0.1nm). The drawback of on-chip lasers over their off-chip counterparts is their intrinsically lower efficiency and their higher sensitivity to variation in chip activity (since they are physically located above the processing layer). More precisely, each VCSEL is located above a CMOS driver that converts an input electrical data coming from an IP core (represented as a binary voltage) into a current, as illustrated in Figure 9.2(a). The current propagates through a TSV and directly modulates the VCSEL. An optical signal is emitted vertically and is redirected to a horizontal waveguide through a grating. The power of the optical signal injected into the network (OP_{net}) thus depends on i) the intensity of the modulation current I_{laser}, ii) the laser efficiency (η_{laser}) and the taper coupling efficiency ($\eta_{coupling}$, assumed to be 70%). The VCSEL efficiency is highly sensitive to its temperature: it can drop from 15% at 40°C to 4% at 60°C. This rather low efficiency leads to a high dissipated power (P_{laser}) which, together with the power dissipated by the CMOS driver (P_{driver}) and that of the chip part at the source (P_{chip}), influences the on-chip laser temperature. Hence, for a given modulation current, the power of the emitted optical signal (OP_{laser}) depends on the laser temperature, which is influenced by P_{chip}, P_{laser} and P_{driver}, as illustrated in Figure 9.2(b).

In addition, the emitted signal wavelength is dependent on the laser temperature. Ideally, the laser wavelength is designed to match the corresponding MR resonant wavelength at the target. However, the MR resonant wavelength can also drift with its temperature, which is influenced by the power dissipated by the chip part at the target area/surrounding (P'_{chip}), the MR heater (P_{heater}), and also the on-chip laser (P'_{laser}) in the same interface

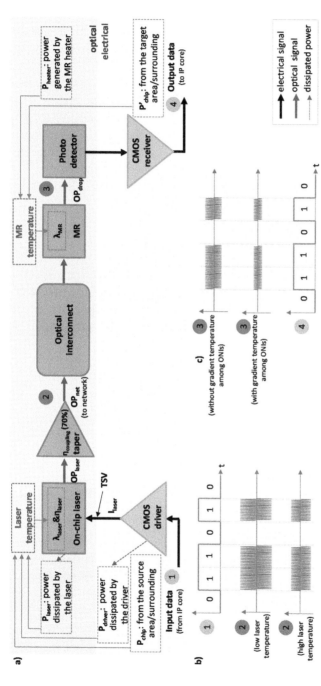

Figure 9.2   a) Principles of communication in an ONoC, considering the thermal effect. The efficiency of the on-chip laser (e.g., VCSEL) and the emitted signal wavelength depend on its temperature, which is influenced by the CMOS driver and the chip activity. The MR resonant wavelength depends on the MR temperature, which is influenced by the laser in the same interface, chip activity and MR heater. b) The signal at the source side. c) The signal at the target side.

(Figure 9.2(a)). Hence, the signal power dropped by the MR at the target side (OP_{drop}) depends on the wavelength alignment among the optical devices, i.e. the gradient temperature, as illustrated in Figure 9.2(c). Thus, a low average temperature and gradient temperature is necessary.

Furthermore, for an increasing activity of the processing layer (which is expected to result in additional communications), either the optical interconnect bandwidth will decrease assuming an identical modulation current (the SNR being lower, data will be re-emitted) or the optical interconnect power consumption will increase (a higher modulation current is required to compensate the reduced efficiency). The modulation current must thus be carefully selected since i) a too small value will lead to low SNR and ii) a too high value will lead to a power-hungry solution.

9.3.4 Contribution

The gradient temperature and average temperature in ONI are critical to design on-chip laser-based optical interconnect: i) **low gradient temperature** within an ONI eases the run-time calibration process and reduces the design complexity; ii) **low average ONI temperature** helps to maintain a reasonable power efficiency of the on-chip laser (e.g., VCSEL).

For the first time, we propose a temperature-aware design methodology allowing the efficient use of CMOS-compatible on-chip lasers (VCSELs). Temperature evaluation and system-level analyses allow the estimation of the SNR in ONoC.

9.4 Proposed Design Methodology

In this section, we describe the methodology for the thermal-aware design of optical interconnect.

9.4.1 Design Methodology Overview

The proposed methodology (illustrated in Figure 9.3) allows design space exploration at both device level and system level. For this purpose, the main characteristics of the optical devices (e.g., lasers, MRs, waveguides, photodetectors) are taken into account in device-level models. Architectural aspects such as interconnect size, topology/layout, and implementation technologies are taken into account at the system-level models ("IP Models" in the figure).

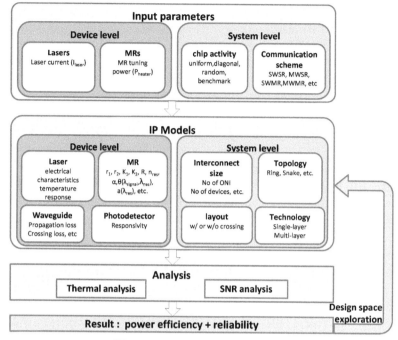

Figure 9.3 Proposed methodology.

Key input parameters such as the laser driver current (I_{laser}), MR tuning power (P_{heater}), chip activity and communication scheme are specified by users. I_{laser} and P_{heater} allow tuning the optical link by aligning signal wavelengths with MRs resonant wavelengths. Different chip activities (e.g., uniform, diagonal, etc.) simulate the power dissipated by the processing layer and communication scheme determines signal paths. From these parameters, thermal simulations are carried out to generate temperature maps of the optical layer. From the estimated temperature of the devices, SNR is evaluated using analytical models, which allows to carry out design space exploration (red arrow in the figure). Hence, energy efficiency and reliability tradeoff solutions can be explored, thus allowing to adapt the interconnect according to the optical interconnect requirements.

9.4.2 Thermal Analysis

In order to perform thermal evaluations, our architecture model is based on the real physical structure of the system. The various components of the system (i.e. package, die, heat sources, and optical devices) are represented as rectangular blocks, defined by their dimension, their position, and a constitutive material. The blocks can be assigned power values, which allow modeling of the heat sources of the system. The Back-End-Of-Line (BEOL) is modeled as a thin layer (10 μm) and the heat sources (cores, cache, router, etc.) are represented as rectangular blocks with power values, situated in the BEOL layer.

IcTherm[1] [32] is a thermal simulator for electronic devices that accurately models their complex structure and provides 3D full-chip temperature maps. IcTherm solves the physical equations that govern the temperature in the chip, using the Finite Volume Method [33], a numerical method for solving partial differential equations. IcTherm was validated against the commercial simulator COMSOL [34]: its maximal error was found to be less than 1% [32]. The structure of the system is discretized into small cubic cells that match the distribution of the materials and the heat sources. Figure 9.4 illustrates the discretization of a section of the system. Because the interfaces contain micro-scale components (e.g. TSVs, VCSELs and CMOS drivers), we use a fine-grain resolution with a cell size of 5 μm × 5 μm to mesh the region containing the interfaces. For the rest of the system, we use a coarser resolution with a cell size of 100 μm × 100 μm for the heat sources and 500 μm × 500 μm for the package.

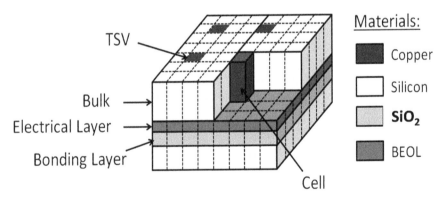

Figure 9.4   IcTherm computes the heat transfer between cells and outputs the temperature value of each cell. This thermal map allows the gradient temperature to be computed between any two points in the system.

---

[1]IcTherm website: http://www.ictherm.com/

9.4.3 SNR Analysis

For a given activity scenario, the thermal map gives the temperature of the lasers and the MRs, from which the gradient temperature of each ONI is extracted. We assume that the gradient temperature within an ONI must remain below 1°C (since we consider MRs with 1.55 nm 3 dB bandwidth, 0.1 nm drift of their resonant wavelength corresponds to at most 6.5% transmission loss). Design space on the MRs heater power can be explored in order to satisfy the 1°C gradient temperature constraint *inside* each ONI. However, the temperature gradient *among* ONIs influences the SNR, as detailed in the following.

9.4.3.1 SNR model

The SNR of one communication channel at the receiver is formalized as follows:

$$SNR = 10 \log_{10} \frac{P_{signal}}{P_{noise}} \tag{9.1}$$

At the receiver, P_{signal} is the expected signal power and P_{noise} is the crosstalk noise power dropped from other unexpected signals by considering the influence of the thermal effect. The signal attenuation and crosstalk power are detailed in the following subsection.

SNR analysis allows the ONoC reliability to be estimated for a given chip activity. This crucial information allows the exploration of the design space and in particular the power consumption of the driver. Indeed, P_{driver} is directly related to the laser modulation current, and it therefore impacts the laser efficiency and the optical signal power. Such exploration is illustrated in the results section.

9.4.3.2 Signal attenuation and crosstalk

When the signal propagates along the waveguide and encounters the MRs for filtering, a portion of the signal is coupled into the MRs. The coupled portion depends on the alignment between the signal wavelength and resonant wavelength of the MR, according to the transmission ratios $\varphi_t(\lambda_{signal}, \lambda_{res})$ and $\varphi_d(\lambda_{signal}, \lambda_{res})$. As an example, two communications $C_{i \to j}$ and $C_{s \to d}$ ($ONI_i \to ONI_j$ and $ONI_s \to ONI_d$ separately) are considered. In Figure 9.5, the signals ($\lambda_{T,i \to j}$ and $\lambda_{T,s \to d}$) and two MRs for filtering (resonant at $\lambda_{R,i \to j}$ and $\lambda_{R,s \to d}$) are shown to illustrate the occurrence of signal attenuation and crosstalk, where $\lambda_{T,i \to j} = \lambda_{R,i \to j}$ and $\lambda_{T,s \to d} = \lambda_{R,s \to d}$ at design time. The integrated MR heater is used to reduce the gradient temperature in ONIs.

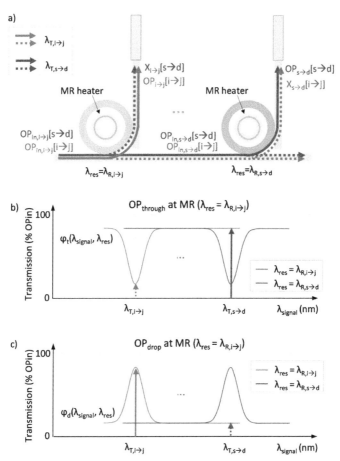

Figure 9.5   a) Signal attenuation and crosstalk along the communication path without temperature influence; b) $OP_{through}$, the transmitted optical power on the MR through port ($\lambda_{res}=\lambda_{R,i\rightarrow j}$) when signal wavelengths are $\lambda_{T,i\rightarrow j}$ and $\lambda_{T,s\rightarrow d}$; c) OP_{drop}, the transmitted signal power at the drop port of MR ($\lambda_{res}=\lambda_{R,i\rightarrow j}$) when signal wavelengths are $\lambda_{T,i\rightarrow j}$ and $\lambda_{T,s\rightarrow d}$.

In Figure 9.5(a), at the MR ($\lambda_{res}=\lambda_{R,i\rightarrow j}$, in blue) for filtering signals of communication $C_{i\rightarrow j}$, without considering the thermal effect, the signal at $\lambda_{T,i\rightarrow j}$ is dropped by the MR connected to the photodetector ($OP_{i\rightarrow j}[i\rightarrow j]$), the value of which is related to the input power ($OP_{in,i\rightarrow j}[i\rightarrow j]$) at the MR. The signal wavelength of communication $C_{s\rightarrow d}$ ($\lambda_{T,s\rightarrow d}$, in red) is designed to be different from the resonant wavelength of the MR ($\lambda_{R,i\rightarrow j}$). In this way, the signal passes through the MR and propagates along the waveguide with a power loss ($X_{i\rightarrow j}[s\rightarrow d]$), resulting in crosstalk to the expected signal of $C_{i\rightarrow j}$. The corresponding transmission of both signals at the through and drop ports of the MR ($\lambda_{res}=\lambda_{R,i\rightarrow j}$) is shown in Figure 9.5(b) and (c) respectively. The signal power $OP_{i\rightarrow j}[i\rightarrow j]$ and crosstalk $X_{i\rightarrow j}[s\rightarrow d]$ are formalized as following:

$$OP_{i\rightarrow j}[i \rightarrow j] = OP_{in,i\rightarrow j}[i \rightarrow j] \times \varphi_d(\lambda_{T,i\rightarrow j}, \lambda_{R,i\rightarrow j}) \qquad (9.2)$$

$$X_{i\rightarrow j}[s \rightarrow d] = OP_{in,i\rightarrow j}[s \rightarrow d] \times \varphi_d(\lambda_{T,s\rightarrow d}, \lambda_{R,i\rightarrow j}) \qquad (9.3)$$

In Figure 9.5(a), at the MR ($\lambda_{res}=\lambda_{R,s\rightarrow d}$) designed for filtering the communication signal $C_{s\rightarrow d}$, the expected signal ($\lambda_{T,s\rightarrow d}$, in red line) is dropped by the MR ($OP_{s\rightarrow d}[s\rightarrow d]$). The signal at another wavelength (in blue, $\lambda_{T,i\rightarrow j}$) propagates over the waveguide, and meanwhile the crosstalk ($X_{s\rightarrow d}[i\rightarrow j]$) to the expected signal ($OP_{s\rightarrow d}[s\rightarrow d]$) occurs. The signal power $OP_{s\rightarrow d}[s\rightarrow d]$ and crosstalk $X_{s\rightarrow d}[i\rightarrow j]$ are calculated with the following equations:

$$OP_{s\rightarrow d}[s \rightarrow d] = OP_{in,s\rightarrow d}[s \rightarrow d] \times \varphi_d(\lambda_{T,s\rightarrow d}, \lambda_{R,s\rightarrow d}) \qquad (9.4)$$

$$X_{s\rightarrow d}[i \rightarrow j] = OP_{in,s\rightarrow d}[i \rightarrow j] \times \varphi_d(\lambda_{T,i\rightarrow j}, \lambda_{R,s\rightarrow d}) \qquad (9.5)$$

In Figure 9.5(a), two communications are considered. The crosstalk to a signal comes from another undesired signal. In the considered network, the crosstalk received at each photodetector is estimated by considering all other signals at undesired wavelengths.

However, when the temperature over the chip changes, there is a drift in wavelength of both the signal emitted by on-chip laser (e.g. $\lambda_{T,i\rightarrow j}$) and MRs for filtering (e.g. $\lambda_{R,i\rightarrow j}$). For instance, the temperature drift of the on-chip laser and MR is $\Delta T_{T,i\rightarrow j}$ and $\Delta T_{R,i\rightarrow j}$ respectively. β is assumed to be the coefficient of the resonant wavelength drift dependent on the temperature. By considering the influence of the temperature drift, the wavelengths ($\lambda_{T,i\rightarrow j}(T)$ and $\lambda_{R,i\rightarrow j}(T)$) can be formalized as following:

$$\lambda_{T,i\rightarrow j}(T) = \lambda_{T,i\rightarrow j} + \Delta T_{T,i\rightarrow j} \times \beta \qquad (9.6)$$

$$\lambda_{R,i\rightarrow j}(T) = \lambda_{R,i\rightarrow j} + \Delta T_{R,i\rightarrow j} \times \beta \qquad (9.7)$$

As a consequence of the temperature drift, for a given signal, the expected signal power (P_{signal}) decreases more seriously and more crosstalk noise (P_{noise}) comes from other wavelengths, compared with the scenario without temperature drift.

9.4.3.3 Transmission principles of MR

Figure 9.6 illustrates the transmission of an optical signal OP_{in} at a wavelength λ_{signal} into an MR with a resonant wavelength λ_{res}. The 3 dB

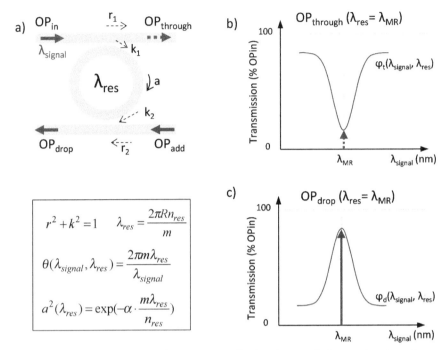

Figure 9.6 MR model: a) device geometry; b) signal transmission at the through port (OP$_{through}$) and c) signal transmission at the drop port (OP$_{drop}$).

bandwidth of the signal is assumed to be small compared to that of the MR (0.1 nm and 1.55 nm [20] respectively). The level of the signal power at the through port (OP$_{through}$) and drop port (OP$_{drop}$) depends on the respective transmission ratios at both ports (i.e. φ_t and φ_d), which rely on the alignment between λ_{signal} and λ_{res} (Figure 9.6(b) and (c)). Specifically, the actual transmission ratios depend on the device geometry (i.e. ring radius, R), the self(cross)-coupling coefficient (i.e. r_1, r_2, k_1 and k_2), the power attenuation coefficient α, and the single-pass phase shift $\theta(\lambda_{signal}, \lambda_{res})$. Table 9.1 summarizes the parameters and Equations (9.1) and (9.2) give the transmission φ_t and φ_d extracted from [35]. For a fixed radius (indicating a fixed resonant wavelength), since the gap between the MR and waveguide changes, the self-coupling coefficients (r_1 and r_2) also change accordingly, as well as the cross-coupling coefficients (k_1 and k_2). This impacts the transmission ratio accordingly, as well as the BW$_{3\,dB}$. We assume symmetric coupling (i.e. $r_1 = r_2$) to maximize the transmission at resonance to the drop port [36].

Table 9.1 Related parameters

Parameter	Description	Unit
r_1, r_2	Self-coupling coefficient	
k_1, k_2	Cross-coupling coefficient	
R	MR radius	um
n_{res}	Effective refractive index of MR (varies with applied voltage, device geometry and ambient temperature)	
m	Resonant mode number of MR	
λ_{res}	MR resonant wavelength	nm
λ_{signal}	Signal wavelength in vacuum	nm
α	Power attenuation coefficient	dB/cm
a (λ_{res})	Single-pass amplitude transmission	
$\theta(\lambda_{signal}, \lambda_{res})$	Single-pass phase shift	
BW_{3dB}	3dB bandwidth of MR	nm
$dn_{res}/d\lambda$	Wavelength dependence of refractive index	nm^{-1}
β	coefficient of resonant wavelength drift dependent on the temperature	nm/°C

In Figure 9.6(b) and (c), the transmission spectra of the MR (resonant at λ_{MR}) are considered to illustrate the relationship clearly. As an example, the signal wavelength λ_{signal} is assumed to be equal to λ_{MR}. In the ideal design, the input signal (OP_{in}) at $\lambda_{signal} = \lambda_{res}$ is totally transmitted to the drop port. However, in the real case, the maximum transmission to the drop port is obtained when $\lambda_{signal} = \lambda_{res}$ (in Figure 9.6(c)), with a portion of signal propagating to the through port in the waveguide (Figure 9.6(b)). When $\lambda_{signal} \neq \lambda_{res}$ (above 1.55 nm), most of the input signal power continues to propagate along the waveguide to the through port. Only a small portion of the signal power is coupled to the drop port. The misalignment of the wavelengths may be required for proper ONoC operation (for wavelength-based routing) or it can be a side effect related to a temperature difference between the ONIs (the laser source and MR for transmitting and receiving the signal separately). By assuming that λ_{signal} equals λ_{res} for an identical temperature of the laser source and the MR during the communication process, at most 50% of the signal will be (wrongly) dropped from the waveguide under a 7.7°C temperature difference (0.77 nm misalignment). This will lead to crosstalk and signal attenuation, which is evaluated with an analytical model.

$$\varphi_t(\lambda_{signal}, \lambda_{res}) = \frac{a^2(\lambda_{res})r_2^2 - 2a(\lambda_{res})r_1 r_2 \cos[\theta(\lambda_{signal}, \lambda_{res})] + r_1^2}{1 - 2a(\lambda_{res})r_1 r_2 \cos[\theta(\lambda_{signal}, \lambda_{res})] + [a(\lambda_{res})r_1 r_2]^2}$$

$$(9.8)$$

$$\varphi_d(\lambda_{signal}, \lambda_{res}) = \frac{a(\lambda_{res})(1 - r_1^2)(1 - r_2^2)}{1 - 2a(\lambda_{res})r_1 r_2 \cos[\theta(\lambda_{signal}, \lambda_{res})] + [a(\lambda_{res})r_1 r_2]^2}$$

(9.9)

Given the transmission ratios at the through and drop ports, the corresponding optical power ($OP_{through}$ and OP_{drop}) can be obtained as the following equations:

$$OP_{through} = OP_{in} \times \varphi_t(\lambda_{signal}, \lambda_{res}) \qquad (9.10)$$

$$OP_{drop} = OP_{in} \times \varphi_d(\lambda_{signal}, \lambda_{res}) \qquad (9.11)$$

According to the resonant condition of MR, when the resonant mode and the radius are fixed, the MR resonant wavelength (λ_{res}) is proportional to the effective refractive index (n_{res}). The effective refractive index is sensitive to the temperature, which will further influence the MR resonant wavelength due to the thermo-optic effect [20]. Thus, the drift of the MR resonant wavelength ($\Delta\lambda_{res}$) can be formalized linearly with the temperature drift ΔT as follows [20]:

$$\Delta\lambda_{res} = \Delta T \times \beta \qquad (9.12)$$

9.5 Case Study

We first detail the considered architecture (i.e., package, electrical layer, optical layer and on-chip laser model). Then, the SNR is analyzed based on the architecture.

9.5.1 System Specification

9.5.1.1 SCC and package

The target system used in our experiments is based on Intel's Single-Chip Cloud Computer (SCC) [37] (Figure 9.7), a 24-tile, 48-core IA-32 45 nm processor with a maximum power dissipation of 125W (for cores operating at 1 GHz). Given the large number of cores, the SCC is a good candidate as the electrical layer for silicon-photonic communications.

We model the same package as the one used by Intel [37]. Figure 9.8 shows the assembly view of the target system, which contains the following components: steel back-plate, motherboard, socket, SCC chip with silicon-photonic links and on-chip laser sources, copper lid and heat sink.

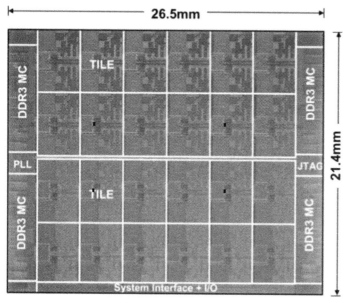

Figure 9.7 SCC chip.

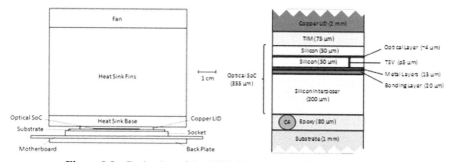

Figure 9.8 Packaging of the SCC chip and the optical interconnect.

9.5.1.2 ORNoC

In ORNoC [3], each IP core communicates with another IP through waveguides forming a ring. The following operations are performed:

- Injection: the IP core injects an optical signal into a waveguide through its output port data. The wavelength of the signal specifies the destination of the IP core;
- Pass through: the incoming signal propagates along the waveguide (i.e., no MR with the same resonant wavelength is located along the waveguide);

- Ejection: the incoming optical signal is ejected from the waveguide and is redirected to the destination IP core. This is achieved by an MR located along the waveguide and with the same resonant wavelength as the signal.

In ORNoC, the same wavelength can be used to realize multiple communications on the same waveguide at the same time. Furthermore, multiple waveguides can be used to interface IP cores. Both clockwise (C) and counter-clockwise (CC) rotation can be considered for signal propagation, where each direction is realized on a separate waveguide.

We assume an ORNoC with 4 waveguides and 4 lasers per waveguide per ONI, and using a serpentine layout. Two waveguides are for C rotation while the other two are for the CC rotation, as shown in the inset of Figure 9.9.

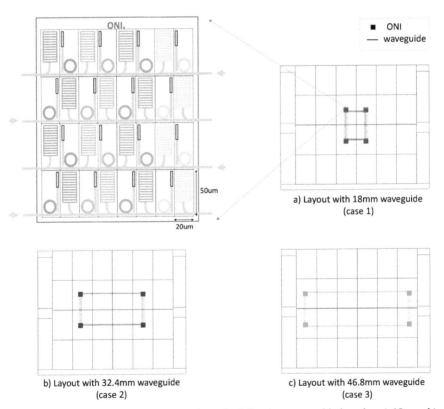

a) Layout with 18mm waveguide (case 1)

b) Layout with 32.4mm waveguide (case 2)

c) Layout with 46.8mm waveguide (case 3)

Figure 9.9 The location of the ONI leads to the following waveguide lengths: a) 18mm, b) 32.4 mm and c) 46.8 mm respectively.

The CMOS drivers and receivers are placed on an empty area of the electrical router of SCC tile. We evaluate the influence of the ONIs location on the SNR. As illustrated in Figure 9.9(a), (b) and (c), we consider 3 scenarios leading to 18 mm (in red), 32.4 mm (in blue) and 46.8 mm (in green) waveguide lengths, based on SCC (Figure 9.7). P_{VCSEL} and P_{heater} are set to 3.6 mW and 1.08 mW respectively (simulations validate that the gradient temperature within each ONI remains below 1°C). OP_{VCSEL} is estimated from the ONI average temperature and the on-chip laser characteristics.

9.5.2 SNR Analysis in the Considered Architecture

Figure 9.10 illustrates a waveguide propagating optical signals injected and ejected by the ONIs. Since we consider ORNoC [3] as the implementation in the optical layer, no arbitration is needed and passive MRs are utilized (i.e. the MRs resonant wavelength is defined at design time but shifts with temperature variation). A communication between a source interface ONI_S and a destination interface ONI_D (denoted as $C_{s \to d}$ and represented by the red line) implies that an optical signal is 1) generated and modulated by a VCSEL-based transmitter ($T_{s \to d}$) and 2) dropped to a photodetector by an MR ($R_{s \to d}$). The quality of the communication $C_{s \to d}$ is influenced

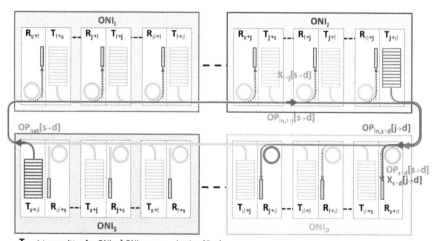

$T_{s \to d}$: transmitter for $ONI_S \to ONI_D$ communication ($C_{s \to d}$)
$R_{s \to d}$: receiver for $ONI_S \to ONI_D$ communication ($C_{s \to d}$)

Figure 9.10 Signal propagation in a waveguide. The quality of a communication depends on the signal attenuation and the crosstalk.

by the signal attenuation along the path and the crosstalk induced by the other communications $C_{i \to j}$, where $i \neq s$ and $j \neq d$ (e.g. the communication represented by the blue line).

Based on the analysis model in Section 9.4.3, the SNR of communication $C_{s \to d}$ at the receiver $R_{s \to d}$ is formalized as follows:

$$SNR = 10 \log_{10} \frac{P_{signal}}{P_{noise}} = 10 \log_{10} \frac{OP_{s \to d}[s \to d]}{\sum_{i=1}^{N} \sum_{j=1}^{N} X_{s \to d}[i \to j]} \quad (9.13)$$

with $i \to \neq j$ and $i \to j \neq s \to d$,

where $OP_{s \to d}[s \to d]$ is the expected signal power of $C_{s \to d}$ received by $R_{s \to d}$ (in ONI_D), and $X_{s \to d}[i \to j]$ is the crosstalk power of the other communications $C_{i \to j}$ received by $R_{s \to d}$.

The losses encountered by the signal $C_{s \to d}$ depend on the waveguide length and the crossed MRs (in $R_{i \to j}$) in the intermediate ONI_I and ONI_J. In the case where the signal wavelength of $C_{s \to d}$ (denoted $\lambda_{T,s \to d}$) is far enough from the MR resonant wavelength $\lambda_{R,i \to j}$ (in $R_{i \to j}$), the losses are expected to be small. This optimistic scenario occurs if the temperature of the intermediate ONIs (ONI_I and ONI_J in the figure) are strictly equal to the temperature of the source ONI_S. A temperature difference among different ONIs will lead to a misalignment of the wavelengths. In this case, part of the signal power will be coupled from the waveguide and reach an undesired photodetector. This leads to additional crosstalk and reduced signal power at intermediate and destination photodetectors respectively (e.g. $X_{i \to j}[s \to d]$ and $OP_{s \to d}[s \to d]$ in the figure). The losses are formulated as follows:

$$X_{i \to j}[s \to d] = OP_{in,i \to j}[s \to d] \times \varphi_d[\lambda_{T,s \to d}(T), \lambda_{R,i \to j}(T)] \quad (9.14)$$

$$OP_{in,ij}[sd] = OP_{net}[s \to d](T)$$
$$\times \prod_{k=s}^{\substack{j-1, if\, j>s \\ j+N-1, if\, j<s}} L_{k \bmod N} \times \prod_{m=s+1}^{\substack{j, if\, j>s \\ j+N, if\, j<s}} \prod_{k=1}^{\substack{i-1, k \neq m \bmod N, if\, m=j \\ N, k \neq m \bmod N, if\, m \neq j}}$$
$$\varphi_t[\lambda_{T,s \to d}(T), \lambda_{R,k \to m \bmod N}(T)] \quad (9.15)$$

$$OP_{net}[s \to d](T) = slop_{W/A}(T) \times [I_{VCSEL} \text{-} I_{th}(T)] \quad (9.16)$$

$$L_k = L_{propagation}^{l_k} \quad (9.17)$$

where $X_{i \to j}[s \to d]$ is the signal power dropped by $R_{i \to j}$, $OP_{in,i \to j}[s \to d]$ is the power of the signal at the $R_{i \to j}$, $\varphi_d(\lambda_{T,s \to d}(T), \lambda_{R,i \to j}(T))$ is the signal ratio dropped by $R_{i \to j}$. $\varphi_t[\lambda_{T,s \to d}(T)$, and $\lambda_{R,k \to mmodN}(T)]$ is the signal ratio passing by $R_{i \to j}$ (in $R_{i \to j}$, $\varphi_t[\lambda_{T,s \to d}(T), \lambda_{R,k \to 0}(T)]$ $=\varphi_t[\lambda_{T,s \to d}(T), \lambda_{R,k \to N}(T)]$). $OP_{net}[s \to d](T)$ is the power injected by the on-chip laser ($T_{s \to d}$), which is temperature-dependent since the laser efficiency is largely influenced by the temperature. In the equation of $OP_{net}[s \to d](T)$, $slop_{W/A}(T)$ is the property of the laser output power according to the laser driver current I_{VCSEL}, while $I_{th}(T)$ is the threshold current of the laser, varying with the laser temperature T. N is the number of ONIs in the optical interconnect. L_k is the propagation loss (in %) along a communication path (e.g. L_1 for $C_{1 \to 2}$ and L_2 for $C_{2 \to 3}$), l_k is the corresponding length of the waveguide and $L_{propagation}$ is the propagation loss per unit length (in % per cm).

9.5.3 Thermal Characteristics of On-Chip Laser

For the on-chip laser, we consider a VCSEL with a $15 \times 30 \ \mu m^2$ footprint size [17, 30]. It relies on mirrors redirecting the vertically generated light into a horizontal waveguide, which allows the thickness of the laser to be reduced below 4 μm (one of the factors making the VCSEL CMOS-compatible). The direct modulation bandwidth is 12 GHz and the 3 dB bandwidth is about 0.1 nm.

Figure 9.11(a) represents a 3D view of the laser extracted from [17]: it is composed of 3 layers of III–V material (0.6 μm, 0.45 μm and 0.4 μm thickness respectively). The laser effect is generated in the central active layer (in red) and the power is dissipated in the adjacent layers (in grey). Two contacts allow driving of the current from the CMOS layer through 5μm-diameter TSVs. The active layers are surrounded by Si/SiO$_2$ lines constituting the mirror structure, which allows light to be coupled into the horizontal waveguide (optical signal shown in blue) using a taper with an estimated 70% efficiency. Figure 9.11(b) gives the laser efficiency according to its temperature and I_{VCSEL}. Figure 9.11(c) gives the influence of P_{VCSEL} and its temperature on the actual emitted light OP_{VCSEL}.

9.6 Results

In this section, the MR heater power is explored through thermal simulations in order to reduce the gradient temperature. Finally, ONoC solutions are compared according to the SNR.

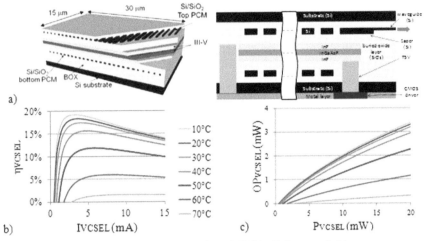

Figure 9.11   VCSEL: a) 3D view extracted from [17], b) efficiency and c) laser output power according to the temperature.

9.6.1 Reduction of the ONI Gradient Temperature

We first evaluate the influence of P_{chip} and P_{VCSEL} on the average and gradient temperature of ONI (e.g., ONI_1 in Figure 9.9). The average temperature is the mean temperature of all the optical devices (i.e., VCSELs and MRs) in the ONI. The gradient temperature is the gap between the maximum and minimum temperatures of the optical devices in the ONI. We run thermal simulations under homogeneous 12.5W, 18.75W, 25W and 31.25W chip activities and we explored P_{VCSEL} ranging from 0 to 6 mW (we assumed $P_{VCSEL}=P_{driver}$, which corresponds to the worst case scenario since the total energy received by the VCSEL is dissipated as heat). Figure 9.12(a) illustrates the average temperature results: a 6W increase of the total chip activity roughly leads to +3.3°C on the average temperature while a 6 mW increase of P_{VCSEL} leads to +11°C. This demonstrates the importance of the calibration of the laser modulation current according to the requirements: a slightly over-sized current will lead to significant power consumption overhead.

Results also show a significant impact of P_{VCSEL} on the gradient temperature between lasers and MRs (1.7°C/mW). Such gradient temperature does not realistically allow the use of the clustering technique for run-time calibration. We thus explore MR heating power values (P_{heater}) to reduce this gradient at design time. As illustrated in Figure 9.12(b), the smallest gradient

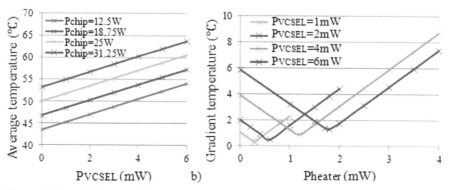

Figure 9.12 a) Influence of P_{VCSEL} and P_{chip} on the average temperature and b) influence of P_{VCSEL} and P_{heater} on the gradient temperature.

is obtained for $P_{heater} = 0.3 \times P_{VCSEL}$. In Figure 9.13, we compare the temperature results with and without MR heaters: for $P_{VCSEL}=1mW$, solutions with and without heaters lead to 0.3°C and 1°C gradient temperature respectively. Significant improvement of the gradient temperature is obtained for higher P_{VCSEL} values: for 6mW, the gradient temperature drops from 5.8°C to 1.3°C (i.e. −4.5°C), which is significant compared to a reasonable 0.8°C increase of the average laser temperature.

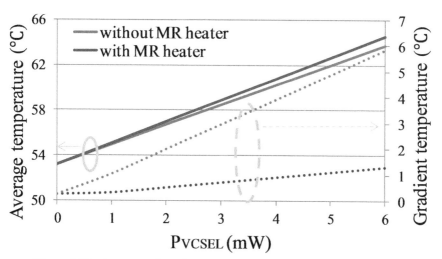

Figure 9.13 Average and gradient temperatures with and without MR heaters.

9.6.2 System Level Estimation of SNR

We evaluate the SNR for the layouts introduced in Figure 9.9. Table 9.2 summarizes the technological parameters that we have considered and Figure 9.14 gives the worst-case SNR results. For a uniform activity, the asymmetric structure of the SCC chip leads to a 3°C difference among the ONIs for 46.8 mm length. The crosstalk is relatively small and the SNR thus only depends on the signal power, which depends on the propagation losses. The SNR drops from 21 dB for 18 mm length to 8 dB for 46.8 mm. For a diagonal activity, the upper-right and bottom-left parts of the chip dissipate 2.6W per tile while the upper-left and bottom-right parts dissipate 5.2W per tile. This leads to heterogeneous temperature of the ONIs (54.62→55.92°C for case 1, 54.33→56.92°C for case 2, 56.16→60.85°C for case 3 respectively). Compared to the uniform activity, the diagonal activity exhibits lower average temperature since upper-right and bottom-left parts have lower power. This leads to higher laser efficiency, further resulting in higher OP_{VCSEL}. However, the temperature gradient *among* the ONIs being higher, additional crosstalk occurs, which results in a lower SNR compared to the uniform activity. A random activity leads to intermediate SNR results.

This analysis validates that the ONoC matches with the receiver sensitivity. Further explorations of the design space allow optimization of the ONoC. For instance, if a given SNR is acceptable, P_{VCSEL} and P_{heater} can be reduced for energy saving. We consider an aggressive scenario for laser efficiency (Figure 9.15(a)), twice that of the conservative case (Figure 9.15(b)). The results of SNR for the various activities are shown in Figure 9.16. The signal power is higher than that in Figure 9.14, since the laser output power OP_{VCSEL} is higher under the same P_{VCSEL} and temperature. Meanwhile, more crosstalk is introduced because of the higher OP_{VCSEL}. For instance, in the 18 mm case with random activity, the signal and crosstalk power figures

Table 9.2   Technological parameters

Parameters	Value
BW_{3dB}	1.55 nm
α	2 dB/cm
n_{MR}	~2.55
m	17
$dn_{res}/d\lambda$	–0.000016 nm^{-1}
β	0.1 nm/°C
$L_{propagation}$	0.5 dB/cm [38]
Wavelength range	C-band
Photodetector sensitivity	–20 dBm (0.01 mW)

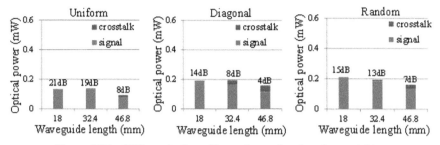

Figure 9.14   SNR results for uniform, diagonal and random activities.

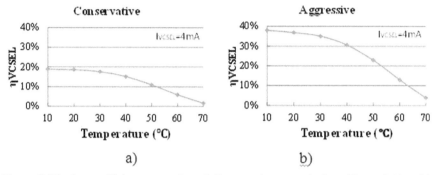

Figure 9.15   Laser efficiency scenarios: a) Conservative scenario from Figure 9.11 and b) Aggressive scenario, twice that of the conservative case.

are 0.515 mW and 0.017 mW respectively in Figure 9.16, compared to 0.205 mW and 0.007 mW in Figure 9.14. However, the SNR results remain almost the same since both signal and crosstalk power increase at the same time. Thus, to meet a given SNR, the energy can be saved by reducing P_{VCSEL}.

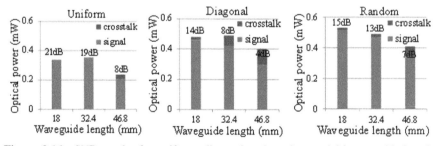

Figure 9.16   SNR results for uniform, diagonal and random activities, considering the aggressive scenario of laser efficiency.

9.7 Conclusion

In this work, we proposed a thermal-aware design methodology for CMOS-compatible on-chip laser-based ONoCs. Thermal simulations allow the exploration of design parameters such as the laser modulation current and heater power in order to reduce the gradient temperature within an ONoC interface. A heater dissipating 30% of the VCSEL power leads to an optimal solution for the considered architecture. SNR analyses allow the comparison of the ONoC for various levels of chip activity.

Acknowledgment

Hui LI is supported by China Scholarship Council (CSC).

References

[1] R. Ho, et al. The Future of Wires. In *Proceedings of the IEEE*, 89(4):490–504, 2001.

[2] I. O'Connor, et al. Reduction Methods for Adapting Optical Network on Chip Topologies to Specific Routing Applications. In *Proceedings of DCIS*, 2008.

[3] S. Le Beux, et al. Layout Guidelines for 3D Architectures including Optical Ring Network-on-Chip (ORNoC). In *19th IFIP/IEEE VLSI-SOC*, 2011.

[4] Luca Ramini, et al. Contrasting wavelength-routed optical NoC topologies for power-efficient 3D-stacked multicore processors using physical-layer analysis. In *Proceedings of DATE*, 2013.

[5] A. Shacham, K. Bergman and L.P. Carloni. Photonic Networks-on-Chip for Future Generations of Chip Multi-Processors. *IEEE Transactions on Computers*, 57(9), pp. 1246–1260, 2008.

[6] Y. Ye, J. Xu, X. Wu, et al. System-level Analysis of Mesh-based Hybrid Optical-Electronic Network-on-Chip. *IEEE International Symposium on Circuits and Systems (ISCAS)*, May 2013.

[7] Y. Ye, et al. A Torus-Based Hierarchical Optical-Electronic Network-on-Chip for Multiprocessor System-on-Chip. *ACM Journal on Emerging Technologies in Computing Systems (JETC)*, 8(1), pp. 5:1–5:26, 2012.

[8] Y. Ye, J. Xu, B. Huang, et al. 3D Mesh-based Optical Network-on-Chip for Multiprocessor System-on-Chip. *IEEE Transactions on Computer-Aided Design of Integrated Circuits and Systems*, 32(4), pp. 584–596, 2013.

[9] M. J. Cianchetti, J. C. Kerekes, and D. H. Albonesi. Phastlane: a rapid transit optical routing network. In *Proceedings of the 36th Annual International Symposium on Computer Architecture (ISCA '09)*, 2009.

[10] L. Zhang, E. E. Regentova, X. Tan. Packet switching optical network-on-chip architectures. *Computers & Electrical Engineering*, 39(2), 697–714, 2013.

[11] M. Heck and J. E. Bowers. Energy Efficient and Energy Proportional Optical Interconnects for Multi-Core Processors: Driving the Need for On-Chip Sources. *IEEE Journal of Selected Topics in Quantum Electronics*, 20(4), 2014.

[12] X. Wu, J. Xu, Y. Ye, and et al. SUOR: Sectioned Undirectional Optical Ring for Chip Multiprocessor. *ACM Journal on Emerging Technologies in Computing Systems (JETC)*, 10 (4), 2014.

[13] Y. Pan, P. Kumar, J. Kim, and et al. Firefly: illuminating future network-on-chip with nanophotonics. In *Proceedings of the 36th Annual International Symposium on Computer Architecture (ISCA '09)*, 2009.

[14] Z. Li, et al. Reliability Modeling and Management of Nanophotonic On-Chip Networks. *IEEE Transactions on Very Large Scale Integration (VLSI) Systems*, Vol. 20, No. 1, pp. 98–111, 2012.

[15] S. Manipatruni, et al. Wide temperature range operation of micrometer-scale silicon electro-optic modulators. *OPTCS LETTERS*, Vol. 33, No. 19, pp. 2185–2187, 2008.

[16] A. Biberman, et al. Thermally Active 4x4 Non-Blocking Switch for Networks-on-Chip. in *IEEE Lasers and Electro-Optics Society*, pp. 370–371, 2008.

[17] E. Timurdogan, C. M. Sorace-Agaskar, J. Sun, E. S. Hosseini, A. Biberman, and M. R. Watts. An ultralow power athermal silicon modulator. *Nature Communications* 5, 4008, 2014.

[18] K. Padmaraju, et al. Thermal stabilization of a microring modulator using feedback control. *Optics Express*, Vol. 20, No. 27, pp. 27999–28008, 2012.

[19] S. S. Djordjevic, et al. CMOS-compatible, athermal silicon ring modulators clad with titanium dioxide. *OPTICS EXPRESS*, Vol. 21, No. 12, 2013.

[20] Y. Ye, et al. System-Level Modeling and Analysis of Thermal Effects in Optical Networks-on-Chip. *IEEE Transactions on Very Large Scale Integration Systems*, Vol. 21, No. 2, pp. 292–305, February 2013.

[21] Y. Ye, Z. Wang, P. Yang, et al. System-Level Modeling and Analysis of Thermal Effects in WDM-Based Optical Networks-on-Chip. *IEEE Transactions on Computer-Aided Design of Integrated Circuits and Systems*, 33(11), 1718–1731, 2014.

[22] Y. Zhang, et al. Power-Efficient Calibration and Reconfiguration for Optical Network-on-Chip. *Journal of Optical Communications and Networking*, 2012.

[23] T. Zhang, et al. Thermal management of Manycore Systems with Silicon-Photonic Networks. In *Proceedings of Design, Automation and Test in Eutope (DATE)* 2014.

[24] C. Chen, et al. Sharing and Placement of On-chip Laser Sources in Silicon-Photonic NoCs. In *Proceedings of International Symposium on Networks-on-Chip*, 2014.

[25] S. Le Beux, et al. Optical Crossbars on Chip, A Comparative Study based on Worst-Case Losses. In *Wiley CCPE*, 2014.

[26] A. Bianco, et al. Optical Interconnection Networks based on Micro-ring Resonators. In *IEEE International Conference on Communications*, 2010.

[27] J. Ahn, et al. Devices and architectures for photonic chip-scale integration. *Applied Physics A: Materials Science & Processing*, 95(4):989–997, 2009.

[28] D. Van Thourhout, T. Spuesens, S. Kumar Selvaraja, et al. Nanophotonic devices for optical interconnect. *IEEE Journal of Selected Topics in Quantum Electronics*, Vol. 16, No. 5, pp. 1363–1375, 2010.

[29] Dana Vantrease, et al. Corona: System Implications of Emerging Nanophotonic Technology. In *Proceedings of ISCA*, 2008.

[30] Markus-Christian Amann and Werner Hofmann. InP-Based Long-Wavelength VCSELs and VCSEL Arrays. *IEEE Journal of Selected Topics in Quantum Electronics*, Vol. 15, No. 3, 2009.

[31] J. Van Campenhout, et al. Electrically pumped InP-based microdisk lasers integrated with a nanophotonic silicon-on-insulator waveguide circuit. *Optics Express*, 15(11), p. 6744–6749, 2007.

[32] A. Fourmigue, et al. Efficient Transient Thermal Simulation of 3D ICs with Liquid-Cooling and Through Silicon Vias. In *Proceedings of DATE* 2014.

[33] S. C. Chapra and R. P. Canale. *Numerical Methods for Engineers*. McGraw-Hill, Inc., New York, NY, USA, 6th edition, 2009.

[34] COMSOL. http://www.comsol.com, July 2014.

[35] Wim Bogaerts, Peter De Heyn, Thomas Van Vaerenbergh, Katrien DeVos, Shankar Kumar Selvaraja, Tom Claes, Pieter Dumon, Peter Bienstman, Dries Van Thourhout, and Roel Baets. Silicon microring resonators. *Laser&Photonics Reviews*, 6(1):47–73, 2012.

[36] Z. Su, E. Timurdogan, J. Sun, M. Moresco, G. Leake, D. Coolbaugh, and M. R. Watts. An on-chip partial drop wavelength selective broadcast network. *Conference on Lasers and Electro-Optics (CLEO)*, 2014.

[37] J. Howard et al. A 48-core IA-32 processor in 45 nm CMOS using on-die message-passing and DVFS for performance and power scaling. *Solid-State Circuits, IEEE Journal of*, 46(1):173–183, 2011.

[38] A. Biberman, et al. Photonic Network-on-Chip Architectures using multilayer deposited silicon materials for high-performance chip multiprocessors. *ACM Journal on Emerging Technologies in Computing Systems*, 7(2) 7:1–7:25.

10

Fault-tolerant Photonic Network-on-Chip

Michael Meyer and Abderazek Ben Abdallah

Department of Computer Science and Engineering, University of Aizu,
Aizu-Wakamatsu, Fukushima, 965-8580 Japan

Abstract

Photonic Networks-on-Chip (PNoCs) promise significant advantages over
their electronic counterparts. In particular, they offer a potentially disruptive
technology solution with fundamentally low power dissipation that remains
independent of capacity while providing ultra-high throughput and minimal access latency. However, the major optical device in PNoC systems,
microring resonators (MRs), are very sensitive to temperature fluctuation and
manufacturing errors. A single MR failure may cause messages to be misdelivered or lost, which results in bandwidth loss or even complete failure of the
whole system. This chapter describes a fault-tolerant PNoC architecture. The
system is based on a fault-tolerant path-configuration and routing algorithm,
a microring fault-resilient photonic router, and uses minimal redundancy to
assure accuracy of the packet transmission even after faulty MRs are detected.

10.1 Introduction

Photonic Network-on-Chip (PNoC) is becoming an attractive solution
enabling ultra-high communication bandwidth in the terabits per second
range, low power, and low communication latency [7, 9–12]. When combined
with Wavelength Division Multiplexing (WDM), multiple parallel optical
streams of data concurrently transfer through a single waveguide, while MRs,
which can be switched as high as 40 GHz, are used to realize wavelength-
selective modulators, and switches [44]. While a single-layer configuration
can provide low-loss waveguides and high-performance photonic devices,
it suffers from limited integration density due to waveguide crossing and

limited real estate. A way to go beyond this limitation is to monolithically stack multiple photonic layers above Si as multilayered electrical interconnections realized in modern electronic circuits [8, 61]. Figure 10.2 shows a high level view of a three-dimensional PNoC (PHENIC) implemented with one electrical control layer and several photonic communication layers [39].

The main components of an PNoC include a laser source, which generates phase-coherent and equally spaced wavelengths, waveguides, which is used as a transmission medium, and modulators and photodetectors, which convert electrical digital data to and from photonic signals [32]. Figure 10.1 shows a typical on-chip optical link that uses an external laser as a light source. It is expected that the laser source could produce up to 64 wavelengths per waveguide for a Dense Wavelength Division Multiplexing (DWDM) network.

Fault tolerance is crucial when considering mission critical applications where the system must correctly function even when something goes wrong. One such an application is that of space travel, where repair or replacement is not a possible option, and billions of dollars would be wasted.

10.1.1 Design Challenges

The photonic domain is immune to transient faults caused by radiation [29], but is still susceptible to process variation (PV) and thermal variations (TV) as well as aging. The aging typically occurs faster in active components as well as elements that have high TV [26]. In the optical domain, the faults can occur in MRs, waveguides, routers, etc. Active components, such as MRs, have higher failure rates than passive components, e.g. waveguides [26]. A single MR failure can cause messages to be misdelivered or lost, which results is in bandwidth loss or even complete failure of the whole system.

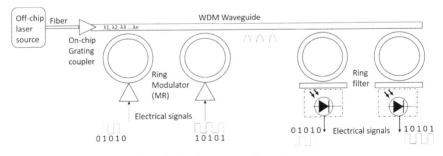

Figure 10.1 Photonic link architecture.

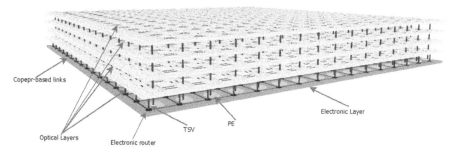

Figure 10.2   3D-Stacked photonic network-on-chip architecture.

Together, fabrication-induced PV and TV effects present enormous perfor-mance and reliability concerns. TV causes a microring to respond to a different wavelength than intended. This can take the form of a passband shift in the MRs. When an MR heats up, it expands, changing its radius, and therefore shifting the wavelengths which it uses to the right [15]. As reported in [44], a change of as little as 1°C can shift the resonance wavelength of a microring by as much as 0.1 nm. This is not permanent and will return when the temperature returns to normal. Therefore, systems' temperature must be kept at a reasonable value in order for the MRs to resonate correctly. This is challenging, especially in large complex computing system, which uses thousands of these components. Trimming technique [4] is generally used to dynamically modify the resonance frequency of a microring to overcome both thermal drift and fabrication inaccuracy. This technique can be accomplished by dynamically increasing the current in the $n+$ region or by heating the ring [4, 22, 48].

PV is the variations of critical physical dimensions, e.g. thickness of wafer, width of waveguides also affect the resonant wavelengths of MRs. This means that not all fabricated MRs can be used due to PV. As a result, network nodes that do not have all working MRs would lose some or all of wavelengths/bandwidth in communication [56]. To solve this problem, Xu et al. [58] proposed a method of flexible wavelength assignment. Because the networks are already built with excess detectors or Modulators for each message, the node with the excess components can compensate and rematch to the components which have been affected by PV.

Over time, all silicon based ICs wear down. We refer to this phenomenon as *aging*. Some of the aging effects only apply to the active components, because of their electrical subcomponents [54], such as the MRs, while other aging affects all parts, even the waveguides.

Recent PNoCs researches (i.e. network topology, router micro-architecture design, and performance and power optimization and analysis) have resulted in several architectures capable of transmitting at a high data bandwidth and low energy dissipation [7, 9–12]. In [8], we proposed an energy-efficient and high-throughput hybrid silicon-photonic network-on-chip based on a smart contention-aware path-configuration algorithm and an energy-efficient non-blocking optical switch to further exploit the low energy proprieties of the PNoC systems. However, little attention has been given to the aspect of fault-tolerance and reliability along the photonic interconnects.

This chapter presents a fault-tolerant PNoC architecture. The system is based on a fault-tolerant path-configuration and routing algorithm, a microring fault-resilient photonic router, and uses minimal redundancy to assure accuracy of the packet transmission even after faulty MRs are detected.

10.1.2 Fault Models

It is worth noting that the light is not sensitive to radiation or electromagnetic fields, the signals which control the optical network can be sensitive to it. The following is a list of actual possible causes that can contribute to the failure of an optical device.

10.1.2.1 PNoC signal strength

Typical NoCs are defined by their power consumption, delay and throughput. PNoCs also have to consider the Signal-to-Noise Ratio at the receiving end. Because they do not buffer and retransmit, the signal gets weaker based off of how many hops it jumps. This does not significantly affect the power the network consumes, but it can lead to a higher sensitivity to noise.

10.1.2.2 Electrostatic discharge

While the waveguides are not electrically conductive, the switches and photo-detectors are. This means that they are sensitive to high currents. One thing which can ruin an IC is electrostatic discharge(ESD). This is when a current enters in through the I/O pins of the control circuit, or it can be caused by an extremely strong magnetic field. This all results in the aforementioned extreme current, and this current causes severe damage to the silicon in the components. Possible points of damage are the dielectric, the PN junctions, and any wiring connecting to the controllers. Because of the scaling, the causing phenomena have become harder to control [24]. This can be prevented by proper packaging to the IC providing ESD protection at the pins.

10.1.2.3 Noise

This is one of the unique things that we categorize as a cause for a fault. The reason is because the noise can be caused simply by poorly matched wavelengths. It can also be caused by creating a path that is too long, or a path that crosses too many intersections. These paths tend to be caused by rerouting or non-minimalistic routing, but other factors can contribute and cause more noise. The most common factors are listed in the following subsections.

10.1.2.4 Aging

Over time, all silicon based ICs wear down. Some of the aging effects only apply to the active components, because of their electrical subcomponents, while other aging affects the optical properties of the components.

Electromigration – This mainly affects the wires which control the ring resonators. It does not affect the waveguides in any way. It originally causes a delay in the wire, and can eventually lead to an open, or to a short to a nearby wire. It achieves this by thinning out the thinnest portion of the wire due to higher current density at the bottleneck [30].

Laser Degradation – After the lasers have been on for several hundred hours, they start to show signs of degradation. This shows in the form of either missing wavelengths, which can cause a channel fault, or general weakening of the original laser signal. In each of these cases, it does not become a true problem until the signal falls to a level where the worst case scenario's Signal-to-Noise ratio is too weak to receive an understandable signal [37].

Photodetector Degradation – Various studies have been done for different types of photodetectors showing that they degrade overtime, particularly from being exposed to thermal conditions or UV light. It is reasonable to assume that no matter what material photo detectors are made out of, they all seem to be vulnerable to degradation due to thermal variation, which is present in all networks [26, 54].

A lot of work has been done to combat the effects of aging. Some examples are Agarwal [2], Keane [30], and Kim [31]. These are mainly focused on the electrical side, but the fact that these do exist show the hope for a future where optical aging can be researched and prevented. Many parameters such as the wavelengths and laser strength can possibly be modified throughout the life of a chip to counteract the aging effects in a similar manner to what Mintarno does for Electrical networks [40].

10.1.2.5 Process variability

This can affect both the active and inactive components of the optical network. The variability accounts for material impurities, doping concentrations, and size and geometries of structures [47]. One single dimple in a particular point in the coupling region of a ring resonator can greatly affect the coupling properties and thus cause problems for the switch, or maybe just the channel. A poor geometry can also cause a certain component to be more sensitive to aging or ESD. Obviously if a variation gets bad enough, an entire link can be rendered useless. This would be considered an early permanent fault, and should be detected before a device is released. The impurities in a waveguide can cause such a block, or cause there to be a change in the reflectivity of the material, and that causes a higher amount of insertion loss, resulting in a lower signal-to-noise ratio. Other similar chains-of-events can occur from bad doping of the photodetectors. Minimizing this process variability can greatly increase the reliability of the system, even without implementing fancier and area or energy heavy redundancies. The unfortunate truth is that with recent advances in scaling, the variability continues to increase [33, 50].

10.1.2.6 Temperature variation

For electrical components, temperature variation can cause changes in properties such as resistivity and cause more power consumption or delay, but in the optical domain, it is quite different. Ring resonators are tuned by heating up the ring, causing them to expand, which changes their passband wavelength. If the chip heats up to a point beyond the tuning, then certain channels just disappear as a whole. The increase in temperature also causes the photodetectors to degrade as mentioned in the previous section. These temperature variations also tend to speed up other forms of aging as well.

Table 10.1 summarizes the physical causes and their effects. Many of these will need to be researched further, and only time will tell exactly how reliable optical is with some other phenomena, but for now, this is a comprehensive list of all physical sources of failures within an optical network. We separated the pure optical from the hybrid components so that it can show exactly how resilient the photons and waveguides really are, when compared with wires, but no Optical Network-on-Chip is completely free of wires.

Table 10.1   Overview of fault causes and effects (ABPV is Accelerated by Process Variation, OOHC is Optical Or Hybrid Components)

	Physical Cause	ABPV	Fault Class	Bursty	OOHC
Noise	Internal reflectance in waveguide	Y	Intermittent Logic Fault	N	Optical
	Waveguide Crossing	N	Intermittent Logic Fault	Y	Optical
	Intrachannel Noise	N	Intermittent Logic Fault	Y	Both
ESD	Build up of Static energy which is quickly released	Y	Permanent Logic Fault	N	Hybrid
Aging	Electron Migration	Y	Intermittent—>Permanent Delay or Logic Fault	N	Hybrid
	Laser Degradation	Y	Intermittent—>Permanent Logic Fault	N	Hybrid
	Photodetector Degradation	Y	Intermittent—>Permanent Delay or Logic Fault	N	Hybrid
	Waveguide Degradation	Y	Intermittent—>Permanent Logic Fault	N	Optical
	MR Degradation	Y	Intermittent—>Permanent Logic Fault	N	Hybrid
Temp. Variation	Variation in wear-out effects due to temp. differences	Y	Intermittent and Permanent delay, stuck-at and misrouting	Y/N	Hybrid
	Performance variation due to temp differences	Y	Intermittent delay and misrouting	Y	Hybrid
	Variation in wear-out effects due to temp. differences	Y	Intermittent and Permanent delay, stuck-at and misrouting	Y/N	Optical

10.2 Fault-tolerant Photonic Network-on-Chip Architecture

The Fault-tolerant Photonic Network-on-Chip (FT-PHENIC) system, shown in Figure 10.3, is a mesh-based topology and uses minimal redundancy to assure accuracy of the packet transmission even after faulty MRs are detected. The system uses Stall-Go mechanism for flow-control, and a Matrix-arbiter as a scheduling technique [3, 13, 14, 39]. FT-PHENIC is also based on a microring fault-resilient photonic router (FTTDOR) [39] and an adaptive path-configuration and routing algorithm. As illustrated in Figure 10.3, the proposed system consists of a Photonic Communication Network (PCN), used for data communication, and an Electronic Control Network (ECN), used for path configuration and routing. Each PE (Processing Element) is connected to a local electrical router and also connected to the corresponding gateway (modulator/detector) in the PCN [8]. Messages generated by the PEs are separated into control signals and payload signals. Control signals are routed in the ECN and used for path configuration and routing. The payloads are converted to optical data and transmitted on the PCN.

10.2.1 Microring Fault-resilient Photonic Router

The block diagram of the Microring Fault-resilient Photonic Router (FTTDOR) is shown in Figure 10.4. It consists of a non-blocking fault tolerant photonic switch (Figure 10.4(a)) and a light weight control router (Figure 10.4(b)). Redundant MRs are carefully placed at special locations on the switch to assure fault tolerance even if one of the MRs on the backup path has a fault. The backup route for the NEWS (North-East-West-South) directions is to actually use the waveguide connected to the core ports as a master backup; therefore, the redundant MRs are all chosen at the locations which connect the NSEW ports to the core.

For a majority of faults, the design of the switch allows for an alternate, slightly less power efficient route. In fact, the backup route is less power-efficient because the packets travel across more waveguide distance, go through more active MRs, and cross more waveguides. However, the switch still maintains all of its functionality. Because backup routes are only intended for use in the switches in which faults have occurred, the extra loss will have minimal effect on the message' signal strength across the whole network.

The FTTDOR was designed to require no MRs from East-West and North-South traffic. Since this kind of traffic accounts for a majority of the traffic of the PCN [39], such design will save on power and continue to

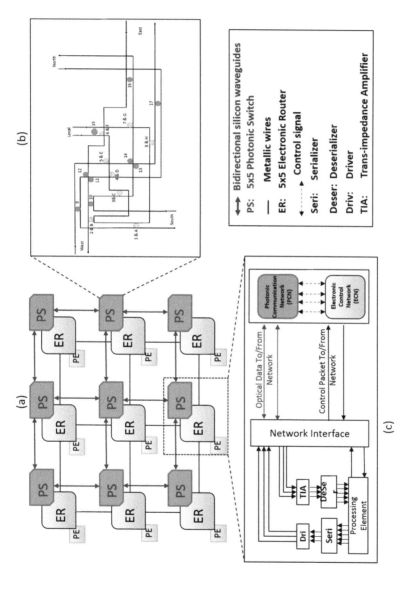

Figure 10.3 FT-PHENIC system architecture. (a) 3x3 mesh-based system, (b) 5x5 non-blocking photonic switch, (c) Unified tile including PE, NI and control modules.

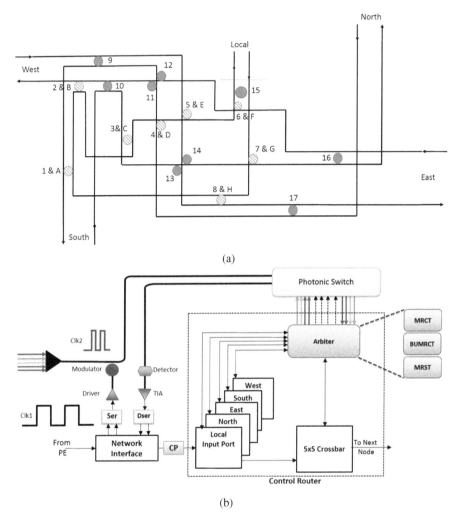

Figure 10.4 Microring fault-resilient photonic router (FTTDOR): (a) Non-blocking fault tolerant photonic switch, (b) Light-weight control router.

function in the case of any MR fails. Assuming that a single location of redundant MRs does not fail all together, the switch is able to maintain all functionality at slowed speeds.

Figure 10.5 shows a reconfiguration example of how MR 9 can be backed up by MRs 5, 15 and 1. Additionally, the MRs which connect parallel waveguides are replaced with racetracks [41]. This allows for a wider pass-band

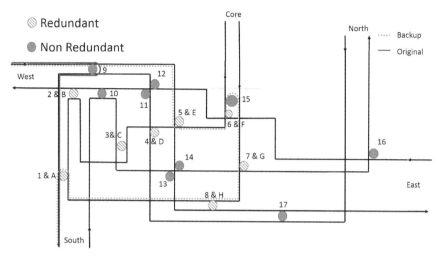

Figure 10.5 Example of how a non redundant MR's functionality can be mimicked by redundant ones.

of light frequencies, makes them less sensitive to physical faults, such as reduced sensitivity to thermally-caused passband shifting. Racetracks also have a larger Mean Time Between Failures (MTBF) [41].

The original form of FTTDOR switch is a five-port non-blocking switch, meaning that it allows for routing from any available port to any other available port. Once a fault is detected, the switch recovers, but there is a chance that it may turn into a blocking switch; however, it should be able to maintain all functionality as long as none of the redundant MRs fail. Because the redundant MRs lie dormant, they do not require much power other than the boost in signal strength required to compensate for the signal loss, caused by passing by an inactive MR, which is minimal. As all rerouting in the switch occurs on the core waveguide, traffic certainly increases on this one waveguides as too many faults occur, which is why it should be treated as a node failure after a threshold of failed MRs is reached.

In addition to tolerating faults, FTTDOR is able to handle the *ACK* signals and the resulting regeneration process of the *Tear-down* signal at each hop. To accomplish this goal, a hybrid switching policy is used: *Spacial-switching* for the data signals by manipulating the state of the broadband switching elements and a *Wavelength-selective switching* for the *Tear-down* signals by using detectors and modulators. Moreover, since the *Tear-down* signals should be checked and regenerated at each hop, it is crucial that their manipulation be automatic and not interfere with data signals, nor cause a blockage inside the switch. When the *Tear-down* is generated at the source NI (Network Interface), it is first sent to the control router. Then, the *Photonic Switch Controller* releases the corresponding MRs and generate another *Tear-down* which is sent to the output-port modulator in the PCN where it continues its path in a hop-by-hop basis until it reaches its destination. At the destination node, the *Tear-down* is detected in the input-port and sent to the *Photonic Switch Controller* in the corresponding electronic router. In this fashion, we can omit the overhead of an additional gateway which becomes significant when we increase the number of cores.

Table 10.2 shows the MRs configuration for data transmission, where 16 MRs are used in a non-blocking fashion. Table 10.3 shows the backup paths for each transmission.

We use the first six wavelengths in the optical spectrum starting from 1550 nm, with a wavelength spacing equal to 0.8 nm to maintain a low cross-talk as reported in [46]. For the acknowledgment signals, we use the first five wavelengths in the optical spectrum starting from 1550 nm: four wavelengths for the *Tear-down* signal where each one is dedicated for each port except the local one. In addition, a single wavelength is used for the *ACK*. The remaining available wavelengths are used for data transmission. The five wavelengths used to control the *ACK* and *Tear-down* signals are notably constant regardless of the network size, in contrast with the fully optical where the number of wavelength used for control and arbitration grows with the network size. Thus, cutting these wavelengths from the available spectrum to be used for control, would not degrade the system bandwidth. These five wavelengths will be negligible especially when DWDM is used providing up to 128 wavelengths per waveguide [16]. The wavelength assignment for each port is shown in Table 10.4.

Should the *Tear-down* signals enter the switch, they need to be redirected to the corresponding electronic router. Since these signals are coming from

Table 10.2   Microring configuration for normal data transmission

Output/Input	Core	North	East	South	West
Core	-	4	6	3	5
North	7	-	16	None	14
East	8	17	-	13	None
South	1	None	12	-	9
West	2	11	None	10	-

Table 10.3 Microring backup configuration for data transmission

Output/Input	Core	North	East	South	West
Core	15	D	F	C	E
North	G	-	6,15,7	None	5,15,7
East	H	4,15,8	-	3,15,8	None
South	A	None	6,15,1	-	5,15,1
West	B	4,15,2	None	3,15,2	-

different ports, and are modulated with different wavelengths, detectors capable of switching all of the four wavelengths are placed in front of the input-ports to intercept the signals. The converted optical signal will be redirected to the electronic router to be processed. According to the included information, the corresponding MRs will be released. For the *ACK*, when the PSCP reaches the destination, 1-bit optical signal is modulated starting from the output port (i.e., opposite direction) and travels back to the source.

With this smart hybrid switching mechanism, we take advantage of the low-power consumption of the optical link by using optical pulses modulated with the adequate wavelength instead of propagating the acknowledgment signals in the ECN. Second, we take advantage of the WDM proprieties by separating the acknowledgment packets and the data signals and let them coexist in the same medium without interfering with each other. This contrasts with the electronic domain where these acknowledgment packets travel for a several hops consequently blocking (preventing) the waiting cores from sending their PSCP packets. Finally, we are able to tolerate faults due to the arrangement of the MRs, and allowing for redundancy at critical locations.

As a primary comparison, we performed a study on the routers, and the loss that they would each have on average, and in their worst case. The results can be seen in Table 10.5. As expected, the Crux [59] performs the best, as its only design goal was to minimize loss and noise, sacrificing a lot of functionality. Values for the calculation were taken from various authors, and can be seen in Table 10.6.

Table 10.4 Wavelength assignment for acknowledgment signal (Mod: Modulator, and Det: Photo-detector)

	Core	North	East	South	West
Input	Mod_{λ_0}	Det_{λ_3}	Det_{λ_2}	Det_{λ_1}	Det_{λ_4}
Output	Det_{λ_0}	Mod_{λ_1}	Mod_{λ_4}	Mod_{λ_3}	Mod_{λ_2}

Table 10.5   Various switches and their estimated losses. AL: Average Loss, WL: Worst Loss

Router	Cros.	MRs	Termi.	AL(dB)	WL(dB)	WL(faulty)(dB)
Crossbar	25	25	10	1.12	1.60	∞
Crux	9	12	2	.657	1.11	∞
PHENIC	27	18	0	1.315	1.615	∞
FT-PHENIC	19	16+9	0	.965	1.115	2.215

10.2.2 Light-weight Electronic Control Router

Figure 10.4(b) illustrates the control router architecture, which is is based upon OASIS-NoC router [5, 6, 13, 14]. As shown in the above figure, the arbiter receives the detected *Tear-down* from the above switch (colored arrows). According to the information encoded in this signal, the corresponding MRs are released and a new *Tear-down* is generated for the next hop until it reaches its final destination and all MRs involved in this communication are released. The figure shows also the connection between the network interface (NI) and the local port, where a configuration packet (CP) is sent from the NI to the local port. The CP could be a setup packet or a path blocked packet. The NI is connected also to the data switch (i.e., PCN). When the source node receives the ACK, the payload is processed by a serializer bank (if needed), a high speed driver, and a modulator to convert the electrical signal to an optical one. At the source node, the optical data leaves the data switch and go through a detection step, a high speed Trans-Impedance-Amplification step, and a deserialization step. At the end the NI's receiver, receives the payload data with its original clock speed.

10.2.3 Fault-tolerant Path-configuration and Routing

The key feature of the Fault-tolerant Photonic Path-configuration algorithm (FTPP) is that it can handle faulty MRs within the photonic switches. When a fault occurs, the algorithm checks for the secondary MRs on the list, and

Table 10.6   Insertion loss parameters for 22 nm process

Parameter	Value
Through Ring Loss	.5 dB [59]
Pass By Ring Loss	.005 dB [19]
Bending Loss	.005 dB [19]
Crossing Loss	.12 dB [59]
Terminator	.01 dB [19]

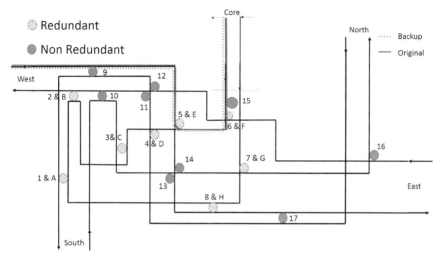

Figure 10.6 Example of how a redundant MR's functionality can be mimicked by its redundancy.

checks their status. The backup MR table can be very simple in the cases of a redundant MR failing, where it is simply replaced by its redundancy, or it can be slightly more complicated, as seen in Figure 10.5.

The FTPP algorithm must meet certain requirements to work with the FT-PHENIC system. It should be also able to remove the dependency between the ECN and PCN which causes a significant latency overhead in conventional hybrid-PNoC systems. In addition, the latency caused by the path blocking, which requires several cycles for the path dropping and the new path setup packet generation is considerably decreased. Another key feature of the configuration algorithm is the efficiency of the ECN resources' utilization. By moving the acknowledgment signals to the upper layer, we can reduce the buffer depth to only 2 slots, since half of the network traffic is eliminated. This reduction is a key factor to design a light-weight router, highly optimized for latency and energy.

Figure 10.7(a) shows an example of a successful path-setup process where all the necessary resources between a given source-destination pair are reserved. The corresponding pseudo code is given in Algorithm 1. Before optical data transmission, the source node issues a *Path-setup-Control-Packet* (PSCP) which is routed in the ECN and includes information about the destination and source addresses. In addition to the source and destination

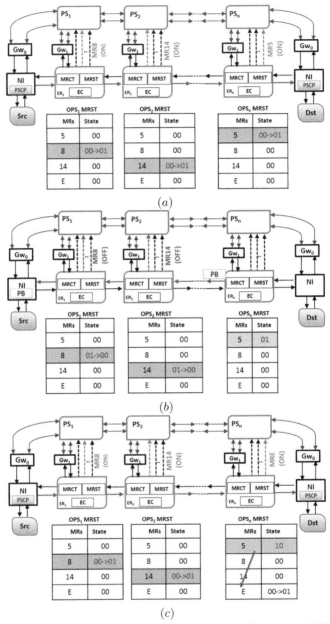

Figure 10.7   Microring fault-resilient path-configuration: a) Path-setup, (b) Path-blocked, (c) Faulty MR with Recovery. GW_0: Gateway for data, GW_1: Gateway for acknowledgment signals, PS: photonic Switch, MRCT: Micro Ring Configuration Table, MRST: Micro Ring State Table. 00=Not faulty,Not blocked, 01=Not faulty, Blocked, 10=Faulty.

Algorithm 1: Fault-tolerant path-configuration algorithm.

```
   // Path Setup Control Packet for communication i, PSCPi
   // Path Blocked Packet for communication i, PBi
   Input: Si, Di
   // From ACK detector
   Input: DetcACKs
   // To ACK modulator
   Output: ModACKs
   // From Teardown detector
   Input: TeardModi
   // To Teardown modulator
   Output: TeardModi
   // To Microring resonator
   Output: MRsj=0...n
   // Buffer writing and routing computation stages
```

1 initialization;
2 **while** *(Path-Setup-Control-Packet (PSCP) !=0)* **do**
3   DestAdd ← PSCPi;
4   PortIn ← PSCPi;
5   **if** *(resource are available)* **then**   /* check MRs state */
6   **if** *MR is not faulty* **then**
7              $Grant_i$ ← Arbiter;
8   **else if** *Backup MR is Not Faulty* **then**   /* check backup MRs state */
9              $GrantBackup_i$ ← Arbiter;
10   **else**   /* no possible path */
11             $Blocked_i$ ← Arbiter;
12   FaultyNodeList ← Node;
13   **end**
14   **else**   /* generate path blocked */
15         $Blocked_i$ ← Arbiter;
16   **end**
17 **end**
```
   // Path blocked
```
18 initialization;
19 **while** *(PB !=0)* **do**   /* Path blocked arrives */
20   **if** *(MRsi state is **reserved**)* **then**   /* release reserved MRs */
21   release ← MRsi;
22 **end**
```
   // Generate ACK
```
23 initialization;
24 **while** *(NI receiver ← PSCPi)* **do**   /* PSCP arrives to Dest */
25   **if** *(PSCP arrives to NI)* **then**   /* generate ACK to Src */
26   ACK_i ← To modulator ACK (λ0);
27 **end**
```
   // Receives ACK
```
28 initialization;
29 **while** *(NI receiver ← ACKi (λ0))* **do**   /* ACK arrives to Src λ0 */
30   **if** *(ACK arrives to the NIsender)* **then**   /* modulate the data */
31         $Data_i$ ← To Data's Modulator;
32 **end**
```
   // Identify and Generate Teardowni
```

(Continued)

Algorithm 2: Continued

---

33 initialization;
34 **while** *(From detector signal =$Teardown_i$ with λi)* **do**
35 | $findInport \leftarrow \lambda i$;   /* find In-port according to the wavelength */
36 | free $\leftarrow$ MRsi;   /* Free involved MRs */
37 | $Teardwon_i \leftarrow$ To modulator λi; /* generate new Tear-down according to λi */
38 **end**

---

addresses, other information is included. For example, one-bit is used for the Packet-type field. This field can be "0" for a PSCP and "1" when this configuration packet is a Path-blocked. Other information to ensure Quality-of-Service and fault-tolerance, such as Message-ID, Fault-status, Error-Detection-Code, can also be included. For each electrical router, the output-port is calculated according to Dimension-Order routing [6]. Every time the PSCP progresses to the next router, the optical waveguides between the previous and current routers are reserved. Depending on the output port of the electrical router, the corresponding photonic router is configured by switching ON/OFF one or more MRs using the MRs configuration table shown in Table 10.2. In the example shown in Figure 10.7(a), the packet is entering the local input-port attached to the Network Interface (NI) and requesting the east output-port. According to Table 10.2, MR 8 is required and its availability is checked in the (Micro Ring State Table) MRST. In this table, the MR's state is "00" (free and not faulty). Therefore, the switch controller reserves the MR and changes its states from "00" (free and not faulty) to "01" (not free and not faulty). After this successful reservation (hop based), the PSCP continues its path to the next hop and the same procedure is repeated until all necessary MRs are reserved for the complete path. This process is illustrated in $lines$ 1–10 of Algorithm 1. In a case where the requested MRs at a given optical switch along the path are not available, blocking occurs. This can be seen in Figure 10.7(b) where MR 5, which is necessary for the ejection to the local output-port from the west input-port, is used by another communication. In this case, the *PSCP* is converted into a *Path_blocked* packet (PB). The PB, then, travels back to the source node and releases the already reserved resources. The release is done by re-updating the corresponding entries in the MRST to "00" and by sending an electrical "OFF" signal to the corresponding MRs in the PCN. This process is illustrated in $lines$ 11–15 of Algorithm 1.

If a fault is encountered along the way, denoted by a state of "10", seen in Figure 10.7(c), then the switch attempts to use its backup route within the switch to maintain the intended port-to-port communication. This allows for recovery without requiring the whole system to change the route of a packet, and can save on costly retransmission and multiple attempts at setting up the path. Assuming that the backup path is being used for a recovery path, then the algorithm proceeds with sending the standard path blocked packet. When the *PSCP* arrives successfully at the destination node, the NI modulates one-bit acknowledgment (ACK) signal to travel back to the source via the PCN. This can be seen in lines 16–20 of Algorithm 1. Upon the arrival of this *ACK* signal, the source node modulates the payload through the data modulators and sends it to the destination node via the PCN. Lines 21–25 of Algorithm 1 depicts this data/payload transfer phase. The last process of the proposed path-configuration algorithm is the *Tear–down* step as shown in lines 26–31 of Algorithm 1. When the entire payload is transmitted, it is necessary to release the reserved optical resources. This is handled by the source node which sends a $Tear-down$ packet to the destination after predetermined number of cycles depending on the source-destination addresses, transmission bandwidth and message size.

The source's NI sends the electronic $Tear-down$ packet (TD) to the first electronic router ER_1. The Electronic Controller (EC) in this router indexes the MRCT with input-output ports information and determines the MRs that need to be released. As we can see in this figure, the state of MR 8, previously reserved in the path-setup process, is reset to *Free* (state="00") and electrical "OFF" signals are sent to the MR.

After the MRs are deactivated, a new optical Tear-down signal is generated according to the used wavelength. It is sent through the PCN to the next hop where it is converted back to electrical and redirected to the EC in the corresponding electronic router to be processed. After this process, the MRs are released and a new optical Tear-down signal is generated. This process is repeated until the *Tear-down* reaches the destination and all optical resources are released. It is important to mention that the path-setup and path-blocked processes of the proposed algorithm are very similar to the conventional ones [1, 7, 10, 18, 25, 52]. The main difference is that the MRST in our proposal contains only two states: *Free* and *Active*. The MRs are set "ON" as soon as the PSCP succeeds to reserve them. In the conventional mechanisms, three states are necessary: *Free*, *Reserved*, and *Active*. When the PSCP finds the requested MRs *Free*, it updates their states in the MRST to *Reserved* without turning them "ON". When the complete path-setup process is completed, the ACK signal travels back to the source node and sets the corresponding MRs "ON" by updating their states in the MRST to *Active*. With the proposed algorithm, some portions of the reserved path might be set "ON" and then "OFF" due to the unavailability of the resources. However, it enables the fast ACK transmission in the PCN.

In conventional path-configuration algorithms, the ACK and Tear-down packets are transmitted in the ECN and have to go through all the buffering, routing computation, and arbitration stages. With the proposed algorithm, they are carried via the PCN. As a consequence, the ETE latency can be significantly reduced in addition to the dynamic energy saving that can be achieved. Additionally, conventional path-configuration algorithms do not check for faulty MRs. This will allow the system to tolerate more MR failures, and take advantage of the fault tolerant switch.

10.3 Evaluation

We evaluate the FT-PHENIC system using a modified version of PhoenixSim which is developed in the OMNeT++ simulation environment [19]. The simulator incorporates detailed physical models of basic photonic building blocks such as waveguides, modulators, photodetectors, and switches. Electronic energy performance is based on the ORION simulator [27]. We evaluate the bandwidth performance and energy consumption for 16, 64 and 256 cores systems.

We compare the performance of the FT-PHENIC systems with the baseline PHENIC [8], and the system using the algorithm proposed by Xiang et al. [56]. Xiang's network was chosen over other typical systems [17, 45, 53, 55], because it uses some form of fault tolerance, and most of their results would mimic the baseline PHENIC.

For the fault related data, we disabled a certain number of MRs at random, and recorded the data. To get better results, we would run each system at each fault rate 10 times, and then averaged each test's total energy, average bandwidth, and average latency. Currently, the MR is disabled for the whole test, and thus models either a permanent or intermittent fault. Dealing with passband shift or temporary overheating of an MR is outside of the scope of this paper, beyond redundancy as a solution. The fault rates were chosen to span from 0 to 30% due to the fact that at this point, all of the tested networks were in deadlock.

Table 10.7 Configuration parameters

Network Configuration	Value
Process technology	32 nm
Number of tiles	256,64,16
Chip area (equally divided amongst tiles)	400 mm^2
Core frequency	2.5 GHz
Electronic Control frequency	1 GHz
Power Model	Orion 2.0
Buffer Depth	2
Message size	2 kilobytes
Simulation time	10 ms (25 10^8 cycles)

10.3.1 Complexity Evaluation

The complexity evaluation considers the number of used rings and the resulting static thermal tuning. The number of used MRs is given by Equation (10.1), where $Mod/Detc_{(ring)}$ is the number of rings required to modulate/detect the payload signal. $Switch_{(ring)}$ is the number of ring required for the photonic switch to route the optical data. Finally, the $ACKs(ring)$ is the number required to handle the acknowledgment signal.

$$Total_{(ring)} = Mod/Detc_{(ring)} + Switch_{(ring)} + ACKs(ring)$$

$$(10.1)$$

Tables 10.9 and 10.10 show the comparison results for 64 and 256 cores system, respectively. We can see that the optimized networks have the lowest number of rings. In fact, this kind of network is even more sensitive to MR faults as each MR is critical for the functionality of the node. In addition, with minimal number of rings, the resulting insertion loss is lower than the fault

Table 10.8 Photonic communication network energy parameters

Network Configuration	Value
Datarate (per wavelength)	2.5 GB/s
MRs dynamic energy	375 fJ/bit
MRs static energy	400 μW
Modulators dynamic energy	25 fJ/bit
Modulators static energy	30 μW
Photodetector energy	50 fJ/bit
MRs static thermal tuning	1 μW/ring

Table 10.9   MR requirement comparison results for 64 cores systems

	FT-PHENIC	PHENIC	Xiang
Mod/Detc	64	64	64
Switch	1152	1152	1600
ACKs	640	640	-
Redundant MRs	384	-	-
Total	2240	1856	1664
Sta. Power(mW)	44	37	33

tolerant design. For the proposed FT-PHENIC system, it has an additional rings used for acknowledgment signal, compared to the other networks, as well as for fault-tolerance. This increase can reach 33% when compared to the optimized crossbar and PHENIC systems. We also observe the same behavior when evaluating the required static thermal tuning, which is required to maintain the functionality of the ring, under 20K temperature with $1\mu W$ for each ring.

10.3.2 Latency and Bandwidth Evaluation

Figures 10.8(a) and (b) show the overall average latency and the average latency near the saturation region, respectively. We can see that for zero-load latency, all networks behave in the same way. Near saturation, PHENIC shows more flexibility and scalability in 256 cores when compared to the other networks. For the 64 cores configuration, the crossbar-based system slightly outperforms both PHENIC systems in terms of latency. This can be explained by the use of Optical-to-Electronic conversion of the *Teardown* which affects the overall latency of small networks.

The latency is heavily affected by the failure rate of MRs, and as the systems fail more, the latency increases until the whole system fails.

Table 10.10   MRs requirement comparison results for 256-core systems

	FT-PHENIC	PHENIC	Xiang
Mod/Detc	256	256	256
Switch	4608	4608	6400
ACKs	2560	2560	-
Redundant MRs	1536	-	-
Total	8960	7424	6656
Sta. Power(mW)	179	149	133

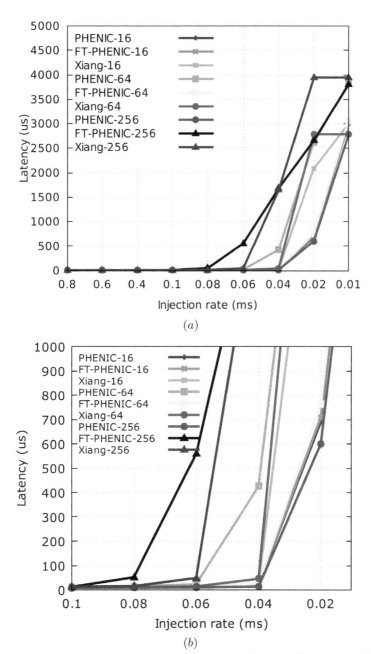

Figure 10.8 Latency comparison results under random uniform traffic: (a) Overall Latency, (b) Latency near-saturation.

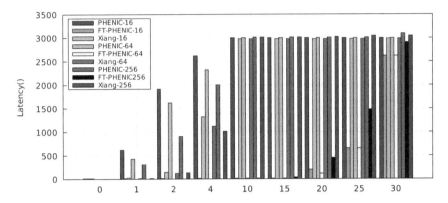

Figure 10.9 Latency results of each system as faults are introduced.

This has a lot to do with failed path setup. Figure 10.9 shows the results of the latency test when adding in varying amounts of MR failures. The FT-PHENIC demonstrates its ability to withstand MR failures over all other systems.

For the achieved bandwidth, Figure 10.10 shows that the bandwidth is increased by about 51% when compared to Xiang' system, for both 64 and 256 cores configurations. When compared to the crossbar, torus and PHENIC systems, we see that the four systems behave similarly. While the torus

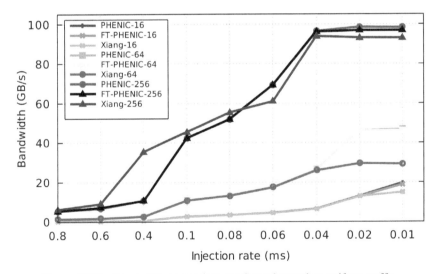

Figure 10.10 Bandwidth comparison results under random uniform traffic.

system has the capability of setting the path with less hop count, the FT-PHENIC system can achieve the same performance without the need for an extra network access which is required for the torus. This behavior is observed for 16, 64 and 256 core systems.

The latency increase caused by failed MRs will in turn cause the bandwidth to decrease. The effects of the failures on the bandwidth can be seen in Figure 10.11. As with the latency, only FT-PHENIC and Xiang show any tolerance to faults, with FT-PHENIC outperforming Xiang.

10.3.3 Energy Evaluation

Figure 10.12 shows the total energy and the energy efficiency comparison results for 16, 64 and 256 cores systems. For the 256 cores configuration, the proposed system outperforms all other networks. This is illustrated by an improvement in terms of energy efficiency reaching 26% when compared the crossbar-based (non blocking). When compared to the torus-based architecture, FT-PHENIC improves the energy efficiency by upwards of 70%. The torus-based architecture offers high bandwidth thanks to the connection between edges leading to short communications. On the other hand, it comes at high energy cost. This can be explained by the fact that the additional input-ports, required for the edge connections established in the torus-based system, incur increased area and consequently an energy overhead.

Figure 10.13 shows the total energy and energy efficiency of the systems when 4% of their MRs have failed. Some systems were not able to complete

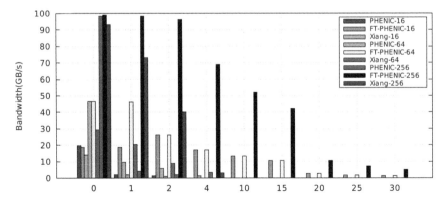

Figure 10.11 Bandwidth comparison results as faults are introduced.

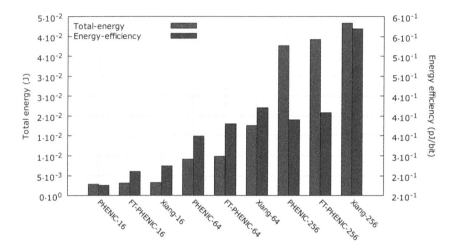

Figure 10.12   Total energy and energy efficiency comparison results under random uniform traffic near-saturation.

simulation, and so their energy is marked as 0J, and an efficiency of 0pJ/bit, just so the functioning ones remain visible. The extra energy comes from the extra run time. It is important to notice how much the scale has changed for the energy efficiency between the fault-free and 4% fault results.

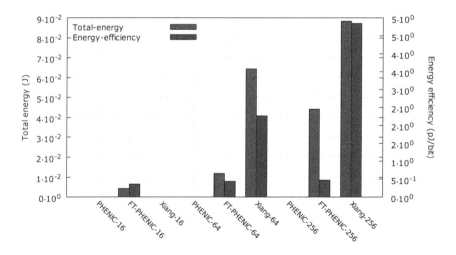

Figure 10.13   Total energy and energy efficiency comparison results under random uniform traffic with 4% of MRs acting faulty.

From these results, we can see that FT-PHENIC outperforms systems with either non-blocking or blocking switches. In addition, it provides heightened energy efficiency, far greater than the torus-based which can offer the same bandwidth as the proposed system. We conclude that the obtained improvement by FT-PHENIC is the result of the association of three main factors together: (1) the non-blocking switch supporting optical acknowledgment signals, (2) the light-weight router with reduced buffer size, (3) and the path setup algorithm to adopt hybrid switching inside the photonic switch.

10.4 Related Literature

There are three main types of optical fault tolerance that we were able to find. The first one is various methods of adaptive routing. The second one is techniques involving redundancy, which is commonly implemented in the network interface by using WDM as a redundancy technique. The third one involves buffering, checking, and proceeding like a standard electronic NoC.

Adaptive routing [36, 49, 57] is the most common method for fault-tolerance in mesh based architectures because of the large amount of possible minimal paths. It does require some extra logic in the routing decision, but this is minimal compared to an extra interconnect at each location. For it to truly support multiple faults, it must also support non-minimal routing in order to avoid a non-reserved deadlock situation. It should also be noted that implementing fault tolerance on a deadlock free algorithm can negate that feature. This is not troublesome to optical networks as deadlock is a non-issue due to the fact that end-to-end is reserved before the transmission can start, and is only an issue during path setup.

Ramesh et al. proposed a method [49] of determining and using backup routes. The algorithm determines the least cost path. This path will be used unless there is a fault detected, in which case the backup path is used. Ramesh proposed using a set of probe packets. When the destination receives one of the probe packets, it then sends a PACK signal for each probe packet. If a packet is dropped due to faults, then a NACK signal is sent. This is a solution of off-chip optical networks though.

Loh breaks his algorithm [36] into a similar fashion to Ramesh. It has a Default Routing algorithm and a backup routing method. His two methods are called Logical Route and Adaptive Route. The Logical Route in his paper is a few sets of dimension order routing. The adaptive algorithm determines which of the deterministic routings to use. This method simply checks for faults along the way, and if it can be detected, then it tries to switch to the other form of dimension order routing. This is an attempt to shift from X to Y when a problem is found in the X direction. This results in a routing algorithm which is minimal and adaptive, deadlock-free, and livelock-free.

Fault Regions [57] is a form of adaptive routing where each node keeps track of the permanent faults of its neighbors. This then allows for the path making decision to be educated with respect to faults up to a certain distance away. It can then guarantee that no old permanent faults are going to cause problems with the transmission. One such an algorithm is proposed by Xingyun [57]. He proposed a quite interesting optical network. It comes in the form of a torus which only allows data in two directions. This allows for some unique fault tolerance ideas. While they may not be minimalistic routing it will switch directions, go under the chip and come back from the top and reroute to avoid a bad crossing. This could possibly cause large amounts of insertion loss from routing around the network's length multiple times. This loss would translate to high power cost, and not yield any true benefits to converting to optical. This is still only monitoring its own outputs though.

Look Ahead Routing [56] is another type of adaptive routing which is most interesting to implement in a nanophotonic setting. This is where a node has knowledge of its neighbors' faulty links, and possibly its neighbors' neighbors' links. With this data at hand, the routing can protect a path and guarantee its success. The only issue would be implementing one of the detection algorithms mentioned at the beginning of this section. Although it hasn't been implemented in a photonic chip yet, there is no obvious reason preventing it from being translated over. Xiang's method [56] uses a Minus-first routing algorithm as a basis. The author does not detail how to detect a faulty link, but once a faulty link is discovered, it runs a Minus-first algorithm, checking each step along the way. This method attempts to find all paths from the source to the destination from the problematic node, and then determines which one requires the least amount of time. This switch shows that only the links are optical, and the switches themselves are electrical. This also allows for the implementation of buffers, which allow for a few more fault tolerance options which can be detailed in Radetzki's paper [47].

Modular redundancy uses WDM (Wavelength Division Multiplexing) as a fault tolerance tool [38, 51, 60]. The general idea is that if a certain wavelength is causing problems, either through noise or a manufacturing defect, and this problem can be detected, then certain wavelengths can be disabled and enabled. This is highly effective for modulator and photo-detector based faults. These focus on permanent and intermittent faults, because a transient fault would occur far too late for a wavelength to be switched. *Noise* has been a large source of faults within optical networks. Currently, there are many different forms of optical switches which are used in networks. The main goal of these switch designs is to reduce the area, when compared to the crossbar switch. We will only focus on the non-blocking switches because of their performance benefits. Three examples of optical switches can be seen in Figure 10.14. The first is an example of a typical optimized switch, which reduces crossings and MRs. The second is the five-port crossbar switch, which uses the maximum number of MRs, crossings, and terminators, but is a simplistic non-blocking design. The last, Crux by Ye et al. [59], is a switch which is optimized for XY-deterministic routing. This allows it to drop some extra MRs, but it no longer maintains the functionality to travel from the Y-direction to X-direction, such as North to East. This does greatly reduce the noise, when compared to other switches which can perform all network routing operations. Many other switches and networks were proposed to improve the SNR [20, 21, 28, 41, 43]. The reason this noise is so heavily researched is explained by Nikdast et al. [42].

Additionally, various authors have looked into the affect of *thermal variance*, and how to combat it [34, 35]. There are various ways to combat it, but the most common way is to cool down the ring to normal temperatures, which can be done by keeping it inactive, or by thermal tuning [35]. Trimming [4] was also one solution, which was mentioned in the introduction, and appears to be a promising answer to the problem. To the best of our knowledge, none of the existing solutions proposed so far take advantage of switch structure to provide fault tolerance. The focus of all other research has been on the

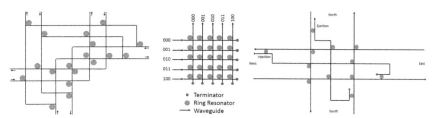

Figure 10.14 Example of photonic switches. From left to right: PHENIC's original [8], Crossbar, and Crux [59].

routing algorithms or different locations to provide modular redundancy, or noise reduction.

10.5 Chapter Summary and Discussion

This chapter presents a fault-tolerant Photonic Network-on-Chip architecture, which uses minimal redundancy to assure accuracy of the packet transmission even after faulty microrings (MRs) are detected. The system is based on a fault-tolerant path-configuration and routing algorithm, and a microring fault-resilient photonic router. Simulation results show that FT-PHENIC enjoys about 50% increase in bandwidth and about 60% decrease in energy related to the typical crossbar unit, versus other reported architectures. Additionally, the FT-PHENIC tolerates MR faults quite well up until around when 20% of the MRs have failed. These encouraging results highlight the potential of using photonic on chip and the FT-PHENIC hybrid PNoC architecture to meet the design and performance challenges of future generations of many-core systems.

One key thing holding back the reliability of optical switches is the reliability of the basic MR unit. This reliability is based off of the physical parameters that are used when designing each unit. We would like to explore the physical properties of the MRs themselves to improve the reliability. As we have previously said in the paper, making small changes to the shape, such as using racetracks [41] has led to an improvement in the reliability of MRs, and a reduction in the sensitivity to thermal variation. This means without changing the bending radius or waveguide thickness or material, they were able to improve reliability, and we would like to continue with such research.

Another item that would greatly aid the development of optical routers is the ability to buffer. Even more so the ability to read the data in multiple locations. Currently, splitting the data to be read will cause a large amount of insertion loss. Buffering is currently limited to causing a delay by creating optical coils, and can only delay it a very minimal amount of time, and cause a large amount of propagation loss [23]. Being able to read the data at multiple locations could allow for error correcting codes to not only fix some of the errors, but also aid in the fault diagnosis schemes.

This research mainly focused on improving the fault-tolerance of the network. We attempted to address the process variation problems of optical switches, but thermal variation is still a large problem. The temperature fluctuation can temporarily cause an MR to respond to an improper wavelength, which can result in larger problems.

References

[1] Cisse Ahmadou Dit Adi, Hiroki Matsutani, Michihirio Koibuchi, Hidetsugu Irie, Takefumi Miyoshi, and Tsutomu Yoshinaga. An efficient path setup for a photonic network-on-chip. In *2010 First International Conference on Networking and Computing*, pages 156–161, Nov 2010.

[2] Mridul Agarwal, Bipul C Paul, Ming Zhang, and Subhasish Mitra. Circuit failure prediction and its application to transistor aging. In *25th IEEE VLSI Test Symposium (VTS'07)*, pages 277–286. IEEE, 2007.

[3] Akram Ben Ahmed. *High-throughput Architecture and Routing Algorithms Towards the Design of Reliable Mesh-based Many-Core Network-on-Chip Systems*. PhD thesis, Graduate School of Computer Science and Engineering, University of Aizu, March 2015.

[4] Jung Ho Ahn, Marco Fiorentino, Raymond G. Beausoleil, Nathan Binkert, Al Davis, D. Fattal, Norman P. Jouppi, Moray McLaren, C.M. Santori, Robert S. Schreiber, S.M. Spillane, Dana Vantrease, and Q. Xu. Devices and architectures for photonic chip-scale integration. *Applied Physics A*, 95(4):989–997.

[5] Abderazek Ben Abdallah. *Multicore Systems-on-Chip: Practical Hardware/Software Design, 2nd Edition*. Atlantis, 2013.

[6] Abderazek Ben Abdallah and Masahiro Sowa. Basic Network-on-Chip Interconnection for Future Gigascale MCSoCs Applications: Communication and Computation Orthogonalization. In *JASSST2006*, 2006.

[7] Achraf Ben Ahmed and Abderazek Ben Abdallah. Phenic: silicon photonic 3d-network-on-chip architecture for high-performance heterogeneous many-core system-on-chip. In *Sciences and Techniques of Automatic Control and Computer Engineering (STA), 2013 14th International Conference on*, pages 1–9, Dec 2013.

[8] Achraf Ben Ahmed and Abderazek Ben Abdallah. Hybrid silicon-photonic network-on-chip for future generations of high-performance many-core systems. *The Journal of Supercomputing*, DOI: 10.1007/s11227-015-1539-0, 2015.

[9] Achraf Ben Ahmed, Meyer Meyer, Yuichi Okuyama, and Abderazek Ben Abdallah. Hybrid photonic noc based on non-blocking photonic switch and light-weight electronic router. In *2015 IEEE International Conference on Systems, Man and Cybernetics (SMC)*, October 2015.

[10] Achraf Ben Ahmed, Michael Meyer, Yuichi Okuyama, and Abderazek Ben Abdallah. Efficient router architecture, design and performance exploration for many-core hybrid photonic network-on-chip

(2d-phenic). In *Information Science and Control Engineering (ICISCE), 2015 2nd International Conference on*, pages 202–206, April 2015.

[11] Achraf Ben Ahmed, Yuichi Okuyama, and Abderazek Ben Abdallah. Contention-free routing for hybrid photonic mesh-based network-on-chip systems. In *The 9th IEEE International Symposium on Embedded Multicore/Manycore SoCs (MCSoc)*, pages 235–242, September 2015.

[12] Achraf Ben Ahmed, Yuichi Okuyama, and Abderazek Ben Abdallah. Non-blocking electro-optic network-on-chip router for high-throughput and low-power many-core systems. In *The World Congress on Information Technology and Computer Applications 2015*, June 2015.

[13] Akram Ben Ahmed and Abderazek Ben Abdallah. Architecture and design of high-throughput, low-latency, and fault-tolerant routing algorithm for 3d-network-on-chip (3d-noc). *J. Supercomput.*, 66(3):1507–1532, dec 2013.

[14] Akram Ben Ahmed and Abderazek Ben Abdallah. Graceful deadlock-free fault-tolerant routing algorithm for 3d network-on-chip architectures. *J. Parallel Distrib. Comput.*, 74(4):2229–2240, April 2014.

[15] Wim Bogaerts, Peter De Heyn, Thomas Van Vaerenbergh, Katrien De Vos, Shankar Kumar Selvaraja, Tom Claes, Pieter Dumon, Peter Bienstman, Dries Van Thourhout, and Roel Baets. Silicon microring resonators. *Laser & Photonics Reviews*, 6(1):47–73, 2012.

[16] Lars Brusberg, Henning Schrder, Marco Queisser, and Klaus-Dieter Lang. Single-mode glass waveguide platform for dwdm chip-to-chip interconnects. In *Electronic Components and Technology Conference (ECTC), 2012 IEEE 62nd*, pages 1532–1539, May 2012.

[17] Johnnie Chan and Keren Bergman. Photonic interconnection network architectures using wavelength-selective spatial routing for chip-scale communications. *Optical Communications and Networking, IEEE/OSA Journal of*, 4(3), March 2012.

[18] Johnnie Chan, Gilbert Hendry, Keren Bergman, and Luca P. Carloni. Physical-layer modeling and system-level design of chip-scale photonic interconnection networks. *Computer-Aided Design of Integrated Circuits and Systems, IEEE Transactions on*, 30(10):1507–1520, Oct 2011.

[19] Johnnie Chan, Gilbert Hendry, Aleksandr Biberman, Keren Bergman, and Luca P Carloni. Phoenixsim: A simulator for physical-layer analysis of chip-scale photonic interconnection networks. In *Proceedings of the Conference on Design, Automation and Test in Europe*, pages 691–696. European Design and Automation Association, 2010.

[20] Sai Vineel Reddy Chittamuru and Sudeep Pasricha. Crosstalk mitigation for high-radix and low-diameter photonic noc architectures. *Design Test, IEEE*, 32(3):29–39, June 2015.

[21] Sai Vineel Reddy Chittamuru and Sudeep Pasricha. Improving crosstalk resilience with wavelength spacing in photonic crossbar-based network-on-chip architectures. In *Circuits and Systems (MWSCAS), 2015 IEEE 58th International Midwest Symposium on*, pages 1–4. IEEE, 2015.

[22] Sai T. Chu, Wugen Pan, Shinya Sato, Taro Kaneko, Brent E. Little, and Yasuo Kokubun. Wavelength trimming of a microring resonator filter by means of a uv sensitive polymer overlay. *Photonics Technology Letters, IEEE*, 11(6):688–690, June 1999.

[23] Sasan Fathpour and Nabeel A. Riza. Silicon-photonics-based wideband radar beamforming: basic design. *Optical Engineering*, 49(1):018201–018201–7, 2010.

[24] Sheng guang Yang, Li Li, Yu ang Zhang, Bing Zhang, and Yi Xu. A power-aware adaptive routing scheme for network on a chip. In *ASIC, 2007. ASICON '07. 7th International Conference on*, pages 1301–1304, Oct 2007.

[25] Gilbert Hendry, Eric Robinson, Vitaliy Gleyzer, Johnnie Chan, Luca P. Carloni, Nadya Bliss, and Keren Bergman. Circuit-switched memory access in photonic interconnection networks for high-performance embedded computing. In *High Performance Computing, Networking, Storage and Analysis (SC), 2010 International Conference for*, pages 1–12, Nov 2010.

[26] Zhan-Shuo Hu, Fei-Yi Hung, Kuan-Jen Chen, Shoou-Jinn Chang, Wei-Kang Hsieh, and Tsai-Yu Liao. Improvement in thermal degradation of zno photodetector by embedding silver oxide nanoparticles. *Functional Materials Letters*, 6(01):1350001, 2013.

[27] Andrew B. Kahng, Bin Li, Li-Shiuan Peh, and Kambiz Samadi. Orion 2.0: A power-area simulator for interconnection networks. *Very Large Scale Integration (VLSI) Systems, IEEE Transactions on*, 20(1):191–196, Jan 2012.

[28] Pradheep Khanna Kaliraj. Reliability-performance trade-offs in photonic noc architectures. 2013.

[29] Roman Kappeler. Radiation testing of micro photonic components. Stagiaire Project Report. ESA/ESTEC. September 29, 2004. Ref. No.: EWP 2263.

[30] John Keane and Chris H Kim. An odometer for cpus: Microprocessors don't normally show wear and tear, but wear they do. *IEEE SPECTRUM*, 48(5):26–31, 2011.

[31] John Keane, Tae-Hyoung Kim, and Chris H Kim. An on-chip nbti sensor for measuring pmos threshold voltage degradation. *IEEE Transactions on Very Large Scale Integration (VLSI) Systems*, 18(6):947–956, 2010.

[32] Brian R. Koch, Alexander W. Fang, Oded Cohen, and John E. Bowers. Mode-locked silicon evanescent lasers. *Optics Express*, 18(15), 2007.

[33] Kelin Kuhn, Chris Kenyon, Avner Kornfeld, Mark Liu, Atul Maheshwari, Wei-kai Shih, Sam Sivakumar, Greg Taylor, Peter VanDerVoorn, and Keith Zawadzki. Managing process variation in intel's 45nm cmos technology. *Intel Technology Journal*, 12(2), 2008.

[34] Hui Li, Alain Fourmigue, Sébastien Le Beux, Xavier Letartre, Ian O'Connor, and Gabriela Nicolescu. Thermal aware design method for vcsel-based on-chip optical interconnect. In *Proceedings of the 2015 Design, Automation & Test in Europe Conference & Exhibition*, pages 1120–1125. EDA Consortium, 2015.

[35] Zheng Li, Moustafa Mohamed, Xi Chen, Eric Dudley, Ke Meng, Li Shang, Alan R Mickelson, Russ Joseph, Manish Vachharajani, Brian Schwartz, et al. Reliability modeling and management of nanophotonic on-chip networks. *Very Large Scale Integration (VLSI) Systems, IEEE Transactions on*, 20(1):98–111, 2012.

[36] Peter KK Loh and Wen-Jing Hsu. Design of a viable fault-tolerant routing strategy for optical-based grids. In *Parallel and Distributed Processing and Applications*, pages 112–126. Springer, 2003.

[37] Serge Luryi, Jimmy Xu, and Alex Zaslavsky. *Future trends in microelectronics: up the nano creek*. John Wiley & Sons, 2007.

[38] Moray McLaren, Nathan Lorenzo Binkert, Alan Lynn Davis, and Marco Florentino. Energy-efficient and fault-tolerant resonator-based modulation and wavelength division multiplexing systems, 22 2014. US Patent 8,705,972.

[39] Michael C. Meyer, Akram Ben Ahmed, Yuichi Okuyama, and Aabderazek Ben Abdallah. Fttdor: Microring fault-resilient optical router for reliable optical network-on-chip systems. In *Embedded Multicore/Many-core Systems-on-Chip (MCSoC), 2015 IEEE 9th International Symposium on*, pages 227–234, Sept 2015.

[40] Evelyn Mintarno, Joëlle Skaf, Rui Zheng, Jyothi Bhaskar Velamala, Yu Cao, Stephen Boyd, Robert W Dutton, and Subhasish Mitra. Self-tuning for maximized lifetime energy-efficiency in the presence of circuit aging. *IEEE Transactions on Computer-Aided Design of Integrated Circuits and Systems*, 30(5):760–773, 2011.

[41] Moustafa Mohamed. *Silicon Nanophotonics for Many-Core On-Chip Networks*. PhD thesis, University of Colorado, 2013.

[42] Mahdi Nikdast and Jiang Xu. On the impact of crosstalk noise in optical networks-on-chip. In *Design Automation Conference (DAC)*, 2014.

[43] Mahdi Nikdast, Jiang Xu, Xiaowen Wu, Wei Zhang, Yaoyao Ye, Xuan Wang, Zhehui Wang, and Zhe Wang. Systematic analysis of crosstalk noise in folded-torus-based optical networks-on-chip. *Computer-Aided Design of Integrated Circuits and Systems, IEEE Transactions on*, 33(3):437–450, 2014.

[44] Christopher J. Nitta, Matthew K. Farrens, and Venkatesh Akella. Resilient microring resonator based photonic networks. In *Proceedings of the 44th Annual IEEE/ACM International Symposium on Micro architecture*, MICRO-44, pages 95–104, New York, NY, USA, 2011. ACM.

[45] Yan Pan, Prabhat Kumar, John Kim, Gokhan Memik, Yu Zhang, and Alok Choudhary. Firefly: illuminating future network-on-chip with nanophotonics. In *ACM SIGARCH Computer Architecture News*, volume 37, pages 429–440. ACM, 2009.

[46] Kyle Preston, Nicolas Sherwood-Droz, Jacob S. Levy, and Michal Lipson. Performance guidelines for wdm interconnects based on silicon microring resonators. In *Lasers and Electro-Optics (CLEO), 2011 Conference on*, pages 1–2, May 2011.

[47] Martin Radetzki, Chaochao Feng, Xueqian Zhao, and Axel Jantsch. Methods for fault tolerance in networks-on-chip. *ACM Computing Surveys (CSUR)*, 46(1):8, 2013.

[48] D. Rafizadeh, J.P. Zhang, S.C. Hagness, A. Taflove, K.A. Stair, S.T. Ho, and R.C. Tiberio. Temperature tuning of microcavity ring and disk resonators at 1.5- mu;m. In *Lasers and Electro-Optics Society Annual Meeting, 1997. LEOS '97 10th Annual Meeting. Conference Proceedings., IEEE*, volume 2, pages 162–163 vol.2, Nov 1997.

[49] Gayatri Ramesh and S SundaraVadivelu. A reliable and fault tolerant routing for optical wdm networks. *arXiv preprint arXiv:0912.0602*, 2009.

[50] Samar K. Saha. Modeling process variability in scaled cmos technology. *IEEE Design & Test of Computers*, 27(2):0008–16, 2010.

[51] Laxman Sahasrabuddhe, Senthil Ramamurthy, and Biswanath Mukherjee. Fault management in ip-over-wdm networks: Wdm protection versus ip restoration. *Selected Areas in Communications, IEEE Journal on*, 20(1):21–33, 2002.

[52] Assaf Shacham, Keren Bergman, and Luca P. Carloni. On the design of a photonic network-on-chip. In *Networks-on-Chip, 2007. NOCS 2007. First International Symposium on*, pages 53–64, May 2007.

[53] Assaf Shacham, Keren Bergman, and Luca P. Carloni. Photonic networks-on-chip for future generations of chip multiprocessors. *Computers, IEEE Transactions on*, 57(9):1246–1260, Sept 2008.

[54] Zhijuan Tu, Zhiping Zhou, and Xingjun Wang. Reliability considerations of high speed germanium waveguide photodetectors. In *SPIE OPTO*, pages 89820W–89820W. International Society for Optics and Photonics, 2014.

[55] Dana Vantrease, Robert Schreiber, Matteo Monchiero, Moray McLaren, Norman P Jouppi, Marco Fiorentino, Al Davis, Nathan Binkert, Raymond G Beausoleil, and Jung Ho Ahn. Corona: System implications of emerging nanophotonic technology. In *ACM SIGARCH Computer Architecture News*, volume 36, pages 153–164. IEEE Computer Society, 2008.

[56] Dong Xiang, Yan Zhang, ShuChang Shan, and Yi Xu. A fault-tolerant routing algorithm design for on-chip optical networks. In *Reliable Distributed Systems (SRDS), 2013 IEEE 32nd International Symposium on*, pages 1–9, Sept 2013.

[57] Qi Xingyun, Feng Quanyou, Chen Yongran, Dou Qiang, and Dou Wenhua. A fault tolerant bufferless optical interconnection network. In *Computer and Information Science, 2009. ICIS 2009. Eighth IEEE/ACIS International Conference on*, pages 249–254. IEEE, 2009.

[58] Yi Xu, Jun Yang, and Rami Melhem. Tolerating process variations in nanophotonic on-chip networks. In *ACM SIGARCH Computer Architecture News*, volume 40, pages 142–152. IEEE Computer Society, 2012.

[59] Yaoyao Ye, Xiaowen Wu, Jiang Xu, Wei Zhang, Mahdi Nikdast, and Xuan Wang. Holistic comparison of optical routers for chip multiprocessors. In *Anti-Counterfeiting, Security and Identification (ASID), 2012 International Conference on*, pages 1–5. IEEE, 2012.

[60] Jing Zhang and B Mukheriee. A review of fault management in wdm mesh networks: basic concepts and research challenges. *Network, IEEE*, 18(2):41–48, 2004.

[61] Shiyang Zhu and Guo-Qiang Lo. Vertically-stacked multilayer photonics on bulk silicon toward three-dimensional integration. *Lightwave Technology, Journal of*, PP(99):1–1, 2015.

11

Techniques for Energy Proportionality in Optical Interconnects

Yigit Demir[1] and Nikos Hardavellas[2]

[1]Intel, USA
[2]Northwestern University, USA

Abstract

The high optical loss of typical photonic components, coupled with the low efficiency of WDM-compatible lasers and the power consumed for microring trimming, dramatically increase the power consumption of silicon-photonic optical interconnects. Unfortunately, the majority of this power is typically wasted. While the full laser power is required to support periods of high interconnect activity, most of it is wasted when activity is low because the laser sources are always on, even when the interconnect transmits no messages. In addition, the power used to tune the network's microrings is affected more by the thermal variations imposed by processor activity, rather than interconnect utilization. Thus, the power drawn by a photonic interconnect is not proportional to interconnect activity.

This chapter reviews some of the recently proposed techniques to achieve energy proportionality in optical interconnects, and provides directions for future research. In particular, Section 11.1 discusses techniques to turn off the laser source (i.e., "power-gate" the laser) when the network is idle in order to minimize the laser's energy consumption. To avoid exposing the laser turn-on delay, these techniques employ mechanisms that predict when communication is imminent and turn on the laser just in time for the transmission. Section 11.2 discusses a technique to minimize the energy consumed to keep the microring resonators within a narrow temperature zone. The technique consists of encapsulating the photonics layer in a porous silicon insulator to keep the thermal energy of the microring heaters from escaping the photonics layer, and employing microfluidics to minimize the impact of high thermal variations from the logic die.

11.1 Laser Power-Gating

11.1.1 Why Lasers Waste Power

The lasers powering silicon photonic interconnects consume a significant amount of power because their output optical power needs to be high enough to compensate for the optical loss of silicon waveguides, optical couplers, and ring resonators. Typical silicon waveguides exhibit optical loss between 0.1–0.3 dB/cm [6], resulting in modest optical loss over short distances. However, replacing global wires with waveguides that traverse the entire chip in a serpentine form can drastically increase the required laser power. Firefly [48] on a 580 mm^2 chip needs a 16cm waveguide, which increases the laser power by 1.5–3x. Aggressive technology can produce low-loss waveguides (0.05 dB/cm [33]), but these are much wider [33, 35]. Their high area occupancy may force the use of narrow data paths (e.g., 2-bit links for an 8×8 array in Oracle MacroChip [33]), which in turn impose significant serialization delays that degrade performance, and ultimately increase power consumption. Moreover, WDM-compatible lasers are only 5–10% [63, 72] energy efficient. Thus, the wall-plug laser power requirement is 10–20x higher than the required laser output power. Manufacturing variations impose additional losses [67, 75], forcing designers to increase the laser power even higher to maintain a safety margin. Most of these costs are hard to avoid and are multiplicative, so the wall-plug laser power can easily grow by 10x when all inefficiencies are factored in.

Unfortunately, the majority of this power is typically wasted. In real workloads the NoC often stays idle for long periods of time: compute-intensive execution phases underutilize the NoC (common in many scientific applications), and servers in the cloud often stay idle or exhibit load imbalances (Google-scale datacenters are typically less than 30% utilized [1]). Figure 11.1 shows that 70% of the messages are more than 8 cycles apart from the previous message in a wide range of compute- and memory-intensive applications. While the full laser power is required to support periods of high interconnect activity, the laser is wasted during idle times because photonic interconnects are always on. In a typical setting, light is continuously injected into the waveguides even if no packets are sent.

11.1.2 The Solution: Laser Power-Gating

Motivated by these observations, recent work proposes laser gating (turning the laser off during idle periods to save power) as an effective technique to

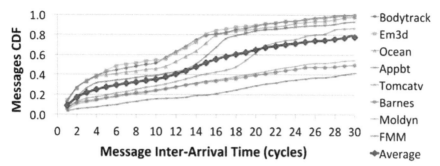

Figure 11.1 CDF of message inter-arrival. 70% of the messages are 8+ cycles apart.

conserve laser power [10, 11, 23, 34]. The laser control scheme proposed for *ATAC+* [34] turns off the lasers when idle, and turns them on in order to transmit. However, turning a laser on can incur significant delays, which may degrade performance and increase the power consumption. Predictive techniques that hide the laser turn-on delay can mitigate this overhead. *EcoLaser* [10] proposes an adaptive mechanism, which leaves the laser on longer than the time required for the current transmission, to allow other senders to opportunistically find the laser on and transmit without incurring a delay. However its design is complicated and still wastes significant laser power compared to an optimal scheme (2x on average).

Proactive Laser (ProLaser) [12] is a state-of-the-art laser control mechanism that mitigates EcoLaser's shortcomings. ProLaser (a) monitors the messages sent in a NoC and correlates them to cache coherence protocol events to predict future messages and turn on the laser proactively, (b) turns on the data portion of the NoC only when it predicts messages that carry data, and (c) employs a Bloom filter at the L2 cache slice of each node in the NoC to predict a cache hit or miss, and provide the laser turn-on request sufficiently early to hide the entire laser turn-on delay. ProLaser saves 49–88% of the laser power and tracks by 94–98% a perfect prediction scheme. Moreover, the power savings of ProLaser allow for providing a higher power budget to the cores, which enables them to run 50–70% faster.

11.1.3 Background

11.1.3.1 Laser primer
Previous work [2, 3, 30, 34, 48] typically uses off-chip lasers due to their ease of deployment and high energy efficiency (up to 30% for Gaussian comb lasers [16]). However, recent work [23] shows that output spectrum

power variations and laser-to-fiber and fiber-to-chip coupling losses add 7–8 dB optical loss, thus off-chip lasers are in reality only 6% efficient. In comparison, on-chip laser sources [31] attain wall-plug efficiencies up to 15%, while enabling wavelength-division multiplexing (WDM). WDM can be implemented by feeding a set of wavelengths generated by an array of single-wavelength lasers into an optical bus. On-chip lasers offer energy efficiency and easy packaging, but their wall-plug power consumption counts against the processor's power budget. In either case, the laser power consumption is a considerable overhead, especially when accounting for realistic optical loss parameters and laser efficiencies, emphasizing the need for power gating the laser source. Power gating on-chip lasers can increase the energy efficiency of a photonic interconnect by up to 4x [23].

Laser power gating has been overlooked due to the high turn-on latency (0.1 μs [23]) of the traditional distributed feedback comb lasers (DFB) which are widely assumed in photonic interconnects [2, 3, 30, 34, 48]. These lasers use diffraction grating to form the optical cavity. Temperature affects the diffraction grating pitch and the active region's refractive index, which alter the diffraction grating's wavelength selection, and hence the laser's emission wavelength. Thus, when the laser turns on it needs time to reach a set temperature and lock at the designated wavelength. This delay hampers laser gating. In contrast, Fabry-Perot (FP) lasers use two discrete mirrors to form the optical cavity, and their emission wavelength depends not on the temperature, but on the n-type doping level and the strain applied during the cavity development. Thus, when FP lasers are turned on (pumped to the lasing threshold), they lase at the designated wavelength within 1.5 ns [5, 29, 37], without requiring time for temperature stabilization and locking. Hence, FP lasers are suitable for laser gating.

The turn-on delay of Fabry-Perot lasers is highly tunable by design parameters, and nanosecond laser turn-on delays are both theoretically predicted [24, 25, 50 pp. 80–82] and achieved in real implementations [5, 19, 32, 37, 49]. To turn the laser on, a supply current is applied to the laser. When the carrier density exceeds the threshold density, laser oscillation starts and light output increases drastically (laser turn-on). The time it takes from the current injection to the laser turn-on is the "laser turn-on delay" which is governed by the carrier life time and is in the order of ns ([50] pp. 80–82). For example, InP-based diode FP lasers [32] have been shown to emit light with a 2ns electrical pulse excitation. InP-lasers have high peak power, their emission wavelength is tunable in a wide range, and have high temperature stability, which makes them WDM compatible. Moreover, InP-lasers can be integrated on Si [19, 49] so they can be used as an on-chip laser source. Similarly, Ge-based FP on-chip lasers have been manufactured [37] and their turn-on delay was measured at 1.5 ns [5, 29, 37]. Ge-lasers can be built within a 1 μm waveguide and occupy only 7.68×10^{-3} mm^2 area per laser, operate in room temperature, and are WDM-compatible as they exhibit gain spectrum over 200 nm [5].

It is important to note that vertical-cavity surface-emitting lasers (VCSELs) are unsuitable for on-chip WDM applications. VCSELs emit significant heat, and their operating wavelengths are defined by the epitaxial growth [23], which challenges the implementation of a multi-wavelength VCSEL array on chip for WDM. It is also hard to protect the integrity of messages with direct laser modulation due to chirping and the pattern effect [50].

11.1.3.2 Nanophotonic interconnect topologies

In Single-Writer-Multiple-Reader (SWMR) [30] crossbars, each router has its own dedicated data channel that delivers messages to all other routers (Figure 11.2(a)). R-SWMR [34, 48] crossbars add a reservation channel to SWMR. A sender in R-SWMR first broadcasts on its reservation channel a flit with the receiver's ID (in Figure 11.2(a), router R1 broadcasts on RCH1 a flit with ID=2). Upon receiving a reservation flit, the receiver (R2) turns on its demodulators to receive the message from the sender's data channel (CH1), which is now dedicated to transfer data from the sender to the receiver. Reservation channels are narrow because reservation flits only carry the receiver ID and message type information. However, the laser power required to broadcast increases exponentially with the number of readers, making it impractical to broadcast at high-radix crossbars (e.g., radix-64). Instead of having a single broadcast link with many readers, slicing [3] spreads the readers across multiple waveguides and enables high-radix R-SWMR crossbars.

11.1.4 Laser Control Schemes

A brief description of the laser control schemes we model appears in Table 11.1. The table also includes a brief description of the no-control (No-Ctrl) and the power-equivalent (Power_Eq) networks for completeness, even though there is no laser control applied to these designs. We consider both on-chip and off-chip laser sources. Because the interconnects we model are based on R-SWMR buses, each sender employs its own private sets of lasers

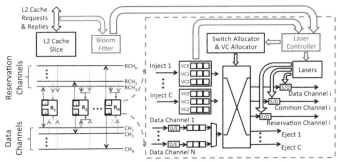

(a) R-SWMR Crossbar with Laser Control.

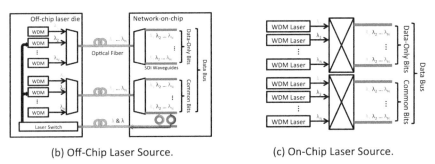

(b) Off-Chip Laser Source.   (c) On-Chip Laser Source.

Figure 11.2 R-SWMR with laser control, off-chip and on-chip laser configurations.

in both cases. When on-chip lasers are used, the lasers are placed next to each sender (Figure 11.2(a). When off-chip lasers are used, each sender still has its own private sets of lasers which couple onto the chip through optical fibers. Each sender controls its lasers by employing an additional optical link that couples from the chip to the laser sources through an additional optical fiber (Figure 11.2(c). This additional path increases the delay to control the lasers. All performance and energy overheads of the additional link are included in the evaluation.

11.1.4.1 Proactive laser control

ProLaser [12] predicts when a message transmission on the NoC is imminent and turns on the corresponding lasers just enough cycles early to hide the turn-on delay. It also segregates the data and control busses, similar to techniques proposed for electrical NoCs [26, 39], and predicts if the data-portion of the optical bus is required. ProLaser makes these predictions both

Table 11.1 Brief description of modeled laser control schemes

Name	Description
No-Ctrl	Baseline scheme; the lasers are not controlled and they are always on.
Power_Eq	An optical NoC with no laser control, similar to No-Ctrl. However, the network's width is scaled down to approximate the average power consumption of ProLaser [12].
EcoLaser	EcoLaser [10] keeps the laser on for a few more cycles (adapting at runtime by monitoring the traffic) after the last transmission to allow senders to transmit opportunistically without incurring a laser turn-on delay. EcoLaser still wastes laser power because it turns on the entire optical bus, even when sending data-less coherence messages.
Simple	EcoLaser with independent laser control for the data and the control bits of the optical bus. The data segment of the bus is turned on only when data are transmitted.
ATAC+	Proposed in [34]. It turns off the lasers when idle, and turns them on with enough power for one receiver to transmit (unicast mode). The unicast mode of ATAC+ is identical in operation to R-SWMR [54], which is the optical bus technology we model in this work. ATAC+ does not employ prediction and does not segregate the data and control busses. We only model the laser control scheme of ATAC+ and employ it on the same optical networks we use for all the other control schemes.
ProLaser	ProLaser [12]. monitors coherence events and cache hits using Bloom filters to predict when communication is imminent, and turns on the lasers just in time for communication. The control and data portions of the optical bus are independently controlled.
Perfect	Oracle scheme with full knowledge of future interconnect requests. The lasers turn on and stabilize exactly when a packet is ready for transmission, and turn off immediately afterwards, unless the next request arrives before the laser can turn off and back on again.
xx-OffChip	Schemes with the suffix "-OffChip" refer to the corresponding laser control scheme for off-chip lasers. Schemes without a suffix imply their application to on-chip lasers.

for the cache miss events that initiate a coherence transaction, as well as for the intermediate messages generated by the cache coherence protocol (e.g., invalidations, acks).

ProLaser accurately predicts the initial cache miss events that initiate a coherence transaction by employing a small Bloom filter [54] next to the L2 cache slice of each node in the NoC. A request to the L2 cache is sent in parallel to the Bloom filter. A lookup in the Bloom filter takes 1 cycle, when an L2 hit takes 14 cycles. A Bloom filter lookup consumes 1.09pJ at 16 nm technology (over-estimated using CACTI [74]). When the Bloom filter predicts an L2 miss, ProLaser does not turn on the whole bus, thereby avoiding energy waste, but turns on only the common bits. When the Bloom filter predicts an L2 hit, the whole data bus is turned on 1.5 ns before the L2 hit ends. ProLaser employs a 1KB counting Bloom filter per L2 cache slice, with 4K 2-bit counters and a hash function that takes the lowest 22 bits of the address above the block offset. This configuration produces 1.64–1.98% false positives. False positives do not degrade performance but waste laser energy.

  ProLaser accurately predicts future messages generated by the cache coherence protocol by monitoring the message types. For example, in a directory-based cache coherence protocol, every data message is generated upon receipt of a read, write, or directory-forwarded request. When a node receives a forwarded request (i.e., the node is the owner or a sharer), ProLaser turns on the whole data bus, anticipating that this node will send out a data reply. When a node receives a read or write request, a lookup in the local L2 cache slice's Bloom filter decides the type of the reply: an L2 miss generates another read request (to the tile with the memory controller), while an L2 hit generates a data reply. ProLaser turns the lasers on proactively for both of these messages, but it turns on the data-only portion only upon predicting an L2 hit. ProLaser also sends out acknowledgement messages quickly by turning on the control bits right after a reply or an invalidation. ProLaser does not predict L1 misses. However these requests do not incur high latency overhead (1–2 cycles), as they only use the common bits, which are frequently active. Other non-critical messages, such as write-backs, are also not predicted by ProLaser, as they do not have a significant impact on the overall performance.

  Figure 11.2.a shows the microarchitecture of ProLaser. ProLaser adds a Bloom filter to the L2 cache slice and a laser controller which makes predictions by monitoring the Bloom filter and the message types in the injection buffers. ProLaser keeps the lasers on until all of the messages queued in the injection buffers are transmitted. Unlike prior schemes, ProLaser is applicable to off-chip laser sources as well.

11.1.4.2 Controlling an off-chip laser source

Off-chip lasers are less efficient than on-chip ones, but their power consumption is not counted against the processor's power budget. In contrast, on-chip lasers do not suffer from additional coupling inefficiencies, but their power consumption reduces the power budget available to cores and caches, may cause overheating, and may degrade the overall performance. To achieve the best of both worlds, [23] proposes to implement an off-chip laser source using an array of single-wavelength lasers similar to the on-chip lasers assumed in this work. This arrangement still requires a fiber-to-chip coupling and higher packaging costs, but avoids comb laser losses and thermal concerns. With such an implementation, a feedback signal from the laser control in the processor die can turn on/off the lasers in the array, thereby improving energy efficiency.

Laser gating techniques can thus be extended to off-chip WDM laser arrays [12, 23] by adding an additional waveguide, fiber, and a few microrings. The green and red wavelengths in Figure 11.2(c) control the data-only and common-bits lasers separately and they are always on. When a node wants to turn on a laser source, it redirects these wavelengths back to the off-chip source using microrings at the cost of 1 cycle. Signaling the laser source takes 0.4 ns (2 cm waveguide plus 4 cm fiber), i.e., 2 cycles at 5 GHz. Upon receiving a signal, the laser turns on after the turn-on delay, the emitted light travels back to the node within 2 cycles, and the message is sent out.

11.1.5 Experimental Methodology

11.1.5.1 Interconnect modeling

We model interconnects using Booksim 2.0 [9], a cycle-accurate NoC simulator. Optimal bitrates have been calculated to be 4–10 Gbps in the near term [23], thus we model an optical NoC transmitting at 10 Gbps per wavelength. The optical links transmit at both edges of the clock [23, 27, 48, 65]; hence we model a 5 GHz on-chip clock, similar to [3, 27, 44, 45, 47, 48, 65, 70]. To provide high bandwidth, photonic links offer wide busses, which can send a data message in one cycle. In a typical configuration a data message is 600 bits and contains a 64-byte cache block, 64-bit address, 20-bit ID, and a 4-bit message type. Messages carrying data are transmitted in one cycle on a 300-bit optical bus [27, 48, 65] comprising 5 waveguides at 64 DWDM, where the optical links transmit at both edges of the clock. Small coherence messages are transmitted in two 44-bit data-less flits, thus 256 bits of the optical bus remain idle. Thus, 44 bits of the optical bus are activated for all messages (common bits), and the remaining 256 bits are used only for data. Figure 11.2(b) illustrates the separation of the data bus into control bits and data-only bits. Laser power-gating techniques with the ability to control the two sections independently require an additional set of lasers for the data-only bus. This does not increase the total laser power, as the optical link loss and the total number of wavelengths remain the same. We model 1-cycle routers with 1-cycle electrical to optical and optical to electrical bit conversions on a 480 mm^2 chip. The chip employs a 10 cm waveguide with 5 cycles round trip time at 5 GHz. The link latency (1–5 cycles) is calculated based on the traversed waveguide length. The buffers are 20-flits deep, with a flit size of 300 bits. We model 1.5 ns turn-on delay.

11.1.5.2 Modeling optical and electrical multicore NoC

To evaluate the impact of laser gating on a realistic multicore, we model a 64-core processor on a full-system cycle-accurate simulator based on Flexus 4.0 [22, 66] integrated with Booksim 2.0 [9] and DRAMsim 2.0 [52]. Table 11.2 details the architectural modeling parameters. We model a shared physically-distributed L2 cache and directories. The memory controllers are uniformly distributed on the chip, and they use the same interconnect with virtual channels to avoid deadlock. All messages below L1 traverse the interconnect. We calculate the power consumption of the electrical interconnect using DSENT [59]. We model a 16 nm node according to ITRS projections [18]. The simulated system executes a selection of workloads from PARSEC, SPLASH-2, and other scientific applications.

All systems we model employ a throttling mechanism to keep the chip within a safe operational temperature below 90°C. We assume a DVFS (dynamic voltage and frequency scaling) mechanism that scales voltage and frequency aggressively to maximize performance while staying within 90°C. Thus, core frequencies stay at a reasonable 1.6–3.8 GHz range for all applications (3–3.2 GHz on average). To decouple the impact of laser gating from the exact design of DVFS, we assume a fine-grain scheme that can switch at 200 MHz steps within that range.

Table 11.2   Architectural parameters

CMP Size	64 Cores, 480 mm^2
Cores	UltraSparc III ISA, OoO, 4-wide dispatch/retire, 96-entry ROB
Clock Freq.	1.6–3.8 GHz for chips under realistic physical constraints (90°C max temperature). DVFS sets the core frequency. The average across all applications is 3 GHz
L1-I/D Caches	64 KB, 2-way, 64-byte block, 2-cycle hit, 2 ports, 32 MSHRs, 16-entry victim cache
L2 Cache	Shared, 512 KB/core, 16 way, 64-byte block, 14-cycles, 32 MSHRs, 16-entry victim cache
Mem. Control.	One per 4 cores, 1 channel per memory controller, round-robin page interleaving
Main Memory	Optically connected memory [2], 10 ns access
Networks	Radix-16 R-SWMR and Firefly [48]

We collect runtime statistics from full-system simulations and use them to calculate the power consumption of the system using McPAT [36], and the power consumption of the optical NoC using the analytical power model by Joshi *et al.* [27]. We provide these power calculations to HotSpot 5.0 [58] to estimate the temperature of the chip. The estimated temperature is then used to refine the leakage power estimate through McPAT. The new power calculation is fed back to HotSpot to refine the temperature profile, and we iterate this process until temperature stabilizes. We adjust DVFS based on the stable-state power and temperature estimates.

To put ProLaser's performance and energy consumption into perspective, we include in our evaluation a high-performance all-electrical on-chip interconnect: a 4-ary 4-flat flattened butterfly (Flat-Butterfly) [28] derived by combining a 4-level butterfly network with concentration 4. For Flat-Butterfly we model routers with 10 input and output ports and 3-cycle routing delay. Routers are connected through 88-bit bi-directional links (1-flit control, 7-flit data messages) with 1-cycle local, 2-cycle semi-local and 3-cycle global link delays. To show the range of laser gating's impact, we evaluate its application on two optical network topologies that are at the opposite ends of the spectrum: a radix-16 R-SWMR and Firefly [48].

The radix-16 R-SWMR approximates a worst-case scenario for laser gating. It has low power consumption and its high concentration factor (4) creates heavier traffic. The low power consumption and heavy traffic limit the opportunity to power-gate the laser. Firefly connects 16 local clusters (4 routers each) with 4 R-SWMR crossbars. The local clusters use an electrical ring to route packets within the destination cluster, and each of the routers in a local cluster is connected to a different R-SWMR crossbar. The local electrical ring has 150-bit bi-directional links with 1-cycle delay. Firefly has high laser power consumption and a low concentration (1), which results in light traffic, thus giving ample opportunity to conserve laser power.

Using this infrastructure we compare the performance (instructions per sec) and energy per instruction (EPI) of the schemes in Table 11.1. Our design impacts only the on-chip energy consumption, so we calculate EPI accounting for the energy consumed on the processor chip only, without accounting for the energy to access main memory.

11.1.5.3 Laser power modeling

The photonic devices we model include on-chip or off-chip lasers, ring modulators to modulate the light (electrical-to-optical conversion), waveguides to route optical signals to their destination, and resonant demodulators to demodulate the optical signal (optical-to-electrical conversion). By employing DWDM with sufficient separation between wavelengths (e.g., 5 nm), lasers of different wavelengths can be guided in the same waveguide without interfering with each other, which increases the bandwidth density. There are several waveguide technologies that offer low optical loss in the 0.272–0.6 dB/cm range [4, 6, 17, 57, 62]. For ProLaser we choose to employ silicon nitride (Si_3N_4) waveguides because they have high light confinement, low intrinsic optical loss in the C-band (0.4 dB/cm) and can achieve superior reproducibility in a CMOS-compatible platform [17]. Similarly, DWDM-compatible modulators and demodulators using resonant rings have been manufactured and characterized [61], and can handle energy-efficient signal conversions at speeds over 10 GHz.

To keep all designs grounded to reality, we model devices with parameters that have been experimentally demonstrated in recent manufactured prototypes. Table 11.3 presents the optical parameters for the nanophotonic devices we model along with the corresponding parameter sources. The modulation and demodulation energy is 317 fJ/bit at 10 GHz [61]. The laser power per wavelength and total laser power are calculated in Table 11.3 using the analytical models introduced in [27] (we use the analytical models to provide the breakdown; DSENT [59] calculates similar results). The total laser power in Table 11.3 is the wall-plug laser power and accounts for both the data and reservation channels, plus the laser efficiency of 15%.

Table 11.3   Nanophotonic parameters for Radix-16 R-SWMR crossbar

	Per Unit	On-Chip Laser Total (Radix-16 R-SWMR)	Off-Chip Laser Total (Radix-16 R-SWMR)
DWDM		32	32
Splitter [61]	1 dB	3 dB	3 dB
WG Loss [17]	0.4 dB/cm	4 dB	4 dB
Nonlinearity [61]	1 dB	1 dB	1 dB
Modulator Ins. [61]	3 dB	3 dB	3 dB
Ring Through [60]	0.01 dB	5.12 dB	5.12 dB
Filter Drop [60]	1.5 dB	1.5 dB	1.5 dB
Coupler [61]	1.2 dB		2.4 dB
Total Loss		17.62 dB	20.02 dB
Detector [41]		−20 dBm	−20 dBm
Laser Power Per λ		0.578 mW	1.01 mW
Total Laser Power	15% Eff.	18.56 W	32.36 W

11.1.5.4 Resonant ring heater modeling

To calculate the total ring heating power we extend the method by Nitta et al. [44] by additionally accounting for the heating of the photonic die by the operation of the cores. We model the thermal characteristics of a 3D-stacked architecture where the photonic die sits underneath the logic die. We use the 3D-chip extension of HotSpot [58] to model the transient temperature changes in the optical die. After we execute a workload and collect transient temperature traces, we calculate the ring heating power required to maintain the entire photonic die at the constant micro-ring trimming temperature during the entire execution. In addition, we account for the individual ring trimming power required to overcome process variations, as described in [27]. The ring trimming power is less significant when using smaller-radix crossbars.

11.1.5.5 The overheads of laser control

Each R-SWMR bus is powered by two sets of 32 lasers. Each set has one laser per wavelength in 32-way DWDM. The lasers of each set are first muxed and then split into 5 waveguides, thus 64 lasers create a 300-bit optical bus spread over 10 waveguides. This architecture is better than 32 lasers split into 10 waveguides, as the latter requires an additional level of splitters and incurs 26% higher optical loss.

Controlling the common bits of the bus independently introduces an additional set of 44 lasers. Thus, for example ProLaser for radix-16 employs 704 lasers more than a baseline radix-16 crossbar. These additional lasers occupy 5.4 mm^2 [37], increasing the overall area of the photonic devices by 6.2%. Overall, ProLaser on radix-16 R-SWMR crossbars employs a total of 1728 lasers, which occupy 13.25 mm^2 (i.e., 2.8% of the chip area of a 480 mm^2 chip, of which 1.1% of the chip area is ProLaser's overhead). In the case of Firefly, the additional lasers are 2816 and occupy 21.6 mm^2. ProLaser in the Firefly topology requires a total of 6912 lasers, which occupy a total of 53 mm^2 (11% of the 480 mm^2 chip area, of which 4.5% is ProLaser's overhead).

When an off-chip laser is used, 64 wavelengths are muxed into one optical fiber to couple onto the chip. Thus, each bus requires 3 optical fibers (one for the data bits, one for the common bits, and one for off-chip laser control, Figure 11.2(c)). For example, ProLaser for radix-16 uses 48 fibers attached to the chip, and 192 fibers for Firefly. 192 fibers need 48 mm at a generous 250 μm coupling pitch (tapered couplers need just 25 μm), occupying only 54% of the chip circumference.

Controlling the off-chip lasers requires an additional always-on optical link. The additional link increases static laser power by 0.6%, which corresponds to only 0.07% overhead over the chip's peak power because this is only a 2-bit link. On top of that static power consumption, each time the link is used the system goes through an E/O conversion at the L2 tile and an O/E conversion at the off-chip laser source. These operations consume a total of 634fJ. Compared to an L2 access, which consumes 78.35pJ, each laser control command imposes a dynamic energy overhead of 0.8% over an L2 cache access, which corresponds to only 0.009% overhead over the average dynamic energy consumption of the chip across workloads. Our evaluation faithfully models the additional link and all these effects.

ProLaser additionally employs a 1KB counting Bloom filter at each L2 cache slice. The L2 cache slice is 512KB, thus the Bloom filters impose a 0.2% area overhead on the L2 cache (i.e., less than 0.1% area overhead over the total 480 mm^2 chip area). The Bloom filter is an efficient L-CBF [54] structure of 4K 2-bit entries. A Bloom filter lookup consumes 1.09pJ (overestimated using CACTI [74]). For comparison, an access to the 512KB 16-way L2 cache slice consumes 78.35pJ. Thus, the dynamic energy overhead of a Bloom filter is 1.4% of the L2 dynamic energy. All these overheads are faithfully modeled in our evaluation.

Laser gating trades off message latency for energy savings, thus it is expected to achieve lower performance than No-Ctrl. Prior work [12] has analyzed this effect on multicores with on-chip lasers running real-world workloads and found that the schemes in Table 11.1 save 30–46% of the laser energy on average at the cost of 10–17% slowdown, while ProLaser saves 61% with 1.7% slowdown. However, in a realistic setting, the saved power will reduce thermal emergencies and the need for core throttling, and actually increase performance.

11.1.6 Experimental Results

Under realistic thermal and power constraints, DVFS scales voltage and frequency to keep the chip within 90°C. Thus, core frequencies stay at a reasonable 1.6–3.8 GHz range for all applications (3–3.2 GHz on average). DVFS in No-Ctrl throttles the cores frequently to keep them below 90°C. Laser gating, however, reduces the laser power and leads to a cooler chip, with less core throttling and higher performance.

The impact of laser gating depends on the total laser power consumption of the photonic network. Off-chip lasers incur additional coupling losses, and thus consume 2.5x higher power than on-chip lasers. The majority of this power is dissipated away from the multicore chip. However, the power consumption of the off-chip lasers can be as high as the power budget of the multicore itself (due to high coupling losses and low efficiency), so the impact of laser gating energy savings on the total system energy remains significant.

11.1.7 Case Study 1: Radix-16 R-SWMR

The radix-16 crossbar approximates a worst-case scenario for laser gating. It has low power consumption and its high concentration (4) creates heavy traffic. These two effects limit the potential for laser gating. Laser gating schemes save a significant fraction of the laser power, but they expose some of the laser turn-on delay and lead to lower performance under workloads sensitive to NoC latency. ProLaser outperforms all other schemes and achieves high performance (higher than even No-Ctrl, Figure 11.3), high energy savings, and low energy per instruction (EPI, Figure 11.4). ProLaser on a radix-16 crossbar with off-chip lasers consumes 32.36W (Table 11.3), of which only 2.6W are dissipated on chip. Thus, laser control schemes can no longer increase performance by reducing core throttling and are slower than No-Ctrl. However, ProLaser saves a significant portion of the laser energy, which reduces the total energy consumption of the chip (12% lower EPI) at just 2% slowdown.

11.1.8 Case Study 2: Firefly

Firefly [48] consists of 4 radix-16 R-SWMR crossbars with slightly longer waveguides, thus its power consumption is 4x that of a radix-16 R-SWMR

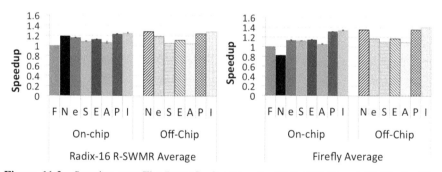

Figure 11.3 Speedup over Flat-Butterfly for (a) radix-16 R-SWMR and (b) Firefly. The evaluated designs are from left to right: Flat-B. (F), No-Ctrl (N), Power_Eq (e), Simple (S), EcoLaser (E), ATAC+ (A), ProLaser (P), and Perfect (I), and their off-chip implementations.

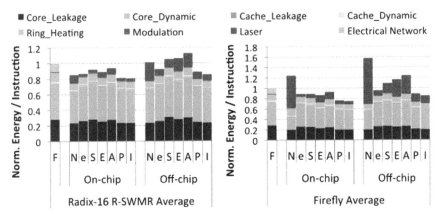

Figure 11.4   EPI for (a) radix-16 R-SWMR and (b) Firefly. The evaluated designs are from left to right: Flat-B. (F), No-Ctrl (N), Power_Eq (e), Simple (S), EcoLaser (E), ATAC+ (A), ProLaser (P), and Perfect (I), and their off-chip implementations.

crossbar. The high laser power and low concentration (1) of Firefly results in light traffic, thus giving ample opportunity for laser gating. The wall-plug power consumption for Firefly with on-chip lasers is 76.7W. All laser control schemes outperform No-Ctrl (Figure 11.3), because the laser energy savings are a considerable fraction of the chip's power budget. Off-chip lasers consume 134.06W for Firefly, of which only 10.96W are dissipated on the chip. On average ProLaser-OffChip is slightly slower than No-Ctrl-OffChip, but has 44% lower EPI (52% max, Figure 11.4).

In all cases, ProLaser's performance and EPI match Perfect's by 94–96%, which indicates that ProLaser is harvesting the majority of the possible laser energy savings.

11.1.9 Related Work

Several other techniques complementary to laser power-gating have been proposed to reduce the power requirement of a NoC. Zhou et al. [70] control active splitters to tune channel bandwidth on a binary tree network and increase channel utilization, which leads to higher energy efficiency. Chen et al. [8] distribute laser power across multiple busses in a multi-bus NoC based on the changes in the bandwidth demand. Flexishare [47] minimizes static power consumption by fully sharing a reduced number of channels across the network. Neel et al. [43] employ runtime power management techniques to reduce the laser power by adapting the width of the network, i.e., scaling the number of channels available for communication. Chen et al. [7] dynamically activate/deactivate L2 cache banks and switch on/off the corresponding silicon-photonic links in the NoC. Nitta et al. [45] show the energy inefficiency of photonic interconnects under low utilization, and propose to improve efficiency by recapturing the energy of photons, which are not used for communication.

11.2 Minimizing Ring Trimming Power

11.2.1 Introduction

Silicon photonics can be manufactured by adding a few new steps in the CMOS manufacturing process [8]. While silicon-photonic devices can be manufactured alongside CMOS logic even on the same die [8], designers typically assume a simplified process where the photonic components are housed within a photonic die, which is 3D-stacked to a logic die that contains cores, caches, and other electronic components. Due to this arrangement, the thermal variations of the logic or memory die [38] directly couple to the photonic devices. These thermal variations may occur rapidly depending on the workload, are both spatial and temporal in nature, and can exceed 30°C [53] difference. These thermal fluctuations may prevent silicon-photonic designs based on microring resonators from functioning. Microrings are tuned to resonate at a particular wavelength when they are at a set temperature, but they are highly thermally sensitive devices. For example, the resonant wavelength of a microring modulator with 5 μm diameter shifts by 0.11 nm/°C [40]. As a typical photodetector requires optical power no less than 3 dB below the peak to operate properly, the microrings can withstand no more than 2.8°C of temperature shift assuming 5 nm wavelength separation.

To keep the microrings resonating at their appropriate wavelengths designers employ trimming, a technique that dynamically shifts the microring's resonant wavelength towards the red through heating, or shifts it towards the blue through current injection. Trimming by current injection causes instability and thermal runaways [44], thus microrings are typically maintained at a constant temperature using the heaters only. Modulators with integrated heaters have been shown to produce error-free 10 Gb/s modulation across a 60°C temperature variation range, with comparable tuning efficiencies [73]. Because only the heaters are used, the microrings are tuned to temperatures above the maximum temperature that the microprocessor reaches.

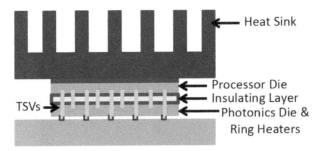

Figure 11.5 Insulating nanophotonics with Parka.

Unfortunately this means that the heaters need to work constantly to keep the microrings tuned, and at the same time the majority of the heating power is wasted as it dissipates through the package to the heat sink. Thus, it is common for microring heaters to consume upwards of 40W [44], most of which is wasted. To make matters worse, the heaters also heat the logic layer close to its thermal limit, which forces the system to throttle the cores, thereby reducing performance. The runaway heat also increases the frequency and magnitude of thermal emergencies, and accelerates the aging of the logic die.

11.2.2 Solution: Photonic Die Insulation with Parka

Recent work proposed a rather simple yet effective solution: thermally decouple the 3D-stacked logic die from the photonics die by introducing an insulating layer between them to maintain higher thermal stability and easier trimming. Parka [13] reduces the wasted energy and the heating of the logic layer by thermally decoupling the 3D-stacked logic die from the photonics die by placing an insulation layer between them (Figure 11.5). The insulation layer increases the thermal resistivity of the heat path from the photonics layer to the heat sink, and (a) allows for easier microring trimming by trapping the heat within the photonics layer, (b) reduces the temperature variation in the photonics layer, and (c) minimizes the heating of the logic die induced by the microring heaters. This allows Parka to reduce the ring heating power by 3.8–5.4x on average [13] over a baseline scheme with no insulation. Moreover, the energy savings allow for providing a higher power budget to the cores, which enables them to run 11–34% faster.

The processor die in a Parka architecture is placed close to the heat sink to allow better cooling, while an oxidized 150μm-thick macro porous Si layer realizes the thermal insulation, as porous Si has 100x lower thermal conductivity than Si [42]. The power delivery and communication between the dies is maintained through high aspect ratio through silicon vias (TSVs) [20, 56, 68]. Parka does not depend on the exact insulator technology used. Insulation can be achieved also by a 5μm-thick air or vacuum cavity etched between layers [68].

Adding the insulation layer is expected to increase the manufacturing cost only marginally. The porous Si insulation layer can be integrated into the CMOS process by oxidizing a plain silicon die through a simple electrochemical process [42]. The porous Si die is not subject to the regular yield-induced costs of dies that implement complex logic and require multiple mask exposures and metal layers, and thus it is significantly cheaper. The addition of the porous Si layer also does not affect the number of TSVs and the number of pins in the package, which together with the logic and photonic dies constitute the dominant cost factors [15, 69]. The thickness of the insulation layer impacts the TSVs' height, but the cost is highly insensitive to it [15, 69]. The additional layer will incur 3D-bonding costs, but these will increase the total cost by less than 1.5% [15, 69].

11.2.3 Experimental Methodology

We evaluate Parka on a realistic multicore with optical NoC using the methodology in Section 11.1.5 and the parameters in Table 11.2 and Table 11.3.

11.2.3.1 Modeling the Ring-Heater Power Consumption

We model a photonic die with microrings tuned to 90°C (363.15°K) following the methodology outlined in Section 11.1.5.4. We model a multicore where 50μm-thick logic and photonic dies are 3D-stacked, and separated by a 150 μm porous Si insulation layer (Figure 11.5). The thermal resistivity is 0.01 m-K/W for Si and 1 m-K/W for porous Si [42]. We evaluate Parka's thermal behavior via a transient thermal analysis at 300 μs time steps using the 3D extension of HotSpot [58]. The ambient temperature is fixed at 45°C (318.15°K).

Our model accounts for the thermal impact of TSVs and also for the individual ring trimming power required to overcome process variations, as described in [27]. We model a design that employs a total of 76,800 microrings, which are driven by one TSV each. We model high-aspect ratio TSVs with 10 μm diameter [56]. All the TSVs together cover a 6 mm^2 area, which corresponds to 1.25% of the chip area and contributes only 0.5% to the total cost [15, 69]. This overhead is due to 3D-stacking the photonic and the logic dies, and it is incurred by both Parka and the baseline system.

11.2.3.2 Modeling cooling solutions

The ring-heating power requirement depends highly on the cooling solution. Aggressive cooling solutions are capable of faster heat removal from the processor stack, which is likely to force the ring heaters to work even harder to keep the photonic layer at the tuned temperature. Therefore, the thermal decoupling that Parka advocates will be more important when better cooling solutions are employed. We model both forced-air cooling (convective thermal resistance $R_{conv} = 0.25$ K/W) and a liquid cooling solution ($R_{conv} = 0.15$ K/W [55]).

For the liquid cooling solution we assume that microchannels facilitate forced convective interlayer cooling with single-phase fluids. We assume water, as other single-phase fluids with higher thermal capacitance are toxic and thus impractical. The TSVs are etched within 100μm-wide microchannel walls as in [53]. We assume uniformly distributed microchannels and equivalent fluid flow rate through each channel in the same layer. Although nonuniform heat flux can induce fluid flow variations, the variations stay below 2% for single-phase flows and have negligible impact on the cooling system's performance [53]. The fluid pump and valve consume 1.3W per 10 ml/min flow, and the power is linear to the volumetric fluid flow [53].

11.2.4 Experimental Results

11.2.4.1 Impact on ring-heating power consumption

We evaluate the thermal shielding effect of the insulating layer by observing the temperature variation in the photonics die resulting from temperature fluctuations in the processor die. We increase the power consumption in the processor layer from idle to its maximum allowed level (130W), and observe the temperature change in the photonics layer (Figure 11.6). The processor die stays at 66°C (339.15°K) when in the idle state, and its temperature reaches 90°C (363.15°K) rapidly when it is turned on (within 18ms). The temperature of the photonics die closely tracks the temperature change of the processor die when there is no insulation. However, for Parka, it takes twice as long for the photonics layer to reach 90°C because of the insulating layer.

Thermally decoupling the photonics layer from the rest of the processor stack allows for trimming with less ring heater power consumption, because it does not allow the heat generated by the ring heaters dissipate through

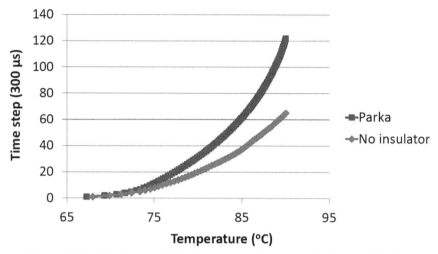

Figure 11.6 Transient analysis of temperature fluctuations in the photonics die.

the heat sink easily. The insulating layer increases the thermal resistance on the heat path to the heat sink, so it traps the heat within the photonics die. Therefore, Parka's ring-heaters can heat the photonics die with less power. Figure 11.7 shows a scenario where we present both the shielding and heat trapping effect of Parka. Figure 11.7(a) shows a snapshot (at time t_0) of the thermal map of the processor die when running a real workload (appbt). We assume that at time t_0 all processors stop, and they only dissipate leakage power until time t_1. We estimate that the processor die leakage power is 30W when idle. Figure 11.7(b) shows the temperature maps of the photonics layer at time t_1. The photonics layer stays at a higher temperature for Parka, as it retains the heat due to the insulating layer.

In the example we assume that the ring heaters are also off until time t_1. At time t_1, the ring heaters are turned on to bring the photonics layer to a stable 90°C (363.15°K), and Figure 11.7(c) shows the power distribution of these ring heaters. We observe that Parka requires less ring-heating power. There are two reasons for this: first, the photonics layer is at a higher temperature at time t_1, so there is a smaller temperature difference (to 90°C) to cover. Second, it is easier to close this temperature difference with Parka because the heat generated by the ring heaters stays within the photonics die.

The amount of ring-heating power required to keep the photonics layer at a stable 90°C highly depends on the power consumption of the processor die. When the processor die is idle, the ring heaters have to work harder.

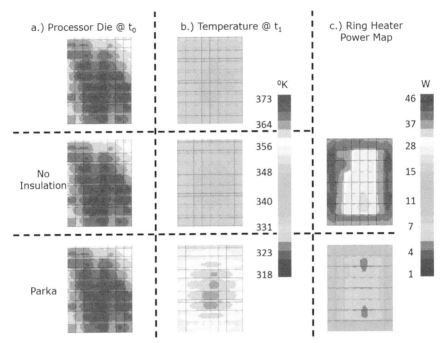

Figure 11.7   Case study: Impact of thermal insulation on the photonics layer temperature and the ring-heating power consumption.

Figure 11.8 shows that for every processor die utilization level, Parka consumes less ring-heating power (up to 3.5x lower) than the ring-heating power required without insulation.

To observe the impact of the cooling solution the cooling solution on Parka, we repeat the same ring-heating power estimation with a liquid cooling solution. We observe that with liquid cooling the operational temperature at the processor layer stays under 90°C when the processor die consumes up to 250W, while forced-air cooling can sustain at best only up to 130W.

We analyze the ring-heating power required on a realistic multicore by running a collection of diverse workloads on our simulated multicore system and calculating the average ring-heating power consumed by each application (Figure 11.9). The temperature fluctuations are higher when running memory-intensive workloads (e.g., bodytrack, em3d, ocean, appbt), hence the ring-heating power consumption is also higher. On average ring heaters consume 16.9W (22.4W max) when there is no insulation. Parka allows for easier trimming by shielding from short fluctuations and trapping the heat, so

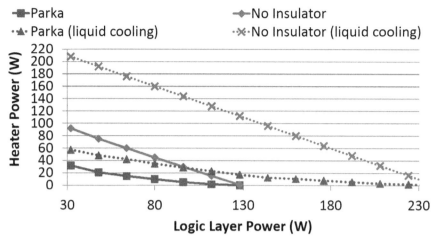

Figure 11.8 Ring-heating power vs. processor die power.

it consumes on average 3.8x less ring heating power (4.4W on average). Parka is essential when using aggressive cooling solutions, because ring heaters have to consume 28.2W on average when there is no insulation, and only 5.2W with liquid cooling.

11.2.4.2 Impact on a realistic multicore

Ring heaters warm up and keep the photonics die at a slightly higher temperature than the maximum operating temperature of the processor [44]. However, while heating the photonics die, the ring heaters also heat the processor die when there is no insulation. Heating the processor die forces it to operate close to its maximum operating temperature, in which case, even a small

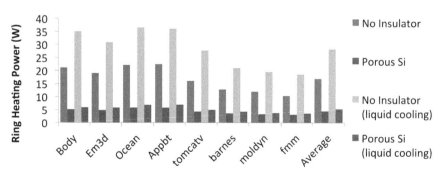

Figure 11.9 Average ring-heating power consumption of real-world applications.

increase in the utilization can cause a temperature spike which pushes the processor out of the safe operating limits causing it to throttle, and reducing performance.

Parka allows ring heaters to consume 3.8x less power on average, and thus the processor layer remains cooler. For example, when the logic layer consumes 90W, the ring heaters consume 36W when there is no insulation, but only 7.2W with Parka. As a result, the logic layer stays at 74°C with Parka, while it reaches 90°C without insulation.

Without insulation, the ring heaters keep the processor die near the limits of safe operating temperature, so any increase in the processor utilization can push the processor into thermal emergencies. As shown in Figure 11.10, the execution activity increases around time steps 12 and 62 push the processor temperature over 90°C without insulation, whereas with Parka the processor stays cooler and avoids the thermal emergencies. The processor runs into thermal emergencies up to 19% of the execution time on real applications (2% on average) when there is no insulation (Figure 11.10). The cores need to be throttled or completely turned off during a thermal emergency to allow for the chip to cool and avoid permanent damage. In contrast, Parka's processor die largely avoids thermal emergencies, and only experiences them for less then 1% of the execution time (Figure 11.10(b)).

Because Parka shields the processor die from the heaters in the photonic layer, it results in a cooler chip with less core throttling and a higher power budget for the cores, which leads to higher performance. As a result, Parka allows the processor to run 11% faster on average (18% maximum) [13]. The ring-heating power consumption is higher when an aggressive cooling solution is employed, so the power savings of Parka are also higher.

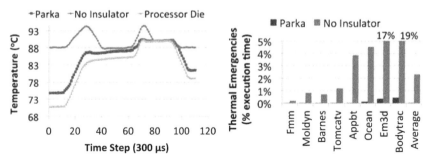

Figure 11.10 Temperature trace (appbt) presenting thermal emergencies in a multicore, and the percentage of execution time spent under thermal emergencies.

With liquid cooling, Parka outperforms the processor without the insulation by 23% on average (34% maximum).

11.2.5 Related Work

There are several techniques that can be used to resolve the thermal challenges of the silicon microring resonator devices. Methods to reduce the thermal dependence of microrings to tolerable levels include athermalization using negative thermo-optic materials or the embedment of the microring in a thermally-balanced interferometric structure. However, it is challenging to integrate the necessary polymer and TiO_2 materials into a CMOS-compatible fabrication process, and the interferometric structure still suffers from susceptibility to fabrication tolerances, increases the footprint of the microring, and it is challenging to adapt the technique to larger microring switch fabrics [46]. Thus, control-based techniques that aim to detect and react to the resonance shift due to thermal fluctuations are preferable, and several prototypes have been shown to withstand thermal variations across a wide temperature range up to $32–60°K$ [73]. Excellent surveys on this topic that are available in the literature [46].

11.3 Future Work

If more efficient lasers were invented, the impact of any laser energy-saving technique would drop. Because ProLaser is only 2–6% away from a perfect oracle scheme, there is limited opportunity left beyond ProLaser for laser gating. Thus, newer techniques would have to rely on a different mechanism, e.g., adjusting the laser power to the minimum level sufficient for reliable transmission of each flit [34], architectures that efficiently share optical components [47], or recapturing the energy of photons not used for communication [45]. Research on improving the energy efficiency of lasers and minimizing the optical loss of photonic components will also go a long way. Extending such techniques to on-board or datacenter-scale networks and processor disintegration or disaggregated architectures will bring a new wave of much-needed innovation in that space [11, 14].

The thermal sensitivity of microrings is a significant impediment to their practical use. Research in materials, athermal ring design, and coupling microrings to MZIs are promising directions to alleviate this constraint [21, 51, 64, 71]. Borrowing from techniques to counteract process variations may also lower the power wasted on microring trimming (e.g., adding supplementary microrings and allowing flexible assignment of wavelengths to network nodes [67]).

Acknowledgements

This work was supported by the National Science Foundation under awards CCF-1218768 and CCF-1453853.

References

[1] L. A. Barroso and U. Holzle. The case for energy-proportional computing. *IEEE Computer*, 40(12):33–37, 2007.

[2] C. Batten, A. Joshi, J. Orcutt, A. Khilo, B. Moss, C. W. Holzwarth, M. A. Popovic, H. Li, H. I. Smith, J. L. Hoyt, F. X. Kartner, R. J. Ram, V. Stojanovic, and K. Asanovic. Building many-core processor-to-DRAM networks with monolithic CMOS silicon photonics. *IEEE Micro*, 29(4):8–21, 2009.

[3] C. Batten, A. Joshi, V. Stojanovic, and K. Asanovic. Designing chip-level nanophotonic interconnection networks. *IEEE Journal on Emerging and Selected Topics in Circuits and Systems*, 2(2):137–153, 2012.

[4] W. Bogaerts and S. K. Selvaraja. Compact single-mode silicon hybrid rib/strip waveguide with adiabatic bends. *IEEE Photonics Journal*, 3(3):422–432, 2011.

[5] R. E. Camacho-Aguilera, Y. Cai, N. Patel, J. T. Bessette, M. Romagnoli, L. C. Kimerling, and J. Michel. An electrically pumped germanium laser. *Optics Express*, 20(10):11316–11320, 2012.

[6] J. Cardenas, C. Poitras, J. Robinson, K. Preston, L. Chen, and M. Lipson. Low loss etchless silicon photonic waveguides. *Optics Express*, 17(6):4752–4757, 2009.

[7] C. Chen, J. L. Abellan, and A. Joshi. Managing laser power in silicon-photonic NoC through cache and NoC reconfiguration. *IEEE Transactions on Computer-Aided Design of Integrated Circuits and Systems*, 34(6):972–985, 2015.

[8] C. Chen and A. Joshi. Runtime management of laser power in silicon-photonic multibus NoC architecture. *IEEE Journal of Selected Topics in Quantum Electronics*, 19(2):3700713–3700713, 2013.

[9] W. J. Dally and T. B. *Principles and Practices of Interconnection Networks*. Morgan Kaufmann Publishing Inc., 2004.

[10] Y. Demir and N. Hardavellas. EcoLaser: An adaptive laser control for energy efficient on-chip photonic interconnects. *Proceedings of the International Symposium on Low-Power Electronics and Design*, 2014.

[11] Y. Demir and N. Hardavellas. SLaC: Stage laser control for a flattened butterfly network. *Proceedings of the IEEE International Symposium on High Performance Computer Architecture*, pp. 321–332, 2016.

[12] Y. Demir and N. Hardavellas. Energy proportional photonic interconnects. *ACM Transactions on Architecture and Code Optimization*, 13(5), 2016.

[13] Y. Demir and N. Hardavellas. Parka: Thermally Insulated Nanophotonic Interconnects. *Proceedings of the 9th International Symposium on Networks-on-Chip*, 2015.

[14] Y. Demir, Y. Pan, S. Song, N. Hardavellas, J. Kim, and G. Memik. Galaxy: A high-performance energy-efficient multi-chip architecture using photonic interconnects. *Proceedings of the 28th ACM International Conference on Supercomputing*, 2014.

[15] X. Dong, J. Zhao, and Y. Xie. Fabrication cost analysis and cost-aware design space exploration for 3-D ICs. *IEEE Transactions on Computer-Aided Design of Integrated Circuits and Systems*, 29(12), 2010.

[16] G.-H. Duan, A. Shen, A. Akrout, F. V. Dijk, F. Lelarge, F. Pommereau, O. LeGouezigou, J.-G. Provost, H. Gariah, F. Blache, F. Mallecot, K. Merghem, A. Martinez, and A. Ramdane. High performance InP-based quantum dash semiconductor mode-locked lasers for optical communications. *Bell Labs Technical Journal*, 14(3):63–84, 2009.

[17] J. P. Epping, M. Hoekman, R. Mateman, A. Leinse, R. G. Heideman, A. van Rees, P. J. van der Slot, C. J. Lee, and K.-J. Boller. High confinement, high yield Si_3N_4 waveguides for nonlinear optical applications. *Optics Express*, 23(2):642–648, 2015.

[18] European, Japan, Korean, Taiwan, and US Semiconductor Industry Associations. The international technology roadmap for semiconductors (ITRS). 2012.

[19] A. W. Fang, H. Park, O. Cohen, R. Jones, M. J. Paniccia, and J. E. Bowers. Electrically pumped hybrid AlGaInAs-silicon evanescent laser. *Optics Express*, 14(20):9203–9210, 2006.

[20] A. C. Fischer, S. J. Bleiker, T. Haraldsson, N. Roxhed, G. Stemme, and F. Niklaus. Very high aspect ratio through-silicon vias (TSVs) fabricated using automated magnetic assembly of nickel wires. *Journal of Micromechanics and Microengineering*, 22(10):105001, 2012.

[21] B. Guha, B. B. C. Kyotoku, and M. Lipson. CMOS-compatible athermal silicon microring resonators. *Optics Express*, 18(4):3487–3493, 2010.

[22] N. Hardavellas, S. Somogyi, T. F. Wenisch, R. E. Wunderlich, S. Chen, J. Kim, B. Falsafi, J. C. Hoe, and A. G. Nowatzyk. SimFlex: a fast, accurate, flexible full-system simulation framework for performance evaluation of server architecture. *SIGMETRICS Performance Evaluation Review, Special Issue on Tools for Computer Architecture Research*, 31(4):31–35, 2004.

[23] M. Heck and J. Bowers. Energy efficient and energy proportional optical interconnects for multi-core processors: Driving the need for on-chip sources. *IEEE Journal of Selected Topics in Quantum Electronics*, 20(4):1–12, 2014.

[24] H. Hisham, G. Mahdiraji, A. Abas, M. Mahdi, and F. Adikan. Characterization of transient response in fiber grating fabry-perot lasers. *IEEE Photonics Journal*, 4(6):2353–2371, 2012.

[25] H. Hisham, G. Mahdiraji, A. Abas, M. Mahdi, and F. Adikan. Characterization of turn-on time delay in fiber grating fabry-perot lasers. *IEEE Photonics Journal*, 4(5):1662–1678, 2012.

[26] J. Huh, C. Kim, H. Shafi, L. Zhang, D. Burger, and S. W. Keckler. A NUCA substrate for flexible CMP cache sharing. *Proceedings of the 19th Annual International Conference on Supercomputing*, pp. 31–40, 2005.

[27] A. Joshi, C. Batten, Y.-J. Kwon, S. Beamer, I. Shamim, K. Asanovic, and V. Stojanovic. Silicon-photonic clos networks for global on-chip communication. *Proceedings of the IEEE International Symposium on Networks-on-Chip*, pp. 124–133, 2009.

[28] J. Kim, W. J. Dally, and D. Abts. Flattened butterfly: A cost-efficient topology for high-radix networks. *Proceedings of the 34th Annual International Symposium on Computer Architecture*, pp. 126–137, 2007.

[29] L. C. Kimerling. Scaling functionality with silicon photonics: Achievement and potential. http://www.orc.soton.ac.uk/fileadmin/seminar_pdf/ UKSP_Showcase_-_Lionel_Kimerling.pdf, November 2013.

[30] N. Kirman, M. Kirman, R. K. Dokania, J. F. Martinez, A. B. Apsel, M. A. Watkins, and D. H. Albonesi. Leveraging optical technology in future bus-based chip multiprocessors. *Proceedings of the 39th IEEE/ACM Annual International Symposium on Microarchitecture*, pp. 492–503, 2006.

[31] B. R. Koch, E. J. Norberg, B. Kim, J. Hutchinson, J.-H. Shin, G. Fish, and A. Fang. Integrated silicon photonic laser sources for telecom

and datacom. *Proceedings of the Optical Fiber Communication Conference/National Fiber Optic Engineers Conference*, page PDP5C.8, 2013.

[32] E. Kotelnikov, A. Katsnelson, K. Patel, and I. Kudryashov. High-power single-mode InGaAsP/InP laser diodes for pulsed operation. *Proceedings of SPIE*, 8277:827715–827715–6, 2012.

[33] A. Krishnamoorthy, R. Ho, X. Zheng, H. Schwetman, J. Lexau, P. Koka, G. Li, I. Shubin, and J. Cunningham. Computer systems based on silicon photonic interconnects. *Proceedings of the IEEE*, 97(7):1337–1361, 2009.

[34] G. Kurian, C. Sun, C.-H. Chen, J. Miller, J. Michel, L. Wei, D. Antoniadis, L.-S. Peh, L. Kimerling, V. Stojanovic, and A. Agarwal. Cross-layer energy and performance evaluation of a nanophotonic manycore processor system using real application workloads. *Proceedings of the 26th IEEE International Parallel Distributed Processing Symposium*, pp. 1117–1130, 2012.

[35] G. Li, J. Yao, H. Thacker, A. Mekis, X. Zheng, I. Shubin, Y. Luo, J. hyoung Lee, K. Raj, J. E. Cunningham, and A. V. Krishnamoorthy. Ultralow-loss, high-density soi optical waveguide routing for macrochip interconnects. *Optics Express*, 20(11):12035–12039, 2012.

[36] S. Li, J. H. Ahn, R. D. Strong, J. B. Brockman, D. M. Tullsen, and N. P. Jouppi. McPAT: an integrated power, area, and timing modeling framework for multicore and manycore architectures. *Proceedings of the 42nd IEEE/ACM Annual International Symposium on Microarchitecture*, pp. 469–480, 2009.

[37] J. Liu, X. Sun, R. Camacho-Aguilera, L. C. Kimerling, and J. Michel. Ge-on-Si laser operating at room temperature. *Optics Letters*, 35(5):679–681, 2010.

[38] S. Liu, B. Leung, A. Neckar, S. O. Memik, G. Memik, and N. Hardavellas. Hardware/software techniques for DRAM thermal management. *Proceedings of the 17th IEEE International Symposium on High Performance Computer Architecture*, pp. 515–525, 2011.

[39] P. Lotfi-Kamran, B. Grot, and B. Falsafi. Noc-out: Microarchitecting a scale-out processor. *Proceedings of the 45th Annual IEEE/ACM International Symposium on Microarchitecture*, pp. 177–187, 2012.

[40] S. Manipatruni, R. K. Dokania, B. Schmidt, N. Sherwood-Droz, C. B. Poitras, A. B. Apsel, and M. Lipson. Wide temperature range operation of micrometer-scale silicon electro-optic modulators. *Optics Letters*, 33(19):2185–2187, 2008.

[41] G. Masini, A. Narasimha, A. Mekis, B. Welch, C. Ogden, C. Bradbury, C. Sohn, D. Song, D. Martinez, D. Foltz, D. Guckenberger, J. Eicher, J. Dong, J. Schramm, J. White, J. Redman, K. Yokoyama, M. Harrison, M. Peterson, M. Saberi, M. Mack, M. Sharp, P. D. Dobbelaere, R. LeBlanc, S. Leap, S. Abdalla, S. Gloeckner, S. Hovey, S. Jackson, S. Sahni, S. Yu, T. Pinguet, W. Xu, and Y. Liang. CMOS photonics for optical engines and interconnects. *Proceedings of the National Fiber Optic Engineers Conference and Optical Fiber Communication Conference and Exposition*, pp. 1–3, 2012.

[42] B. Mondal, P. Basu, B. Reddy, H. Saha, P. Bhattacharya, and C. Roychoudhury. Oxidized macro porous silicon layer as an effective material for thermal insulation in thermal effect microsystems. *Proceedings of the International Conference on Emerging Trends in Electronic and Photonic Devices Systems*, pp. 202–206, 2009.

[43] B. Neel, M. Kennedy, and A. Kodi. Dynamic power reduction techniques in on-chip photonic interconnects. *Proceedings of the 25th Great Lakes Symposium on VLSI*, pp. 249–252, 2015.

[44] C. Nitta, M. Farrens, and V. Akella. Addressing system-level trimming issues in on-chip nanophotonic networks. *Proceedings of the 17th IEEE International Symposium on High Performance Computer Architecture*, pp. 122–131, 2011.

[45] C. Nitta, M. Farrens, and V. Akella. DCOF: An arbitration free directly connected optical fabric. *IEEE Journal on Emerging and Selected Topics in Circuits and Systems*, 2(2):169–182, 2012.

[46] K. Padmaraju and K. Bergman. Resolving the thermal challenges for silicon microring resonator devices. *Nanophotonics*, 3(4-5):269–281, 2013.

[47] Y. Pan, J. Kim, and G. Memik. Flexishare: Channel sharing for an energy-efficient nanophotonic crossbar. *Proceedings of the IEEE International Symposium on High-Performance Computer Architecture*, pp. 1–12, 2010.

[48] Y. Pan, P. Kumar, J. Kim, G. Memik, Y. Zhang, and A. Choudhary. Firefly: Illuminating future network-on-chip with nanophotonics. *Proceedings of the 36th Annual International Symposium on Computer Architecture*, 2009.

[49] M. Paniccia and J. Bowers. First electrically pumped hybrid silicon laser. http://www.intel.com/content/dam/www/public/us/en/documents/technology-briefs/intel-labs-hybrid-silicon-laser-announcement.pdf, 2006.

[50] K. Petermann. *Laser Diode Modulation and Noise, Advances in Opto-electronics,* Volume 3, Springer, 1988.

[51] V. Raghunathan, W. N. Ye, J. Hu, T. Izuhara, J. Michel, and L. Kimerling. Athermal operation of silicon waveguides: spectral, second order and footprint dependencies. *Optics Express*, 18(17):17631–17639, 2010.

[52] P. Rosenfeld, E. Cooper-Balis, and B. Jacob. DRAMsim2: A cycle accurate memory system simulator. *IEEE Computer Architecture Letters*, 10(1):16–19, 2011.

[53] M. M. Sabry, A. K. Coskun, D. Atienza, T. S. Rosing, and T. Brunschwiler. Energy-efficient multiobjective thermal control for liquid-cooled 3-D stacked architectures. *IEEE Transactions on Computer-Aided Design of Integrated Circuits and Systems*, 30(12):1883–1896, 2011.

[54] E. Safi, A. Moshovos, and A. Veneris. L-CBF: A low-power, fast counting bloom filter architecture. *IEEE Transactions on Very Large Scale Integration Systems*, 16(6):628–638, 2008.

[55] K. Sankaranarayanan, B. H. Meyer, W. Huang, R. Ribando, H. Haj-Hariri, M. R. Stan, and K. Skadron. Architectural implications of spatial thermal filtering. *Integration VLSI Journal*, 46(1):44–56, 2013.

[56] T. Sarvey, Y. Zhang, Y. Zhang, H. Oh, and M. Bakir. Thermal and electrical effects of staggered micropin-fin dimensions for cooling of 3D microsystems. *Proceedings of the IEEE Intersociety Conference on Thermal and Thermomechanical Phenomena in Electronic Systems*, 2014.

[57] S. K. Selvaraja, W. Bogaerts, D. V. Thourhout, and R. Baets. Record low-loss hybrid rib/wire waveguides for silicon photonic circuits. *Proceedings of Group IV Photonics*, pp. 1–3, 2010.

[58] K. Skadron, M. R. Stan, W. Huang, S. Velusamy, K. Sankaranarayanan, and D. Tarjan. Temperature-aware microarchitecture. *Proceedings of the Annual International Symposium on Computer Architecture*, pp. 2–13, 2003.

[59] C. Sun, C.-H. O. Chen, G. Kurian, L. Wei, J. Miller, A. Agarwal, L.-S. Peh, and V. Stojanovic. DSENT – a tool connecting emerging photonics with electronics for opto-electronic networks-on-chip modeling. *Proceedings of the 6th IEEE/ACM International Symposium on Networks-on-Chip*, pp. 201–210, 2012.

[60] C. Sun, Y. H. Chen, and V. Stojanovic. Designing processor-memory interfaces with monolithically integrated silicon-photonics. *Proceedings*

of the Conference on Lasers and Electro-Optics Pacific Rim, pp. 1–2, 2013.

[61] C. Sun, M. T. Wade, Y. Lee, J. S. Orcutt, L. Alloatti, M. S. Georgas, A. S. Waterman, J. M. Shainline, R. R. Avizienis, S. Lin, B. R. Moss, R. Kumar, F. Pavanello, A. H. Atabaki, H. M. Cook, A. J. Ou, J. C. Leu, Y.-H. Chen, K. Asanovic, R. J. Ram, M. A. Popovic, and V. M. Stojanovic. Single-chip microprocessor that communicates directly using light. *Nature*, 528(7583):534–538, 2015.

[62] R. Takei, S. Manako, E. Omoda, Y. Sakakibara, M. Mori, and T. Kamei. Sub-1 dB/cm submicrometer-scale amorphous silicon waveguide for backend on-chip optical interconnect. *Optics Express*, 22(4):4779–4788, 2014.

[63] S. Tanaka, S.-H. Jeong, S. Sekiguchi, T. Kurahashi, Y. Tanaka, and K. Morito. Highly-efficient, low-noise Si hybrid laser using flip-chip bonded soa. *Proceedings of the IEEE Optical Interconnects Conference*, pp. 12–13, 2012.

[64] J. Teng, P. Dumon, W. Bogaerts, H. Zhang, X. Jian, X. Han, M. Zhao, G. Morthier, and R. Baets. Athermal silicon-on-insulator ring resonators by overlaying a polymer cladding on narrowed waveguides. *Optics Express*, 17(17):14627–14633, 2009.

[65] D. Vantrease, R. Schreiber, M. Monchiero, M. McLaren, N. P. Jouppi, M. Fiorentino, A. Davis, N. Binkert, R. G. Beausoleil, and J. H. Ahn. Corona: System implications of emerging nanophotonic technology. *Proceedings of the 35th Annual International Symposium on Computer Architecture*, 2008.

[66] T. F. Wenisch, R. E. Wunderlich, M. Ferdman, A. Ailamaki, B. Falsafi, and J. C. Hoe. SimFlex: statistical sampling of computer system simulation. *IEEE Micro*, 26(4):18–31, 2006.

[67] Y. Xu, J. Yang and R. Melhem, Tolerating process variations in nanophotonic on-chip networks. *Proceedings of the 39th Annual International Symposium on Computer Architecture*, pp. 142–152, 2012.

[68] Y. Zhang, H. Oh, and M. Bakir. Within-tier cooling and thermal isolation technologies for heterogeneous 3D ICs. *Proceedings of the IEEE International 3D Systems Integration Conference*, pp. 1–6, 2013.

[69] J. Zhao, X. Dong, and Y. Xie. Cost-aware three-dimensional (3D) many-core multiprocessor design. *Proceedings of the 47th ACM/IEEE Design Automation Conference*, 2010.

[70] L. Zhou and A. Kodi. Probe: Prediction-based optical bandwidth scaling for energy-efficient NoCs. *Proceedings of the 7th IEEE/ACM International Symposium on Networks on Chip*, pp. 1–8, 2013.

[71] L. Zhou, K. Okamoto, and S. J. Ben Yoo. Athermalizing and trimming of slotted silicon microring resonators with UV-sensitive PMMA upper-cladding. *IEEE Photonics Technology Letters*, 21(17):1175–1177, 2009.

[72] A. Zilkie, B. Bijlani, P. Seddighian, D. C. Lee, W. Qian, J. Fong, R. Shafiiha, D. Feng, B. Luff, X. Zheng, J. Cunningham, A. V. Krishnamoorthy, and M. Asghari. High-efficiency hybrid III–V/Si external cavity DBR laser for 3um SOI waveguides. *Proceedings of the 9th IEEE International Conference on Group IV Photonics*, pp. 317–319, 2012.

[73] W. Zortman, A. Lentine, D. Trotter, and M. Watts. Integrated cmos compatible low power 10gbps silicon photonic heater-modulator. *Proceedings of the National Fiber Optic Engineers Conference and Optical Fiber Communication Conference and Exposition*, pp. 1–3, 2012.

[74] N. Muralimanohar, R. Balasubramonian and N. Jouppi, Optimizing NUCA Organizations and Wiring Alternatives for Large Caches with CACTI 6.0. *Proceedings of the 40th Annual IEEE/ACM International Symposium on Microarchitecture*, pp. 3–14, 2007.

[75] Z. Li, M. Mohamed, X. Chen, E. Dudley, K. Meng, L. Shang, A. R. Mickelson, R. Joseph, M. Vachharajani, B. Schwartz and Y. Sun. Reliability Modeling and Management of Nanophotonic On-Chip Networks. *IEEE Transactions on Very Large Scale Integration (VLSI) Systems*, 20(1), pp. 98–111, 2012.

PART IV

On the Impact of Fabrication Non-Uniformity

12

Impact of Fabrication Non-Uniformity on Silicon Photonic Networks-on-Chip

Mahdi Nikdast[1], Gabriela Nicolescu[1], Jelena Trajkovic[2] and Odile Liboiron-Ladouceur[3]

[1]Polytechnique Montréal, Montréal, Canada
[2]Concordia University, Montréal, Canada
[3]McGill University, Montréal, Canada

Abstract

Multiprocessor systems-on-chip (MPSoCs) are rapidly scaling by integrating an increasingly large number of processing cores on a single chip. Consequently, conventional metallic interconnect in such systems imposes a higher power consumption to transmit an even smaller amount of data with higher latency amongst different cores. Optical technology, which has been successful in long distance communications, has indicated promising potentials to realize high performance communication in MPSoCs. As a result, optical network-on-chip (ONoC) is introduced to the communication infrastructure in MPSoCs to realize low power, low latency, and high bandwidth communication. The fundamental optical devices in such systems (e.g., optical filters and switches), however, are extremely sensitive to fabrication non-uniformity (a.k.a. fabrication process variations). In this context, precise analytical models with low computational complexity are required to fully understand the impact of different variations on the performance of ONoCs. Moreover, a comprehensive study on process variations helps determine power penalties required to trim/tune (e.g., thermal tuning) faulty devices in such systems, as well as developing required design techniques to compensate for process variations.

12.1 Introduction

As the technology advances and allows the integration of a large number of processing cores on a single chip, the conventional metallic interconnect in multiprocessor systems-on-chip (MPSoCs) will no longer be able to address the communication in such systems with low power consumption, low latency, and the required bandwidth. Some studies on MPSoCs, which include only tens of cores, have indicated that the metallic interconnect in these systems can consume over 25% of the systems' overall power budget [1]. On the other hand, optical network-on-chip (ONoCs) can boost the communication infrastructure in MPSoCs, addressing the aforementioned issues with metallic interconnect in such systems [2, 3]. Moreover, the bandwidth performance in ONoCs can be further improved by employing wavelength division multiplexing (WDM), which allows simultaneous transmission of multiple optical wavelengths in a single waveguide. Several WDM-based ONoCs have been proposed in which waveguides and microresonators (MRs) are the primary building blocks, integrated to form different optical devices (e.g., optical filters and switches), and various ONoC architectures [4–6]. Realizing a reliable communication in such systems, it is essential to precisely match the central wavelengths (e.g., resonant wavelength in MRs) among different optical devices. Nevertheless, ONoCs are extremely sensitive to fabrication non-uniformity (a.k.a. fabrication process variations), resulting in wavelength mismatches among different devices, and hence performance degradation in the system. Process variations stem from the optical lithography process imperfection, in which variations depend on the resist sensitivity, resist age or thickness, exposure change, and etching [7].

Process variations have been mostly studied at the device-level by exploring different variations either inside a single die (within-die) [8], among multiple dies on the same wafer (within-wafer) [9], among different wafers (wafer-to-wafer) [10], or even among different fabrication runs (run-to-run) [7]. L. Chrostowski, *et al.* studied 371 identical racetrack resonators of 12 μm radius fabricated on a 16×9 mm chip, and indicated a strong linear correlation between their resonant wavelength variations and their physical distances [8]. In [9], S. K. Selvaraja, *et al.* indicated within-wafer silicon thickness non-uniformity of ± 20.68 nm over a 200 mm silicon-on-insulator (SOI) wafer. X. Chen, *et al.* studied process variations in microring resonators, racetrack resonators, and directional couplers identically designed but fabricated on different dies on two different wafers (LETI and IMEC). They reported a process variation of 1.3, 1.3, and 0.33 nm^2/cm for microrings, racetracks, and directional couplers, respectively [10]. In [7], W. A. Zortman *et al.* quantified the source and impact of process induced resonant frequency variation for microdisk resonators across an individual die, entire wafers, and wafer lots for separate process runs. They demonstrated that the primary driver of the resonator non-uniformity is SOI thickness variation, as opposed to diameter variations, which can exceed 10 nm across a wafer.

Some efforts have also been made to study the impact of process variations on ONoCs at the system-level. Y. Xu *et al.* proposed a technique, called BandArb, to mitigate the effects of thermal and process variations on the bandwidth performance of silicon photonic networks by dynamically allocating the bandwidth at run-time based on demand and temperature [11]. In [12], M. Mohamed *et al.* proposed a reliability-aware design flow to improve photonic on-chip interconnects under process and thermal variations. S. V. R. Chittamuru *et al.* proposed a framework, called PICO, for mitigating crosstalk noise in ONoCs with physically trimmed MRs, improving the optical signal-to-noise ratio (OSNR) and reliability in these networks [13].

When designing an optical device, the designer can sweep the device design parameters (a.k.a. corner analysis) in a numerical simulation (e.g., FDTD: Finite-difference time-domain), trying to incorporate in the simulation the impact of different possible variations during the device fabrication (e.g., variations in the top silicon thickness and waveguide width). Nevertheless, when dealing with a large-scale ONoC, which consists of hundreds and thousands of integrated optical devices, employing numerical simulations is not feasible due to their considerable computation cost at such scale. Addressing this issue, this chapter presents an accurate and computationally efficient analytical model to study the impact of different process variations on MR-based optical devices and passive large-scale ONoCs. The proposed analytical study is based on a bottom-up approach: we study the impact of different process variations on the fundamental optical components (i.e., waveguides, component-level), then on the optical devices (i.e., MR-based optical filters and switches, device-level), and finally on a case study of a general passive ONoC that can be applied to any given ONoC architecture (i.e., system-level). Furthermore, our study includes the design, fabrication, and characterization of several MRs (as test-structures) to indicate process variations in silicon photonic fabrications, as well as exploring different characteristics of the variations (e.g., correlations among different devices with respect to their physical distances on the chip) in the fabrication.

The rest of the chapter is organized as follows. In the next section, we study the impact of process variations on optical waveguides and MR-based add-drop filters and switches. Leveraging the analytical models developed at the component-level and device-level, the following section introduces the general passive WDM-based ONoC architecture, considered as a case study, and the study on its optical SNR under fabrication process variations. The discussion then turns to the quantitative simulation results of our proposed models, as well as their evaluations against numerical simulations. The following section then presents our fabrication details and results.

Finally, the last section briefly concludes the chapter and discusses possible future directions.

12.2 Optical Waveguides and MR-Based Devices

We start by studying the impact of process variations on several properties of optical waveguides, such as effective and group indices. The effective index quantifies the overall phase delay per unit length in a waveguide, relative to the phase delay in vacuum. Also, the group index describes the velocity at which the envelope of a propagating pulse travels and is a characteristic of a dispersive waveguide. These properties determine the fundamental characteristics of different optical devices which can be constructed using optical waveguides. For example, the effective index determines the resonant wavelength in MRs. Here, we consider strip waveguides (see Figure 12.1), which are employed for routing in ONoCs as well as constructing passive devices and ONoCs. Our study is based on a two-variable approach to take into

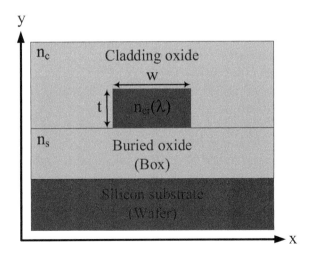

Figure 12.1 Cross section of a strip waveguide.

account the variations in the top silicon thickness (i.e., t) and waveguide width (i.e., w), indicated as the primary concerns in silicon photonics fabrication [7–9, 14].

12.2.1 Strip Waveguides under Process Variations

We use Marcatili's approach, which is an approximate analysis for calculating the propagation modes of rectangular waveguides [15, 16], to study the impact of process variations on the propagation constant as well as the effective and group indices of the fundamental transverse electric (TE) mode in strip waveguides. The analyses for the transverse magnetic (TM) modes can be developed in a similar way, and are not presented in this chapter. Figure 12.1 indicates the cross section of a strip waveguide. As can be seen, the waveguide core, with a refractive index of $n_{cr}(\lambda)$, where λ is the optical wavelength, is buried in a cladding oxide layer at the top and a substrate oxide layer at the bottom. Also, the top cladding oxide and the bottom substrate oxide layers have refractive indices of n_c and n_s, respectively. Note that the refractive index of the waveguide core is larger than that in both the cladding and substrate layers ($n_{cr} > n_{c/s}$), and hence the optical light is confined in the waveguide core. In this chapter, we consider symmetric waveguides in which $n_c = n_s$. Also, we consider the waveguide core to be from silicon while the cladding and substrate are from silicon dioxide (SiO_2). Moreover, the strip waveguide has a width and thickness of w and t, respectively. The refractive index of the waveguide core is a function of the input optical wavelength, considering the impact of silicon dispersion in the core. One can use the Sellmeier equation, which provides an empirical relationship between the refractive index of a medium and the light's wavelength [17], to calculate n_{cr} at a specific λ:

$$n_{cr}^2(\lambda) = 1 + \frac{10.67\lambda^2}{\lambda^2 - 0.31} + \frac{0.003\lambda^2}{\lambda^2 - 1.13} + \frac{1.54\lambda^2}{\lambda^2 - 1104^2}. \qquad (12.1)$$

Maxwell's equations allow us to fully describe the electromagnetic fields (E_x, E_y, H_x, and H_y) in terms of the longitudinal field components (E_z and H_z). For the waveguide core in Figure 12.1, we have [18]:

$$E_{x/y} = \frac{-i}{K^2}\left(\beta\frac{\partial E_z}{\partial x/y} +/- \omega\mu_0\frac{\partial H_z}{\partial y/x}\right), \qquad (12.2)$$

$$H_{x/y} = \frac{-i}{K^2}\left(\beta\frac{\partial H_z}{\partial x/y} -/+ \omega\varepsilon_0 n_{cr}^2(\lambda)\frac{\partial E_z}{\partial y/x}\right), \qquad (12.3)$$

where μ_0 and ε_0 are the permeability and the permittivity of free-space, respectively. Also, K can be defined as:

$$K^2 = n_{cr}^2(\lambda)k_0^2 - \beta^2, \tag{12.4}$$

in which k_0 is the free-space wavenumber that is equal to $\frac{2\pi}{\lambda}$, where λ is the optical wavelength. Also, β is the propagation constant and ω is the angular frequency and it is equal to $\frac{2\pi c}{\lambda}$, in which c is the speed of light in free-space.

Based on Maxwell's equations, one can establish an equation for the longitudinal component of the electric field (E_z), and a similar equation for the longitudinal component of the magnetic field (H_z). These two equations are referred to as the reduced wave equations for strip waveguides [18]:

$$\frac{\partial^2 E_z}{\partial x^2} + \frac{\partial^2 E_z}{\partial y^2} + K^2 E_z = 0, \tag{12.5}$$

$$\frac{\partial^2 H_z}{\partial x^2} + \frac{\partial^2 H_z}{\partial y^2} + K^2 H_z = 0. \tag{12.6}$$

The longitudinal field components of the modal electromagnetic field in the waveguide core can be defined as [16]:

$$E_z = P_1 \sin(k_x(x+\xi))\cos(k_y(y+\eta)), \tag{12.7}$$

$$H_z = P_2 \cos(k_x(x+\xi))\sin(k_y(y+\eta)), \tag{12.8}$$

where P_1 and P_2 are the amplitudes, k_x and k_y are the spatial frequencies, and ξ and η are the spatial shifts. For a symmetric strip waveguide the spatial shifts are small and can be ignored. Applying (12.7) and (12.8) to (12.2) and (12.3), and considering the wave equations in (12.5) and (12.6), the propagation constant in the waveguide core is calculated as:

$$\beta = \sqrt{n_{cr}^2(\lambda)k_0^2 - k_x^2 - k_y^2}. \tag{12.9}$$

Calculating the spatial frequencies k_x and k_y, one can use the following eigenvalue equations obtained by applying boundary conditions (continuity of the fields) at the interfaces between the waveguide core and the cladding/substrate in Figure 12.1 [16, 18]:

$$EV_1(\lambda, W) = \tan(k_x W) - \frac{n_{cr}^2(\lambda)k_x n_{c/s}^2(\gamma_1 + \gamma_2)}{n_{c/s}^4 k_x^2 - n_{cr}^4(\lambda)\gamma_1\gamma_2}, \tag{12.10}$$

$$EV_2(\lambda,T) = \tan(k_y T) - \frac{k_y(\gamma_3 + \gamma_4)}{k_y^2 - \gamma_3\gamma_4}. \qquad (12.11)$$

In these equations, $T = t \pm \rho_t$ and $W = w \pm \rho_w$, in which ρ_t and ρ_w are defined to take into account the variations in the silicon thickness and waveguide width, respectively. Moreover, we have:

$$\gamma_{1/2}^2 = \left(n_{cr}^2(\lambda) - n_{c/s}^2\right)k_0^2 - k_x^2, \qquad (12.12)$$

$$\gamma_{3/4}^2 = \left(n_{cr}^2(\lambda) - n_{c/s}^2\right)k_0^2 - k_y^2. \qquad (12.13)$$

Finally, the effective index of a strip waveguide under silicon thickness and waveguide width variations, and when the optical wavelength is λ is defined as:

$$n_{eff}(T,W,\lambda) = \frac{\beta}{k_0} = \frac{\lambda}{2\pi}\sqrt{n_{cr}^2(\lambda)k_0^2 - k_x^2(T,W,\lambda) - k_y^2(T,W,\lambda)}. \quad (12.14)$$

It is worth mentioning that based on (12.10) and (12.11), k_x and k_y depend on the waveguide thickness and width as well as the optical signal wavelength in (12.14). Considering both the material and waveguide dispersion and the effective index definition in (12.14), the group index, n_g, in a strip waveguide under different variations can be defined as:

$$n_g(T,W,\lambda) = n_{eff}(T,W,\lambda) - \lambda\frac{dn_{eff}(T,W,\lambda)}{d\lambda}. \qquad (12.15)$$

12.2.2 MR-Based Add-Drop Filters and Switches under Process Variations

Leveraging the analytical models in the previous section, this section studies the impact of silicon thickness and waveguide width variations on different properties of MR-based add-drop filters and switches, which are the primary building blocks in passive ONoCs. Figure 12.2 illustrates an add-drop filter, in which an optical signal on the wavelength λ_3 can be added (Figure 12.2(a)) or dropped (Figure 12.2(b)). When the MR is on resonance (i.e., the optical wavelength of the signal on the input port equals the resonant wavelength of the MR), it can switch/drop the optical signal from the input to the drop port, while the optical signal passes the MR towards the through port when the MR is off resonance. Process variations, however, deviate the resonant wavelength of MRs, and hence they cannot switch/drop an optical signal

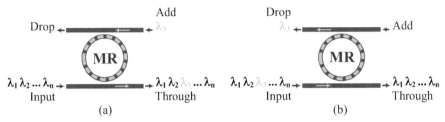

Figure 12.2   Different functions of an MR-based add-drop filter: (a) add an optical wavelength; and, (b) drop an optical wavelength.

at the desired optical wavelength. Consequently, such variations impose power loss and crosstalk in MR-based add-drop filters when switching optical signals, necessitating active tuning of faulty devices that imposes high power penalties.

12.2.2.1 Resonant wavelength shift in MRs

Figure 12.3(a) indicates the structure of an MR-based add-drop filter. As can be seen, a portion of the input optical signal couples to the MR with a cross-over coupling coefficient of κ, and it eventually couples to the drop waveguide through the same coupling coefficient. Similarly, a portion of the optical signal, which is uncoupled, continues propagating towards the

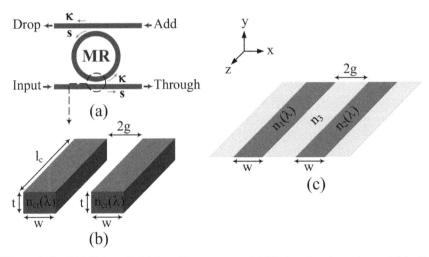

Figure 12.3   (a) MR-based add-drop filter structure; (b) 3D directional coupler model for the coupling region in MR-based add-drop filters; and, (c) 2D approximation of the coupler.

through port and inside the MR with an straight-through coefficient of *s*. Note that the input and drop waveguides are symmetrically coupled to the MR (i.e., symmetric MR-based add-drop filter).

An MR is on resonance when the round-trip optical phase, ϕ_{rtp}, in the MR is an integer multiple of 2π [19]:

$$\phi_{rtp}(T, W, \lambda_{MR}) = \frac{2\pi n_{eff}(T, W, \lambda_{MR}) L_{rtp}(T, W)}{\lambda_{res}} = m2\pi. \qquad (12.16)$$

Here, T and W, respectively, indicate the thickness and width of the MR's waveguide under variations. We assume that the different dimensions as well as the variations in the input waveguide and the MR are the same. Also, m is an integer number that denotes the order of the resonant mode, and λ_{res} is the mth-order resonant wavelength of the MR. Also, L_{rtp} is the round-trip length of the MR and can be defined as $2\pi r(T, W) + 2l_c$, where $r(T, W)$ is the MR's radius under variations and l_c is the coupler length (see Figure 12.3). It is worth mentioning that the effective and group indices of the MR can be calculated using the input waveguide.

As discussed before, process variations deviate the resonant wavelength of MRs. Considering the first order approximation of the waveguide dispersion, (12.15), and (12.16), the resonant wavelength shift in MRs can be defined as [19]:

$$\Delta\lambda_{res}(T, W, \lambda_{res}) = \frac{\Delta_{\rho_{t/w}} n_{eff} \cdot \lambda}{n_g(t, w, \lambda)}. \qquad (12.17)$$

In this equation, λ_{res} is the initial resonant wavelength in the MR. Also, $\Delta_{\rho_{t/w}} n_{eff}$ denotes the effective index changes due to the variations in the thickness or waveguide width, which can be calculated as:

$$\Delta_{\rho_{t/w}} n_{eff} = \left| n_{eff}(T, W, \lambda_{res}) - n_{eff}(t, w, \lambda_{res}) \right|. \qquad (12.18)$$

12.2.3 Optical Spectra of MRs under Process Variations

This section studies the impact of process variations on the optical spectra of MR-based add-drop filters and switches. The variations in the silicon thickness and waveguide width impact the power transmissions from the input port towards the drop and through ports in MR-based add-drop filters and switches. In particular, when variations exist, such devices transmit the input power to the drop and through ports with extra power loss, or, in the worst-case, an optical signal on a neighboring channel (i.e., different wavelength) couples into the MR, whose resonant wavelength is now shifted towards that channel (i.e., crosstalk).

We start by analyzing the coupling coefficients, κ and s in Figure 12.3(a), in MR-based add-drop filters and switches under variations. The coupling region in Figure 12.3(a) is modeled as a 3D directional coupler (DC) in Figure 12.3(b). As can be seen, the coupling region consists of two identical strip waveguides of a length l_c (i.e., the coupler length) which are in proximity, divided with a gap of $2g$. Note that the variations in the waveguide width alter the gap, affecting the coupling coefficients that depend on the coupler length and the gap. Considering Figure 12.3(b), we analytically calculate the fraction of the optical power (i.e., κ^2) that couples from one waveguide to the other one, as well as the rest of that power which remains in the first waveguide (i.e., s^2) based on the supermode theory [20, 21]. Based on this theory, the effective indices of the first two eigenmodes of the coupled waveguides, known as symmetric and antisymmetric modes, determine the coupling coefficients in the MR. Given that the effective index of the symmetric mode is n_e and the effective index of the antisymmetric mode is n_o, the cross-over coupling coefficient is calculated as:

$$\kappa(T, W, \lambda) = \left| sin \left(\frac{\pi}{2L_c(T, W, \lambda)} l_c \right) \right|, \tag{12.19}$$

in which L_c, which is the cross-over length, can be defined as:

$$L_c(T, W, \lambda) = \frac{\lambda}{2(n_e(T, W, \lambda) - n_o(T, W, \lambda))}, \tag{12.20}$$

where $n_e > n_o$. We assume a lossless coupler, and hence $|\kappa|^2 + |s|^2 = 1$, but the optical losses of the coupler are included in the round-trip loss of the entire optical cavity.

Calculating n_e and n_o, one can approximate the 3D DC structure in Figure 12.3(b) with the 2D structure shown in Figure 12.3(c). In this figure, $n_1(\lambda) = n_2(\lambda)$ is the effective index of the slab waveguide with a thickness T in the y direction in Figure 12.3(a). Furthermore, we consider $n_3 = 1.444$ (i.e., refractive index of SiO$_2$ at the optical wavelength of 1550 nm). Finally, using the following eigenvalue equation, we can calculate the effective index of the antisymmetric supermode:

$$EV_3(\lambda, W, g) = tan(2M_1) - \frac{M_1 M_2 \left(1 + coth(2M_2 gW^{-1})\right)}{M_1^2 - M_2^2 coth\left(2M_2 gW^{-1}\right)}, \tag{12.21}$$

in which $2M_1 = k_0 W \sqrt{n_{1/2}^2(\lambda) - (\beta/k_0)^2}$ and $2M_2 = k_0 W \sqrt{(\beta/k_0)^2 - n_3^2}$. Also, replacing coth by tanh gives us the effective index of the symmetric mode [22].

Considering (12.19) and the time-domain coupling theory [23, 24], the transmission from the input port to the through port in MR-based add-drop filters and switches under the variations in the silicon thickness and waveguide width can be calculated as:

$$Tr(T,W,\lambda) = \frac{s(T,W,\lambda) - s^*(T,W,\lambda)\sqrt{A}e^{i\phi_{rtp}}}{1 - \sqrt{A}s^{*2}(T,W,\lambda)e^{i\phi_{rtp}}}, \qquad (12.22)$$

while the power on the drop port is given by:

$$Dr(T,W,\lambda) = \frac{-\kappa^*(T,W,\lambda)\kappa(T,W,\lambda)A^{1/4}e^{\frac{i\phi_{rtp}}{2}}}{1 - \sqrt{A}s^{*2}(T,W,\lambda)e^{i\phi_{rtp}}}. \qquad (12.23)$$

In these equations, $*$ is the complex conjugation. Also, A is the power attenuation that can be calculated as:

$$A(T,W) = e^{-L_p L_{rtp}(T,W)}, \qquad (12.24)$$

in which L_p is the propagation loss of the waveguide in dB/cm. It is worth mentioning that the total round-trip phase, ϕ_{rtp} calculated in (12.16), includes the phase accumulated in the light propagating in the coupler. Moreover, $|\kappa|^2 + |s|^2 = 1$.

12.3 Optical Networks-on-Chip under Process Variations

Leveraging the analytical models in the previous section, we study the impact of process variations on the performance of ONoCs (i.e., system-level). In particular, we study the optical SNR in ONoCs under variations in the silicon thickness and waveguide width. We first explore the impact of different variations on WDM-based optical switches based on MRs, which are widely employed in constructing WDM-based ONoCs. Then, we apply the analytical models for optical switches to study the signal power, crosstalk power, and OSNR in passive WDM-based ONoCs at the system-level.

12.3.1 Process Variations in Optical Switches

Figure 12.4 depicts an MR-based optical switch consisting of n MRs to simultaneously switch n different optical wavelengths. The difference between the

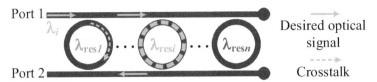

Figure 12.4 An MR-based optical switch consisting of n MRs to simultaneously switch n different wavelengths. In this figure $\lambda_i = \lambda_{resi}$.

resonant wavelengths of two consecutive MRs is called the channel spacing. Also, in Figure 12.4, each MR's radius is slightly different to match the associated resonant wavelength of the MR. As can be seen, the optical signal on the wavelength λ_i enters Port 1 in the optical switch, then it passes the MRs whose resonant wavelengths are different from λ_i, and it finally drops into the MR with the resonant wavelength $\lambda_i = \lambda_{resi}$. All along this path, the optical signal suffers from some power loss and, more importantly, a portion of the optical signal couples into the MRs with different resonant wavelengths (e.g., λ_{res1} in Figure 12.4), which interferes with the desired optical signal on Port 2 as crosstalk. As discussed before, process variations further impacts the power loss and crosstalk in optical switches: MRs whose resonant wavelengths are shifted drop an optical signal with extra power loss, and in the worst-case, fail to drop a desired optical signal and instead drop a signal on a neighboring channel (e.g., λ_i drops into the first MR in Figure 12.4). Here, we consider coherent in-band crosstalk (i.e. when the optical crosstalk noise is at the same wavelength as the desired optical signal) which is of critical concern because it cannot be removed by filtering [25].

Analyzing the power loss and crosstalk in passive optical switches, we can use the analytical models in (12.22) and (12.23). For example, considering the variations in the silicon thickness and waveguide width, and when the optical signal on the wavelength λ_i enters Port 1 in the optical switch in Figure 12.4, the power loss imposed on the optical signal received at Port 2 in the switch can be calculated as [26]:

$$L_{sw}(T,W,\lambda_i) = Tr(T,W,\lambda_i,MR_j)^{2(i'-1)}Dr(T,W,\lambda_i,MR_{i'})L_p^{d(i')}. \quad (12.25)$$

Here, $i = i'$ when there are no variations, while i can be different from i' when variations are introduced (i.e., the optical signal on the wavelength λ_i can drop into $MR_{i'}$). Also, the propagation loss, L_p^d, can be calculated by considering the approximate distance (i.e., d) that the optical signal has traveled in the optical switch. For the optical signal on the wavelength λ_i that passes MRs with λ_{resj}, and then is dropped to the MR with $\lambda_{resi'}$, d can be estimated as $d(i') = 2r_j(2i'+1) + 2i'l_c$ and $1 \leq j \leq i'$, where r_j is the radius of the MR_j. It is worth mentioning that the fourth input parameter in $Tr(T,W,\lambda_i,MR_j)$ indicates the MR number from which the optical signal passes. Similarly, $MR_{i'}$ in $Dr(T,W,\lambda_i,MR_{i'})$ denotes the MR number to which the optical signal drops. The coherent crosstalk noise interfered with the desired optical signal on the wavelength λ_i entered Port 1 in Figure 12.4 can be defined as [27]:

$$C_{sw}(T,W,\lambda_i) = \sum_{k=1}^{i'-1} \left(Tr(T,W,\lambda_i,MR_{1:k-1})^{2(k-1)} Dr(T,W,\lambda_i,MR_k) \right) L_p^{d(k)},$$

(12.26)

in which we consider the worst-case scenario where the coherent crosstalk noise from each MR can be added together [25].

12.3.2 Process Variations at the System-Level in ONoCs

Leveraging the power loss and crosstalk analytical models for optical switches, we model the desired signal power and crosstalk noise power in passive WDM-based ONoCs. At the system-level, we consider a case study of a general passive WDM-based ONoC, indicated in Figure 12.5, which can be applied to different ONoC architectures. The general ONoC has m switching stages where each stage includes n MRs (i.e. $m \times n$ MRs in total), supporting n different wavelengths. Therefore, using the proposed ONoC, n different processing cores can communicate with another n processing cores ($n \times n$). Optical terminators (e.g., waveguide tapers) are employed to avoid the optical signal reflecting back on the waveguides. As indicated in Figure 12.5 and as an example, the source processor core P_a can communicate with the destination processor core P_b using the optical wavelength λ_i. The electrical-optical (E-O) interface is responsible for modulating the optical signal, where we consider a modulator for each processor (n modulators in total, not shown in Figure 12.5 for simplicity). When the optical signal on the wavelength λ_i is equal to the resonant wavelength of the MR with λ_{resi} (i.e., $\lambda_i = \lambda_{resi}$ in Figure 12.5), it couples to that MR and proceeds to the next switching stage (m switching stages in total), while it passes the MRs whose resonant wavelengths are different. Ultimately, it is detected at the photodetector PD_i located at the end of the communication line in the optical-electrical (O-E) interface associated with the destination processor core (n photodetectors in total). Note that the proposed general architecture can be applied to any specific passive ONoC: in any architecture, the optical signal is routed through

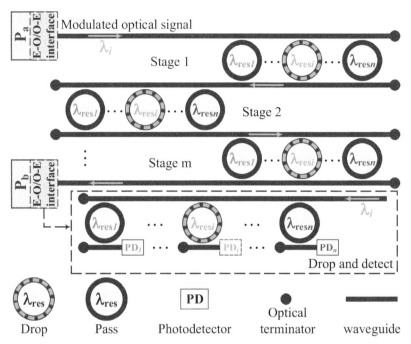

Figure 12.5 An overview of the general passive WDM-based ONoC supporting *n* wavelengths. It has *m* switching stages with *n* MRs at each stage. All the functional devices utilize MRs, including wavelength-selective photodetectors and photonic switches.

a number of MRs ($m \times n$ MRs in our general model), and it is ultimately detected at a photodetector (*n* photodetectors in our general model) located at the end of the communication line. As a result, the models developed in this section can be applied to study process variations in different optical interconnect architectures [4–6, 28–30]. According to Figure 12.5 and for simplicity, the analysis of the power loss and crosstalk for an optical signal starts from the output of the E-O interface at the source processor core (i.e., modulated optical signal in the figure), and ends before the optical signal is detected at a photodetector in the O-E interface at the destination processor core.

Considering the desired signal power received at the photodetectors in Figure 12.5, an optical signal on the wavelength λ_i passes the MRs with $\lambda_i \neq \lambda_{resj}$, couples into the MRs with $\lambda_i = \lambda_{resi'}$, dropped through the MR with $\lambda_i = \lambda_{resi'}$ in the drop and detect section in Figure 12.5, and finally is detected by the photodetector PD_i ($1 \leq i, i', j \leq n$). This communication can be seen as a communication between two processing cores in an MPSoC. Considering (12.25), the desired signal power received at the photodetector PD_i under different variations can be defined as:

$$S_p(T,W,\lambda_i) = P_{in}L_{sw}(T,W,\lambda_i)^m L_{PD}(T,W,\lambda_i), \qquad (12.27)$$

in which P_{in} is the laser output optical power in the E-O interface in Figure 12.5. Also, L_{sw} denotes the switching power loss at each stage, and $L_{PD}(T,W,\lambda_i)$ is the power loss associated with the photodetector PD_i (i.e., the power loss imposed on the optical signal on the wavelength λ_i that passes several MRs and is dropped towards PD_i in the drop and detect section in Figure 12.5), which can be calculated as:

$$L_{PD}(T,W,\lambda_i) = Tr(T,W,\lambda_i,MR_j)^\alpha Dr(T,W,\lambda_i,MR_{i'})L_p^{d'}. \qquad (12.28)$$

In this equation, $\alpha = (m \mod 2)(n-i') + ([m+1] \mod 2)(i'-1)$, and $d' = (\alpha+1)(2r_j+l_c)+2r_j$, which is the estimated optical path length traveled by the optical signal to reach PD_i (see the drop and detect section in Figure 12.5). Also, we have $1 \leq j \leq n$. Similarly, the coherent crosstalk noise power, C_P, interfered with the desired optical signal on the wavelength λ_i and received at the photodetector PD_i under the variations in the silicon thickness and waveguide width can be defined as:

$$C_p(T,W,\lambda_i) = P_{in} \sum_{l=1}^{m} (L_x(T,W,\lambda_i,l)C_{sw}(T,W,\lambda_i,l)) \qquad (12.29)$$

where $L_x(T,W,\lambda_i,l)$ is the power loss associated with the crosstalk signal at the switching stage l ($1 \leq l \leq m$), and it can be calculated similar to (12.25). Also, $C_{sw}(T,W,\lambda_i,l)$ is the crosstalk noise accumulated on the desired optical signal at the switching stage l that can be calculated based on (12.26). Finally, leveraging (12.27) and (12.29), we can define the optical SNR (in dB) as the ratio between the desired optical signal power on the wavelength λ_i and the in-band crosstalk noise power corrupting that signal received at the photodetector PD_i (before detection) as:

$$OSNR(T,W,\lambda_i) = 10\log_{10} \frac{S_p(T,W,\lambda_i)}{C_p(T,W,\lambda_i)}. \qquad (12.30)$$

12.4 Quantitative Simulation Results and Evaluations

This section presents the quantitative simulation results of the analytical models developed in the previous sections at the component-, device-, and system-level. All of the analytical models proposed in this chapter were implemented in MATLAB. Furthermore, we perform numerical simulations in MODE, which is a commercial-grade simulator eigenmode solver and propagator developed by Lumerical [31], to evaluate our proposed method. The original dimensions of the strip waveguides (i.e., before variations) are considered to be $w = 500$ nm and $t = 220$ nm (see Figure 12.1) in this section. Furthermore, we consider the central laser wavelength and the gap (i.e., $2g$ in Figure 12.3(b)) to be 1550 nm and 200 nm, respectively. Also, we assume that $L_p = 2$ dB/cm [32]. The results are obtained for the fundamental TE mode.

12.4.1 Simulation Results at the Component- and Device-level

We first present the quantitative simulation results of the proposed models at the component- and device-level. In this section, a variation range of ± 20 nm (i.e. $\rho_{t/w} \in [-20, 20]$ nm) is considered in our simulations, which is large enough to include possible variations in real fabrications. In order to better indicate the impact of each variation on different properties of strip waveguides and add-drop filters, we first consider applying each variation separately (i.e., ρ_t or ρ_w) in the simulations in this section.

Considering (12.14) and (12.15), Figure 12.6 indicates the effective and group indices of a strip waveguide under silicon thickness and waveguide width variations. In this figure, the waveguide thickness varies from 200 nm to 240 nm, while the waveguide width changes from 480 nm to 520 nm. As can be seen, the x-axis denotes the variation range for both the silicon thickness and waveguide width (i.e., $\rho_{t/w}$). When $\rho_{t/w}$ increases, the effective index in Figure 12.6(a) increases, while the group index shown in Figure 12.6(b) remains almost the same when the thickness variations (ρ_t) increase, and it slightly decreases as the width variations (ρ_w) increase. The group indices changes under the variations in the thickness and waveguide width can be described based on (12.15) and by considering the effective indices variations in Figure 12.6(a). Also, considering Figure 12.6, one can notice that the impact of silicon thickness variations on the effective and group indices is more important than that of waveguide width variations, which is in agreement with the demonstrations in [7–9, 33]. Furthermore, comparing the numerical simulations results with those from our proposed method, we observe a high accuracy of the proposed method with an average error rate smaller than 1%.

Employing (12.17), the resonant wavelength shift in passive MRs under the silicon thickness and waveguide width variations is indicated

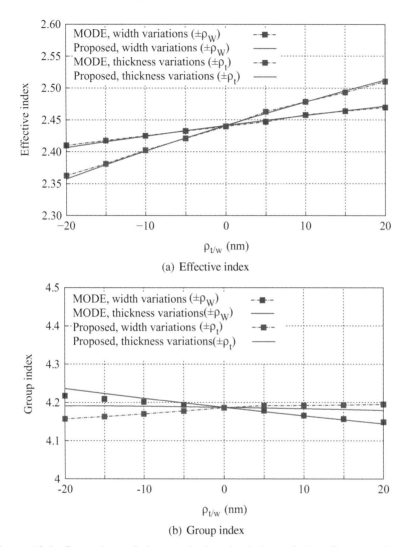

(a) Effective index

(b) Group index

Figure 12.6 Comparison of the quantitative simulations of the effective and group indices in a strip waveguide calculated using MODE (dashed lines with squares) and the proposed method (solid lines) under silicon thickness and waveguide width variations $(\rho_{t/w} \in [-20, 20]$ nm).

in Figure 12.7(a). Note that the resonant wavelength shift in this figure is independent of the MR's radius, and the initial resonant wavelength (i.e., when $\rho_{t/w} = 0$) is at 1550 nm. As can be seen, the resonant wavelength shifts almost linearly with respect to the variations in the silicon

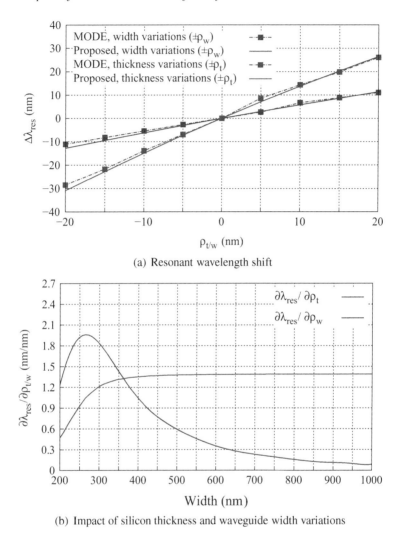

(a) Resonant wavelength shift

(b) Impact of silicon thickness and waveguide width variations

Figure 12.7   (a) Resonant wavelength shift in passive MRs with variations in the silicon thickness and waveguide width; and, (b) the impact of thickness and waveguide width variations on the resonant wavelength in passive MRs when the MR's waveguide width varies from 200 nm to 1000 nm.

thickness or waveguide width. In particular, when $\rho_{t/w} < 0$, there is a blue-shift in the resonant wavelength, while there is a red-shift when $\rho_{t/w} > 0$. Moreover, the impact of silicon thickness variations is more severe compared with that of the waveguide width variations. In the worst-case, when

$\rho_{t/w} \in [-20, 20]$ nm, the silicon thickness variation shifts the resonant wave-length of the MR by approximately 60 nm, while the waveguide width variations shift the resonant wavelength by approximately 20 nm. To better illustrate and distinguish between the impact of each variations on the resonant wavelength of passive MRs, we indicate the resonant wavelength shift slopes (i.e., $\frac{\partial \lambda_{res}}{\partial \rho_t}$ and $\frac{\partial \lambda_{res}}{\partial \rho_w}$) with respect to the variations in the silicon thickness and waveguide width while considering waveguides of different widths (i.e., $200 \le w \le 1000$) in Figure 12.7(b). Note that the waveguide thickness is determined by the SOI thickness on the wafer, which equals 220 nm in this work, and hence we do not consider waveguides of different thicknesses in Figure 12.7(b). As can be seen, at $w = 500$ nm, which is considered as the original waveguide width in this chapter, the impact of silicon thickness variations on the resonant wavelength shift in MRs is much higher than that of the waveguide width variations. Also, one can notice that as the waveguide width increases, the variations in the silicon thickness contributes the most to the resonant wavelength shift in MRs (i.e., wider MRs are more tolerant against waveguide width variations [34]).

Considering (12.19), Figure 12.8(a) indicates the cross-over coupling coefficient in a coupler with a length of 10 μm (i.e., $l_c = 10$ μm) in an MR-based add-drop filter (see Figure 12.3(a)) under variations in the silicon thickness and waveguide width. As can be seen, the cross-over coupling coefficient increases slightly as ρ_w increases (i.e., more power can be coupled as the waveguides get wider), and it slightly decreases as ρ_t increases. As the waveguide width increases, the gap between the input waveguide and MR ($2g$ in Figure 12.3(b)) decreases, and hence more power can be coupled as the waveguides are closer. The gap does not vary with the variations in the silicon thickness. Based on (12.22) and (12.23), Figure 12.8(b) indicates the through and drop ports responses of an MR-based add-drop filter under $\rho_t = \pm 10$ nm, in which $r \approx 9$ μm and $l_c = 4$ μm are used to better indicate the shifts. We excluded the numerical simulation results in the figure for a better illustration. Nevertheless, the results indicate an average error rate smaller than 1%. The initial resonant wavelength, λ_{res}, is 1550 nm, and the free spectral range (FSR) is approximately 9 nm. When $\rho_t > 0$, there is a red-shift in the resonant wavelength, while $\rho_t < 0$ results in a blue-shift in the resonant wavelength. Moreover, as indicated in Figure 12.8(b), the resonant wavelength shift is larger than the FSR (i.e., $\Delta \lambda_{res} >$ FSR). Another important observation is that there is a good agreement between the results in Figure 12.7(a), which is based on (12.17), and the one indicated in Figure 12.8(b).

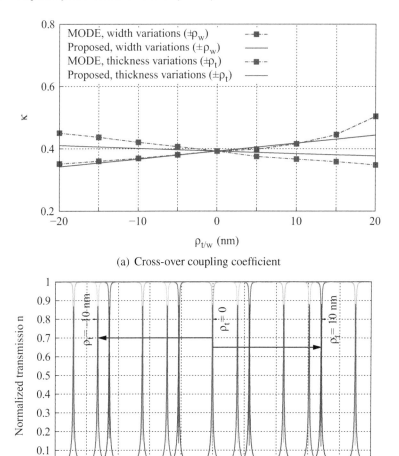

(a) Cross-over coupling coefficient

(b) Optical spectrum of an MR-based add-drop filter when $\rho_w = 0$

Figure 12.8   (a) Cross-over coupling coefficient under the variations in the silicon thickness and waveguide width; and, (b) Optical spectrum of an MR-based add-drop filter with $\rho_t = \pm 10$ nm and $\rho_w = 0$, calculated using the proposed method ($r \approx 9\ \mu$m and $l_c = 4\ \mu$m).

Compared with the numerical simulations results from MODE [31] presented for the component- and device-level in this section, our proposed method demonstrates a high accuracy with an average error rate smaller than 1%. Particularly, one of the major advantages of using the proposed method is its low-computation cost. Performing all the quantitative simulation results in this section (more than 100 simulations), Table 12.1 compares the computation time of our proposed method with that of MODE on a PC computer with 2.66 GHz Intel Core i5 CPU and 8 GB of RAM. While MODE performed the simulations in more than two hours, the proposed method computed the results in less than a minute, hence more than 100 times faster than the numerical simulation. The high-accuracy and low-computation cost of our proposed method enables its application to study large-scale ONoCs under process variations, where employing time-consuming numerical simulations is not feasible.

12.4.2 Simulation Results at the System-level

This section presents the quantitative simulation results of the developed analytical models for the desired signal power, crosstalk power, and OSNR in ONoCs under process variations. We focus on the general ONoC architecture indicated in Figure 12.5. Our system-level simulations consider simultaneous random silicon thickness and waveguide width variations with standard deviations of $\sigma_t = 1$ nm and $\sigma_w = 5$ nm, respectively. Similar to the previous section, the central laser wavelength equals 1550 nm. Also, considering the network in Figure 12.5, we consider MRs of radii 4 μm with couplers with a length of 4 μm, and hence an FSR of approximately 17 nm. Moreover, a channel spacing of 2 nm is considered. It is worth mentioning that each MR's radius is slightly different to cover the whole wavelength range. We assume that the laser output optical power in the E-O interface equals 0 dBm.

Considering (12.27) and (12.29), Figures 12.9(a) and 12.9(b) indicate the desired signal and crosstalk noise power received (i.e., before detection) at each photodetector (x-axis) in the passive WDM-based ONoC in Figure 12.5, in which we consider $m = 4$ and $n = 4$. With no process variations (i.e., $\sigma_{t/w} = 0$), as indicated in Figure 12.9(a), the desired signal power received at different photodetectors is much higher than the crosstalk noise power. On average, the received signal power is –4.2 dBm, while the crosstalk noise power equals –26.5 dBm. When the variations are introduced (i.e., $\sigma_t = 1$ nm

Table 12.1 Computation time comparison between the numerical simulation (MODE [31]) and our proposed method

Method	Computation Time	Average Error Rate
Numerical (MODE)	128 minutes	–
Proposed	54 seconds	1%

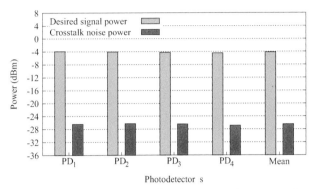

(a) Desired signal and crosstalk noise power when $\sigma_{t/w} = 0$

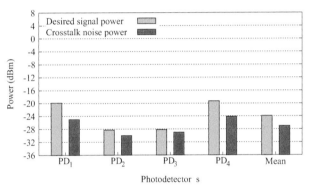

(b) Desired signal and crosstalk noise power when $\sigma_{t/w} \neq 0$

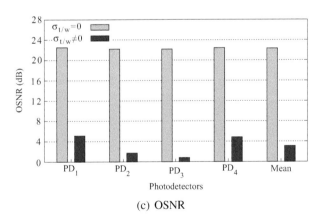

(c) OSNR

Figure 12.9 (a) and (b): Desired signal and crosstalk noise power received at different photodetectors in the passive WDM-based ONoC in Figure 12.5 when $m = 4$ and $n = 4$; and, (c): the OSNR in the ONoC.

and $\sigma_w = 5$ nm), however, both the received desired signal power and crosstalk noise power decrease. When the silicon thickness and waveguide width vary, the resonant wavelengths of the MRs shift. Consequently, the MRs fail to precisely switch optical signals at the desired optical wavelengths at each stage, and hence the received signal power at each photodetector reduces considerably while some crosstalk will accumulate on the desired optical signal through different MRs at different stages. Nevertheless, the accumulated crosstalk on the desired optical signal also deviates, and hence the received crosstalk noise power is slightly reduced at some of the photodetectors. As Figure 12.9(b) indicates, on average, the desired signal power and crosstalk noise power are -23.9 and -27.1 dBm, respectively.

Looking at the OSNR in the system under variations and employing (12.30), Figure 12.9(c) indicates the OSNR of the network when there are no variations (i.e. $\sigma_{t/w} = 0$), and when the variations exist (i.e. $\sigma_{t/w} \neq 0$). As can be seen, the OSNR drops when the variations are introduced. On average, the OSNR is equal to 22.3 dB when $\sigma_{t/w} = 0$, and it reduces to 3.1 dB when $\sigma_t = 1$ nm and $\sigma_w = 5$ nm. Considering the results in Figure 12.9, when $\sigma_{t/w} \neq 0$ and assuming a heater efficiency of 0.8 mW/FSR [35], an average power consumption overhead of 3.2 mW (tuning power) should be considered to align the resonant wavelength of each MR in our ONoC case study. As indicated in this section, fabrication process variations can severely degrade the OSNR of ONoCs.

12.5 Chip Fabrication and Measurement Results

This section presents our fabrication details and its results on process variations in silicon photonic fabrications, as well as exploring different characteristics of the variations (i.e., correlations among different devices with respect to their physical distances on the chip) in the fabrication. An MR-based all-pass filter is designed and considered as a test-structure in this section [36]. The designed MR consists of 220 nm thick SOI strip waveguides with a 500 nm width connected to a TE polarization MR with a 10 μm radius, and a coupler length and gap of 1 μm and 200 nm, respectively. The chip is 8.8×8.8 mm and was fabricated by the Electron Beam (EBeam) Lithography System at the University of Washington.[1] It includes sixty identical copies of the designed MR, all of which occupy 2.1×4.5 mm on

---

[1]The fabrication was performed through Silicon Electronic-Photonic Integrated Circuits program (SiEPIC).

the chip. The MRs were separated by a distance between 60 μm and 4.2 mm. Thirty MRs were placed at $y = 1.5$ mm near the edge of the chip, and the remaining ones were placed at $y = 5.7$ mm near the center of the chip (see Figure 12.10(b)).

Using an automated probe station at the University of British Columbia [37], all the MRs were carefully characterized. The chip was located on a thermal heater to eliminate the impact of thermal variations. The resonant wavelength shift was found to be smaller than the FSR of the MR. Figure 12.10(a) indicates the measured results obtained by automatically testing all the MRs (60 in total). For easier comparison, this figure also indicates the through port response of the MR around the optical wavelength 1550 nm calculated using MODE when $\rho_{t/w} = 0$. As the figure indicates, although all the MRs are identically designed, there is a variation in the resonant wavelengths of the MRs placed at different locations of the chip (see the dashed lines in Figure 12.10(a)). Comparing the fabrication and simulation results, we can see that, in the worst-case, there is a 2.1 nm shift in the resonant wavelength of the MR. Comparing all the MRs within the same distance, we found that the differences in the resonant wavelengths increase with the distance between the MRs. The same conclusion was demonstrated in [8, 10]. Figure 12.10(b) indicates the resonant wavelength shift (i.e., $\Delta\lambda_{resi} = \lambda_{resi} - 1550$ nm for i $\in[1,60]$) versus the physical position of the MRs on the chip (i.e., x and y). As can be seen from our experimental results, compared to the MRs located close to the edge of the chip, the average resonant wavelength shift for the MRs located close to the chip center is smaller by 400 pm [27]. Figure 12.10(b) also indicates the average, μ, and the standard deviation, σ, associated with the resonant wavelength shift: $\mu = 1.6$ nm and $\sigma = 0.2$ nm for the MRs located close to the chip edge, while $\mu = 1.2$ nm and $\sigma = 0.2$ nm for the MRs close to the chip center. This indicates that the top silicon thickness and the waveguide width should be more uniform close to/in the center of our fabricated chip.

Figure 12.10(b) also indicates that the resonant wavelength shift among those MRs which are in proximity (e.g., those located at the edge of the chip or those located at the center) are similar. Further exploring this similarity and indicating that this kind of correlation can exist in other fabricated chips, we compare the resonant wavelength shift differences of each MR pair on another chip, which includes thirty copies of our designed MR, fabricated using the same EBeam facility. In total, 435 comparisons were made (i.e., 30 choose 2, each MR is compared with the other 29 MRs). Figure 12.11 indicates the difference in the resonant wavelengths of each MR pair with

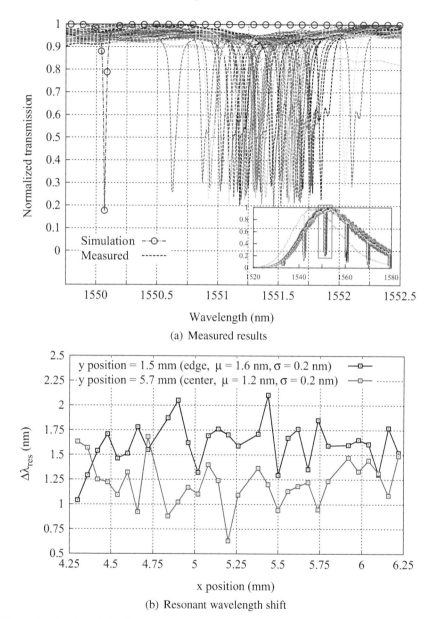

Figure 12.10 (a) Fabrication results obtained by automatically testing all the MRs, as well as the simulation result from MODE when $\rho_{t/w} = 0$; and, (b) resonant wavelength shift versus the physical position of the MRs (x and y) when the original resonant wavelength is at 1550 nm. μ and σ, respectively, denote the average and the standard deviation associated with the resonant wavelength shift.

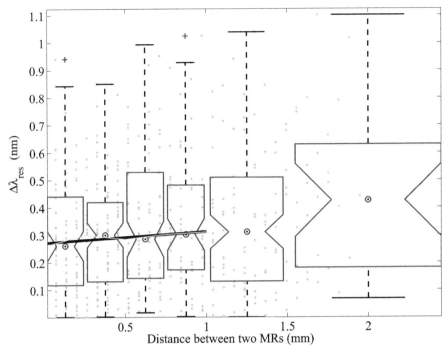

Figure 12.11   Difference in the resonant wavelengths of each MR pair with respect to their distance, where 435 comparisons are made (pink data points).

respect to their distance on the chip.[2] As can be seen, the resonant wavelength shift increases slightly as the physical distance between two MRs increases (i.e., spatial correlation). In particular, the resonant wavelength shift median (the circles in the boxplots) at the considered locations increases with an increase in the distance between two MRs. In other words, those MRs which are in proximity experience almost similar variations, and hence the shift in their resonant wavelength is similar, while this shift is much different for MRs which are not in proximity on the chip. Employing the developed computationally efficient and accurate analytical method in the previous sections and considering a fabricated MR, one can repeatedly simulate the resonant wavelength of the fabricated MR using various values for ρ_t and ρ_w, estimating the silicon thickness and waveguide width variations associated with the measured resonant wavelength.

---

[2]This figure is generated using a MATLAB source code from [8] provided by Dr. Lukas Chrostowski from the University of British Columbia.

12.6 Conclusion

Optical interconnection networks can outperform the inter- and intra-chip communication in MPSoCs by realizing a low-latency and high-bandwidth communication within a reasonable power consumption. Such systems, however, are sensitive to fabrication process variations that cause severe performance degradation in the system. We develop a bottom-up approach to study the impact of fabrication process variations on large-scale passive optical networks-on-chip with high accuracy and computational efficiency. We study the impact of process variations on the fundamental optical components in ONoCs (i.e., waveguides), as well as optical devices (i.e., MR-based optical filters and switches). At the system-level, we consider a general passive WDM-based ONoC, which can be applied to other ONoC architectures, and we study the optical SNR in this system under random process variations. Our results indicate a considerable reduction in the OSNR of the system due to the variations (approximately 20 dB on average). Moreover, our study includes the design, fabrication, and measurement of several MRs (as test-structures) to indicate process variations in silicon photonic fabrications, as well as exploring different characteristics of the variations. For example, we found that the MRs which are in proximity experience similar shift in their resonant wavelengths (i.e., spatial correlation). The high efficiency of the proposed method allows its integration into different silicon photonics design tools, enabling the real-time performance evaluation of ONoCs under process variations.

References

[1] E. Bonetto, L. Chiaraviglio, D. Cuda, G. A. Gavilanes Castillo, and F. Neri. Optical technologies can improve the energy efficiency of networks. In *35th European Conference on Optical Communication*, pages 1–4, September 2009.

[2] Y. Arakawa, T. Nakamura, Y. Urino, and T. Fujita. Silicon photonics for next generation system integration platform. *IEEE Communications Magazine*, 51(3):72–77, March 2013.

[3] S. Rumley, M. Bahadori, R. Polster, S. D. Hammond, D. M. Calhoun, K. Wen, A. Rodrigues, and K. Bergman. Optical interconnects for extreme scale computing systems. *Parallel Computing*, pages 65–80, 2017. doi: 10.1016/j.parco.2017.02.001

[4] X. Wu, J. Xu, Y. Ye, Z. Wang, M. Nikdast, and X. Wang. SUOR: Sectioned undirectional optical ring for chip multiprocessor.

Journal of Emerging Technologies in Computing Systems, 10(4):29:1–29:25, June 2014.

[5] S. Koohi and S. Hessabi. All-optical wavelength-routed architecture for a power-efficient network on chip. *IEEE Transactions on Computers*, 63(3):777–792, March 2014.

[6] E. Fusella and A. Cilardo. H^2ONoC: A hybrid optical-electronic NoC based on hybrid topology. *IEEE Transactions on Very Large Scale Integration Systems*, (99):1–14, 2016.

[7] D. C. Trotter W. A. Zortman and M. R. Watts. Silicon photonics manufacturing. *Optical Express*, 18(23):598–607, 2010.

[8] L. Chrostowski, X. Wang, J. Flueckiger, Y. Wu, Y. Wang, and S. Talebi Fard, "Impact of Fabrication Non-Uniformity on Chip-Scale Silicon Photonic Integrated Circuits," in Optical Fiber Communication Conference, OSA Technical Digest (online) (Optical Society of America, 2014), paper Th2A.37.

[9] S. K. Selvaraja, W. Bogaerts, P. Dumon, D. Van Thourhout, and R. Baets. Subnanometer linewidth uniformity in silicon nanophotonic waveguide devices using CMOS fabrication technology. *IEEE Journal of Selected Topics in Quantum Electronics*, 16(1):316–324, January 2010.

[10] X. Chen, M. Mohamed, Z. Li, L. Shang, and A. R. Mickelson. Process variation in silicon photonic devices. *Applied Optics*, 52(31):7638–7647, November 2013.

[11] Y. Xu, J. Yang, and R. Melhem. Bandarb: Mitigating the effects of thermal and process variations in silicon-photonic network. In *Proceedings of the 12th ACM International Conference on Computing Frontiers*, CF '15, pages 30:1–30:8, New York, NY, USA, 2015. ACM.

[12] M. Mohamed, Z. Li, X. Chen, L. Shang, and A. R. Mickelson. Reliability-aware design flow for silicon photonics on-chip interconnect. *IEEE Transactions on Very Large Scale Integration Systems*, 22(8):1763–1776, August 2014.

[13] S. V. R. Chittamuru, I. G. Thakkar, and S. Pasricha. Pico: Mitigating heterodyne crosstalk due to process variations and intermodulation effects in photonic NoCs. In *Proceedings of the 53rd Annual Design Automation Conference*, DAC '16, pages 39:1–39:6, New York, NY, USA, 2016. ACM.

[14] M. Fiorentino, Z. Peng, N. Binkert, and R. G. Beausoleil. Devices and architectures for large scale integrated silicon photonics circuits. In *IEEE Winter Topicals 2011*, pages 131–132, January 2011.

[15] E. A. J. Marcatili. Dielectric rectangular waveguide and directional coupler for integrated optics. *Bell System Technical Journal*, 48(7):2071–2102, 1969.

[16] W. J. Westerveld, S. M. Leinders, K. W. A. van Dongen, H. P. Urbach, and M. Yousefi. Extension of Marcatili's analytical approach for rectangular silicon optical waveguides. *Journal of Lightwave Technology*, 30(14):2388–2401, July 2012.

[17] D. F. Edwards and E. Ochoa. Infrared refractive index of silicon. *Applied Optics*, 19(24):4130–4131, December 1980.

[18] A. Yariv and P. Yeh. *Photonics: optical electronics in modern communications*. Oxford University Press, 2007.

[19] W. Bogaerts, P. De Heyn, T. Van Vaerenbergh, K. De Vos, S. Kumar Selvaraja, T. Claes, P. Dumon, P. Bienstman, D. Van Thourhout, and R. Baets. Silicon microring resonators. *Laser and Photonics Reviews*, 6(1):47–73, January 2012.

[20] L. Chrostowski and M. Hochberg. *Silicon photonics design from devices to systems*. Cambridge University Press, March 2015.

[21] W. S. C. Chang. *Principles of optics for engineers*. Cambridge University Press, 2015.

[22] K. S. Chiang. Integrated optic waveguides. *Wiley Encyclopedia of Electrical and Electronics Engineering*, 2001.

[23] H. A. Haus, W. P. Haung, S. Kawakami, and N. A. Whitaker. Coupled-mode theory of optical waveguides. *Journal of Lightwave Technology*, LT-5:16–23, 1987.

[24] B. E. Little, S. T. Chu, H. A. Haus, J. Foresi, and J. P. Laine. Microring resonator channel dropping filters. *Journal of Lightwave Technology*, 15(6):998–1005, June 1997.

[25] M. Nikdast, J. Xu, L. H. K. Duong, X. Wu, X. Wang, Z. Wang, Z. Wang, P. Yang, Y. Ye, and Q. Hao. Crosstalk noise in WDM-based optical networks-on-chip: A formal study and comparison. *IEEE Transactions on Very Large Scale Integration (VLSI) Systems*, 23(11):2552–2565, Nov 2015.

[26] M. Nikdast, G. Nicolescu, J. Trajkovic, and O. Liboiron-Ladouceur. Modeling fabrication non-uniformity in chip-scale silicon photonic interconnects. In *Design, Automation Test in Europe Conference Exhibition (DATE)*, pages 115–120, March 2016.

[27] M. Nikdast, G. Nicolescu, J. Trajkovic, and O. Liboiron-Ladouceur. Chip-scale silicon photonic interconnects: A formal study on fabrication non-uniformity. *Journal of Lightwave Technology*, 34(16):3682–3695, August 2016.

[28] D. Vantrease, R. Schreiber, M. Monchiero, M. McLaren, N. P. Jouppi, M. Fiorentino, A. Davis, N. Binkert, R. G. Beausoleil, and J. H. Ahn. Corona: System implications of emerging nanophotonic technology. In *International Symposium on Computer Architecture*, pages 153–164, June 2008.

[29] Y. Pan, J. Kim, and G. Memik. Flexishare: Channel sharing for an energy-efficient nanophotonic crossbar. In *The Sixteenth International Symposium on High-Performance Computer Architecture*, pages 1–12, Jan 2010.

[30] S. Le Beux, J. Trajkovic, I. O'Connor, G. Nicolescu, G. Bois, and P. Paulin. Optical ring network-on-chip (ORNoC): Architecture and design methodology. In *Design, Automation Test in Europe*, pages 1–6, March 2011.

[31] Lumerical Solutions, Inc. http://www.lumerical.com/tcad-products/mode/

[32] T. Alasaarela, D. Korn, L. Alloatti, A. Säynätjoki, A. Tervonen, R. Palmer, J. Leuthold, W. Freude, and S. Honkanen. Reduced propagation loss in silicon strip and slot waveguides coated by atomic layer deposition. *Optics Express*, 19(12):11529–11538, June 2011.

[33] Y. Luo, X. Zheng, S. Lin, J. Yao, H. Thacker, I. Shubin, J. E. Cunningham, J. H. Lee, S. S. Djordjevic, J. Bovington, D. Y. Lee, K. Raj, and A. V. Krishnamoorthy. A process-tolerant ring modulator based on multi-mode waveguides. *IEEE Photonics Technology Letters*, 28(13):1391–1394, July 2016.

[34] C. J. Oton, C. Manganelli, F. Bontempi, M. Fournier, D. Fowler, and C. Kopp. Silicon photonic waveguide metrology using Mach-Zehnder interferometers. *Optics Express*, 24(6):6265–6270, March 2016.

[35] T. Y. Liow, J. Song, X. Tu, A. E. J. Lim, Q. Fang, N. Duan, M. Yu, and G. Q. Lo. Silicon optical interconnect device technologies for 40 Gb/s and beyond. *IEEE Journal of Selected Topics in Quantum Electronics*, 19(2):8200312–8200312, March 2013.

[36] M. Nikdast, G. Nicolescu, J. Trajkovic, and O. Liboiron-Ladouceur. Photonic integrated circuits: a study on process variations. In *Optical Fiber Communication Conference*, page W2A.22. Optical Society of America, 2016.

[37] UBC automated probe station. http://www.siepic.ubc.ca/probestation. Accessed: 2015-02-01.

13

Enhancing Process Variation Resilience in Photonic NoC Architectures

Sai Vineel Reddy Chittamuru,
Ishan G. Thakkar and Sudeep Pasricha

Colorado State University, Forty Collins, Colorado, United States

Abstract

Photonic network-on-chip (PNoC) architectures are a potential candidate for communications in future chip multi-processors as they can attain higher bandwidth with lower power dissipation than electrical NoCs. PNoCs typically use dense wavelength division multiplexing (DWDM) for high bandwidth transfers. Unfortunately, DWDM increases crosstalk noise and decreases optical signal to noise ratio (OSNR) in microring resonators (MRs) threatening the reliability of data communication. Additionally, process variations induce variations in the width and thickness of MRs causing shifts in resonance wavelengths of MRs, which further reduces signal integrity, leading to communication errors and loss in bandwidth. In this chapter, we propose a novel encoding mechanism that intelligently adapts to on-chip process variations, and improves worst-case OSNR by reducing crosstalk noise in MRs used within DWDM-based PNoCs. Experimental results on the Corona PNoC architecture indicate that the proposed approach improves worst-case OSNR by up to 44%.

13.1 Introduction

The ever increasing demand for higher performance computing and technology scaling have driven the trend of integrating a steadily increasing number of processing cores on a single die. With an increase in number of cores, electrical networks-on-chip (ENoCs) are projected to suffer

385

from cripplingly high power dissipation and severely reduced performance [1–3]. Recent developments in the area of silicon photonic device fabrication have enabled the integration of on-chip photonic interconnects with CMOS circuits, and led to the introduction of concept of photonic networks-on-chip (PNoCs) that can offer ultra-high bandwidth, reduced power dissipation, and lower latency than ENoCs.

Several crossbar topology based PNoC architectures have been proposed to date (e.g., [4–6]). These architectures are built using silicon photonic devices such as microring resonators (MRs) and silicon waveguides, and use dense-wavelength-division-multiplexing (DWDM), where a large number of wavelengths are multiplexed in a waveguide to enable high bandwidth parallel data transfers. Un-fortunately, the deleterious effects of fabrication process variations (PV) in MRs and silicon waveguides can reduce reliability in PNoCs. The PV-induced variations in the width and thickness of MRs cause considerable resonance wavelength shifts in MRs [7–9]. An MR couples light of a specific resonance wavelength to/from the waveguide, enabling electrical-to-photonic (modulation) and photonic-to-electrical (detection) operations. PV-induced resonance shifts deteriorate the optical signal to noise ratio (OSNR) in MRs, as they decrease the signal power and increases the crosstalk noise power. This in turn increases the bit-error-rate (BER) in a photonic interconnect. For example, a previous study shows that in a DWDM-based photonic interconnect, when PV-induced resonance shift is over 1/3 of the channel gap, BER of photonic transmission increases from 10^{-12} to 10^{-6} [10]. The increase in BER lowers the reliability of data communication.

The most intuitive way to counteract PV-induced resonance shifts in MRs is to realign the resonant wavelengths by using localized trimming [9] and thermal tuning [7] mechanisms. The first approach causes the wavelength to shift towards the blue end and the latter towards the red end of the resonance spectrum. These mechanisms improve OSNR and decrease BER in DWDM based photonic interconnects. However, realigning the PV-induced red shifts in MRs' resonance wavelengths requires injecting extra free carriers in MRs [9]. The resultant increase of the free carrier concentrations in the MRs increases the absorption related optical loss in the MRs due to the free carrier absorption effect (FCA) [23, 29]. Our analysis in this chapter shows that this additional signal loss decreases Q-factor of MRs, which in turn increases crosstalk noise in MR detectors and reduces OSNR. Thus, the use of trimming to remedy PV is not a viable option, strongly motivating new crosstalk mitigation techniques in PV-affected PNoCs.

We notice that while transmitting data in DWDM-based PNoCs, the impact of PV remedial techniques (such as localized trimming) on the crosstalk noise in MRs depends not only on the amount of PV-induced resonance shifts but also on the characteristics of data values propagating in the network. This implies that the harmful effects of localized trimming on crosstalk noise can be reduced by controlling the relative occurrences of some data values. Therefore, in this chapter, we present a novel technique that intelligently reduces undesirable data value occurrences in a photonic interconnect based on the process variation profile of MRs in the detecting nodes. This technique is easily implementable in any existing DWDM-based crossbar PNoC without requiring major modifications to the architectures. Further, the technique presented in this chapter is lightweight and possesses low overhead. Our main contributions in this chapter are summarized below:

- We present a device-level analytical model that captures the deleterious effects of localized trimming in MRs. Moreover, we extend this model for system-level crosstalk analysis;
- We propose a double bit crosstalk mitigation mechanism (DBCTM) that improves the worst-case OSNR of the PV affected detecting node by encoding specific portions of data to avoid undesirable data occurrences;
- We evaluate our proposed DBCTM technique by implementing it on the well-known crossbar PNoC Corona architecture [4, 11], and compared it with two previously proposed encoding based crosstalk mitigation mechanisms from [13] for real-world multi-threaded PARSEC [12] benchmarks.

13.2 Related Work

DWDM-based PNoCs use photonic devices such as microring resonators (MRs) as modulators and detectors, photonic waveguides, splitters, and trans-impedance amplifiers (TIAs). The reader is directed to [13] for more information on photonic devices. Each constituent photonic device in a PNoC contributes to some type of optical signal loss, the combined effect of which negatively affects the system-level OSNR. In addition to the optical signal loss, the crosstalk noise of the constituent MRs also deteriorates the OSNR. Crosstalk is an intrinsic property of every MR, so both the modulators, detectors, and switches are susceptible to crosstalk noise in DWDM-based PNoCs. Figure 13.1 shows crosstalk noise (as dotted/dashed lines) in modulator and

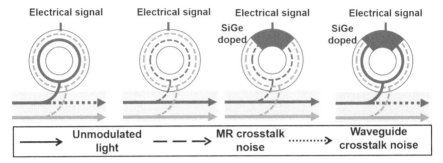

Figure 13.1 MR operation phases in DWDM-based photonic interconnects (note: MR shown has green resonant wavelength): (a) modulator modulating in resonance wavelength (b) modulator in passing (through) mode (c) detector in passing mode (d) detector in detecting mode.

detector MRs during typical modulation/detection phases in the DWDM-based photonic interconnect. Whenever a modulator modulate s/writes a '0' or a detector detects/reads a '1' from a wavelength by removing the light pulse, there is crosstalk generated in the photonic interconnect, as shown in Figure 13.1(a) and (d).

One of the key challenges for the widespread adoption of DWDM-based PNoC architectures is to mitigate the crosstalk noise in their MR detectors. The effect of crosstalk noise on OSNR is negligible in DWDM systems presented in [14] and [15], as these systems use only four DWDM wavelengths per waveguide. On the other hand, in DWDM-based crossbar architectures such as Corona [4] that use 64 wavelength DWDM, there exists significant crosstalk noise. The quantitative results in [16] demonstrate the damaging impact of crosstalk noise in Corona, where the worst-case OSNR is estimated to be about 14 dB in data waveguides, which is insufficient for reliability of data communication. To mitigate the impacts of crosstalk noise in DWDM based PNoCs, two encoding techniques PCTM5B and PCTM6B are presented in [13]. However, none of these works considers process variations and their impact on crosstalk noise in DWDM-based PNoCs.

A few previous works have explored the impact of process variations on DWDM-based photonic links [10, 25, 28]. In [10], the authors discuss a run-time hardware-software management solution that optimizes the performance and reliability of photonic data communication to compensate for PV effects. In [25], a methodology to salvage network-bandwidth loss due to process-variation-drifts is proposed, which reorders MRs and trims them to nearby wavelengths available in the waveguide. Moreover, Mohamed et al. in [28] present power-efficient techniques based on inter-channel hopping and variation-aware routing to compensate for PV effects at run time. All of these PV-remedial techniques are network specific and ignore the harmful effects of PV on crosstalk noise. In contrast, this chapter considers the deleterious effects of PV-remedial techniques and proposes a generalized technique for crosstalk noise mitigation with minimal overhead, to improve OSNR and communication reliability in DWDM-based photonic crossbar PNoCs.

13.3 Analytical Model for PV-Aware Crosstalk Analysis

13.3.1 Impact of Localized Trimming on Crosstalk

As discussed earlier, the localized trimming method is essential to deal with PV-induced resonance red shifts in MRs. However, the use of trimming method in an MR alters its intrinsic optical properties, which leads to increased crosstalk noise and degraded performance in PNoCs using these MRs [23, 29]. In this section, we discuss the effects of the localized trimming method and present analytical models to capture these effects in MRs. Further, this chapter extends these models to generate system-level models for the Corona PNoC architecture [4]. The system-level models enable quantification of signal and noise powers in the constituent MRs and DWDM waveguides of the Corona architecture.

An MR can be considered to be a circular photonic waveguide with a small diameter (Figure 13.1), not to be confused with the larger photonic waveguide in a DWDM-based photonic interconnect for which MRs serve as modulators/writers and detectors/readers. The localized trimming method injects extra free carriers in the circular MR waveguide to counteract the PV-induced resonance red shifts in it. The introduction of extra free carriers reduces the refractive index of the MR waveguide, which in turn induces a blue shift in resonance to counteract the PV-induced red shifts. However, the extra free carriers increase the absorption related optical loss in the MR due to the free carrier absorption effect (FCA) [23]. The increase in the intrinsic optical loss results in a decrease in Q-factor of MRs, which in turn increases MR insertion loss and crosstalk.

We use a process variation map (discussed in Section 13.3.3) to estimate PV-induced shifts in the resonance wavelengths of all the MRs across a chip. Then, for each MR device, we calculate the amount of change in refractive index (Δn_{si}) required to counteract this wavelength shift using the following equation [22]:

$$\Delta n_{si} = \frac{\Delta \lambda_r * n_g}{\Gamma * \lambda_r} \tag{13.1}$$

where, $\Delta \lambda_r$ is the PV-induced resonance shift that need to be compensated for, λ_r is the target resonance wavelength of the MR, n_g is the group refractive index of the MR waveguide, and Γ is the confinement factor describing the overlap of the optical mode with the MR waveguide's silicon core. We assume that the MR waveguides used in this study are similar to those reported in [23], fabricated using standard Si-SiO$_2$ material with a cross section of 450 nm$\times$250 nm. The values of Γ and n_g, for the considered MR waveguides, are set to 0.7 and 4.2 respectively for 1550 nm wavelength [23].

The required change in the free carrier concentration to induce the refractive index change of Δn_{si} at around 1.55 µm wavelength can be quantified using the following equation [22]:

$$\Delta n_{si} = -8.8 \times 10^{-22} \Delta N_e - 8.5 \times 10^{-18} (\Delta N_h)^{0.8} \tag{13.2}$$

where, ΔN_e and ΔN_h are the MR's change in free electron concentration and the change in free hole concentration respectively. The change in the absorption loss coefficient ($\Delta \alpha_{si}$) due to the change in free carrier concentration (owing to the FCA effect) can be quantified using the following equation [22]:

$$\Delta \alpha_{si} = 8.5 \times 10^{-18} \Delta N_e + 6.0 \times 10^{-18} \Delta N_h \tag{13.3}$$

In Equation (13.2) and Equation (13.3), positive values of ΔN_e and ΔN_h correspond to increase in free electron and free hole concentrations respectively, which in turn correspond to decrease in refractive index (negative Δn_{Si} in Equation (13.2)) and increase in absorption loss coefficient (positive $\Delta \alpha_{Si}$ in Equation (13.3)). The Q-factor of an MR depends on the absorption loss coefficient. The relation between the Q-factor and $\Delta \alpha_{si}$, assuming critical coupling of MRs, is given by the following equation [23], here Q' is the loaded Q-factor of MR:

$$Q' = Q + \Delta Q = \frac{\pi n_g}{\lambda_r (\alpha + \Delta \alpha_{si})} \tag{13.4}$$

where, ΔQ is the change in Q-factor and α is the original loss coefficient, which is a sum of three components: *(i)* intrinsic loss coefficient due to material loss and surface roughness; *(ii)* bending loss coefficient, which is a result of the curvature in the MR; and *(iii)* the absorption effect factor that depends on the original free carrier concentration in the MR waveguide core.

Typically, the localized trimming method injects excess concentration of free carriers (electrons and holes) into the MR, which leads to an increase in the absorption loss coefficient (positive $\Delta\alpha_{si}$ corresponding to positive ΔN_e and ΔN_h from Equation (13.3)). As evident from Equation (13.4), a positive value of $\Delta\alpha_{si}$ results in decrease of Q-factor. This causes a broadening of the MR passband, which results in increased insertion loss and crosstalk power penalties.

13.3.2 Crosstalk Modeling for Corona PNoC

In this chapter, we characterize crosstalk in DWDM-based photonic interconnects for the well-known Corona PNoC enhanced with token-slot arbitration [4, 11]. An overview of Corona PNoC is presented in Figure 13.2. In Corona's DWDM-based interconnects, data transmission requires modulating light using a group of MR modulators equal to the number of wavelengths supported by DWDM. Similarly, data detection at the receiver requires a group of detector MRs equal to the number of DWDM wavelengths. Corona PNoC employs ON-OFF keying data modulation at an operating frequency of 5 GHz [4]. This chapter presents analytical equations that model worst-case crosstalk noise power, maximum power loss, and worst-case OSNR in detector MR groups (similar equations are applicable to modulator MR groups).

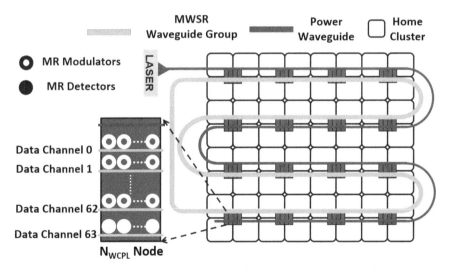

Figure 13.2 Overview of Corona PNoC with MWSR data waveguides.

Before presenting actual analytical equations for Corona, we provide notations for different parameters used in the analytical equations in Tables 13.1 and 13.2. The Corona PNoC is designed for a 256 core single-chip platform, where cores are grouped into 64 clusters, with 4 cores in each cluster. A photonic crossbar topology with 64 data channels is used for data communication between clusters. Each channel consists of 4 multiple-write-single-read (MWSR) waveguides with 64-wavelength DWDM in each waveguide. As modulation occurs on both positive and negative edges of the clock in Corona, such that 512 bits (cache-line size) can be modulated and inserted on 4 MWSR waveguides in a single cycle by a sender.

A data channel starts at a cluster called 'home-cluster', traverses other clusters (where modulators can modulate light and detectors can detect this light), and finally ends at the home-cluster again, at a set of detectors (optical termination). A power waveguide in Corona supplies optical power from an off-chip laser to each of the 64 data channels at its home-cluster, through a series of 1×2 splitters. In each of the 64 home-clusters, optical power is distributed among 4 MWSR waveguides equally using a 1×4 splitter with splitting factor R_{S14}. As all 1×2 splitters are present before the last (64th) channel, this channel suffers the highest signal power loss. Thus, the worst-case signal and crosstalk noise exists in the detector group of the 64th cluster node, and this node is defined as the worst-case power loss node (N_{WCPL}) in the PNoC as shown in Figure 13.2.

For this N_{WCPL} node, the signal power ($P_{signal}(j)$) and crosstalk noise power ($P_{noise}(j)$) received at each detector *j* are shown in Equations (13.5) and (13.6) [14]. $P_S(i,j)$ in Equation (13.7) is the signal power of the *i*th wavelength

Table 13.1   Notations for photonic power loss, crosstalk coefficients [16]

Notation	Parameter Type	Parameter Value (in dB)
L_P	Propagation loss	-0.274 per cm
L_B	Bending loss	-0.005 per $90°$
L_{MI}	Inactive modulator through loss	-0.005
L_{MA}	Active modulator power loss	-0.6
L_{DP}	Passing detector through loss	-0.005
L_{DD}	Detecting detector power loss	-1.6
L_{S12}	1×2 splitter power loss	-0.2
L_{S14}	1×4 splitter power loss	-0.2
L_{S15}	1×5 splitter power loss	-0.2
L_{S16}	1×6 splitter power loss	-0.2
X_{MA}	Active modulator crosstalk factor	-16
X_{DD}	Detecting detector crosstalk factor	-16

Table 13.2   Other model parameter notations

Notation	Crosstalk Coefficient	Parameter Value
Q	Q-factor	9000
FSR	Free spectral range	51.2 nm
L	Photonic path length in cm	
B	Number of bends in photonic path	
λ_j	Resonance wavelength of MR	
R_{S12}	Splitting factor for 1×2 splitter	
R_{S14}	Splitting factor for 1×4 splitter	
R_{S15}	Splitting factor for 1×5 splitter	
R_{S16}	Splitting factor for 1×6 splitter	

received before the *j*th detector. Similarly in Equation (13.9), $P_N(i,j)$ is the incoherent crosstalk noise power of the *i*th wavelength before the *j*th detector. K_S and K_N in Equation (13.10) and (11) represent signal and crosstalk noise power losses before the detector group of N_{WCPL}. $\psi(i,j)$ in Equation (13.8a) represents signal power loss of the *i*th wavelength before the *j*th detector within the detector group of N_{WCPL}; and $\Phi(i,j)$ in Equation (13.8b) is the crosstalk coupling factor of the *i*th wavelength and the *j*th detector.

$$P_{signal}(j) = L_{DD} \, P_S(j,j) \tag{13.5}$$

$$P_{noise}(j) = L_{DD} \, P_N(j,j) + \sum_{i=1}^{n} \Phi(i,j) \, (P_S(i,j)$$
$$+ \, P_N(i,j)) \, (i \neq j) \tag{13.6}$$

$$P_S(i,j) = K_S \, \psi(i,j) \, P_{in}(i) \tag{13.7}$$

$$\psi(i,j) = \begin{cases} X_{DD}(L_{DP})^{j-1}, & \text{If } j-1 \geq i \text{ and } D_B = 1 \\ (L_{DP})^{j-1}, & \text{if } j-1 < i \text{ and } D_B = 1 \\ X_{MA}X_{DD}(L_{DP})^{j-1}, & \text{If } j-1 \geq i \text{ and } D_B = 0 \\ X_{MA}(L_{DP})^{j-1}, & \text{If } j-1 < i \text{ and } D_B = 0 \end{cases} \tag{13.8a}$$

$$\Phi(i,j) = \frac{\delta^2}{\left((i-j)\frac{FSR}{n}\right)^2} + \delta^2, \; Here \; \delta = \frac{\lambda_j}{2Q'} \tag{13.8b}$$

$$P_N(i,j) = \begin{cases} 0, & \text{If } j > i \text{ and } D_B = 1 \\ K_N(L_{DP})^j P_{in}(i), & \text{if } j \leq i \text{ and } D_B = 1 \\ 0, & \text{If } j > i \text{ and } D_B = 0 \\ X_{MA}K_N(L_{DP})^j P_{in}(i), & \text{if } j \leq i \text{ and } D_B = 0 \end{cases} \tag{13.9}$$

$$K_S = \begin{cases} (R_{S14})(L_{S14})(L_P)^L \ (L_B)^B \ (L_{MI})^{64\times 63}, & If \ D_B = 1 \\ (R_{S14})(L_{S14})(L_P)^L \ (L_B)^B \ (L_{MI})^{64\times 62+63}, & If \ D_B = 0 \end{cases}$$

$$(13.10)$$

$$K_N = \begin{cases} (R_{S14})(L_{S14}) \ (L_P)^L \ (L_B)^B \ X_{MA} \ (L_{MI})^{64\times 62+63}, & If \ D_B = 1 \\ (R_{S14})(L_{S14}) \ (L_P)^L \ (L_B)^B \ X_{MA}(L_{MI})^{64\times 62+62}, & If \ D_B = 0 \end{cases}$$

$$(13.11)$$

Due to the use of trimming to remedy PV, crosstalk coupling factor Φ increases with decrease in loaded Q-factor (Q', Equation (13.4)), which in turn increases crosstalk noise in the detectors. For instance, realignment of PV-induced resonance shift of 0.05 nm ($\Delta\lambda_r$, Equation (13.1)) using the trimming method increases the absorption loss coefficient ($\alpha + \Delta\alpha_{si}$) to 11.5cm^{-1} from 9.5cm^{-1} (α, corresponding to the original Q-factor of 9000 given in Table 13.2). This increase in α reduces the Q-factor to 7400 from 9000 (Equation (13.4)), which in turn increases δ (Equation (13.8b)) by 21.6% resulting in about 15× increase in the coupling factor Φ (Equation (13.8b)). As evident from Equation (13.6), this increase in the coupling factor (Φ) results in the increase of crosstalk noise in MR detectors.

We can define OSNR(j) of the *j*th detector of N_{WCPL} as the ratio of $P_{\text{signal}}(j)$ to $P_{\text{noise}}(j)$, as shown in Equation (13.12). These equations are sufficient to analyze signal and crosstalk noise power during the detection of ones (D_B='1') and zeros (D_B='0') in the data waveguide. Section 13.5 uses these models to explain how crosstalk mitigation techniques impact OSNR.

$$OSNR(j) = \frac{P_{signal}(j)}{P_{Noise}(j)} \tag{13.12}$$

13.3.3 Modeling PV of MR Devices in Corona PNoC

We adapt the VARIUS tool [24] to model die-to-die (D2D) as well as within-die (WID) process variations in MRs for the Corona PNoC. This work considers photonic devices with a silicon-dioxide (SiO_2) cladding and silicon (Si) core for our analysis. VARIUS uses a normal distribution to characterize on-chip D2D and WID process variations. The key parameters are mean (μ), variance (σ^2), and density (α) of a variable that follows the normal distribution. As wavelength variations are approximately linear to dimension variations of MRs, this work assumes they also follow the same uniform distribution. The mean (μ) of wavelength variation of an MR is its nominal resonance wavelength. This chapter considers a DWDM wavelength range in the C and L bands [13], with a starting wavelength of 1550 nm and a channel spacing of 0.8 nm. Hence, those wavelengths are the means for each MR modeled. The variance (σ^2) of wavelength variation is determined based on laboratory fabrication data [7] and our target die size. This work considers a 256-core chip with die size 400 mm^2 at 22 nm CMOS technology node. For this die size this work considers a WID standard deviation (σ_{WID}) as 0.61 nm [25] and D2D standard deviation (σ_{D2D}) as 1.01 nm [25]. Further, we also consider a density (α) of 0.5 [25] for this die size. With these parameters, this work uses VARIUS to generate 100 process variation maps. Each process variation map contains over one million points indicating the PV-induced resonance shift of MRs. The total numbers of points picked from these process variation maps are equal to the number of MRs in the Corona PNoC architecture.

13.4 Double-Bit Crosstalk Mitigation (DBCTM)

13.4.1 Overview

In this section, we present our proposed double-bit crosstalk mitigation (DBCTM) technique, which is illustrated in Figure 13.3. As evident from Equation (13.6) presented in Section 13.3.2, crosstalk noise in MR detectors

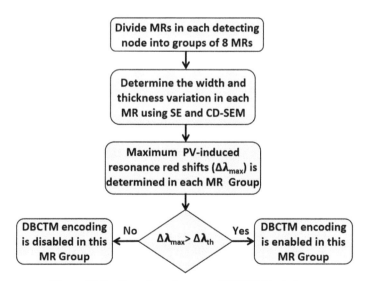

Figure 13.3   Overview of proposed DBCTM technique.

of DWDM-based PNoCs increases with increase in coupling factor (Φ) and increase in signal strength of immediate non-resonating wavelength (i.e., wavelength that is adjacent to MR resonance wavelength). This indicates that the crosstalk noise in a detector can be reduced by reducing the signal strength of immediate non-resonating wavelengths. Therefore, our proposed DBCTM encoding mechanism decreases the signal strength of immediate non-resonant wavelength by modulating a zero on it, which results in reduced crosstalk noise in the detector. For that, the DBCTM technique first divides detecting MRs in the detector bank into groups of 8 MRs each. We can either decrease or increase the number of MRs within a particular group. However, decrease in number of MRs within a group increases the complexity of electrical circuitry for DBCTM encoding mechanism and increases its power dissipation overhead. Though increase in number of MRs within a cluster decreases DBCTM power dissipation overhead because of decrease in complexity of DBCTM's electrical circuitry, but with lower number of MRs the DBCTM lacks the ability to adapt to the PVs within the PNoCs and ultimately reduces its OSNR gains. Thus to balance power and reliability we choose 8 MRs within each group for the proposed DBCTM technique. Then, it determines the maximum PV-induced resonance red shift ($\Delta\lambda_{\max}$) in each MR group. The PV-induced resonance shifts in MRs can be determined by measuring the variations in the thickness and width values of MRs. This is because there is a linear dependency between thickness and width variations of MRs and the resonance wavelength drift of MRs, as discussed in [7] and [10]. For example, a 1 nm variation in width and height of an MR leads to $0.58\sim1$ nm [7, 10] and ~2 nm [7] resonance shifts in MRs respectively. In a real system, spectroscopic ellipsometry (SE) and critical dimension scanning electron microscopy (CD-SEM) can be employed to measure these thickness and width variations in actual fabricated MR devices in an auto-mated manner [7]. In our experimental analysis, we model and estimate PV in MRs using the VARIUS tool [24], a description of which is given in Section 13.3.3.

Once PV-induced resonance red shifts of MRs are determined, we store information about whether to enable or disable encoding for each MR detect-ing group in a read-only memory (ROM) at the modulating node, based on the maximum PV-induced resonance red shift ($\Delta\lambda_{\max}$) value for the group. If this PV-induced resonance red shift value is greater than a threshold red shift value ($\Delta\lambda_{\text{th}}$) for an MR group, we store a '1' to enable DBCTM, else we store a '0' to disable DBCTM for this MR group. MR groups with $\Delta\lambda_{\max} < \Delta\lambda_{\text{th}}$ are thus not impacted. Only MR-groups with $\Delta\lambda_{\max} > \Delta\lambda_{\text{th}}$ employ DBCTM encoding.

13.4.2 DBCTM Sensitivity Analysis with Corona PNoC

Our DBCTM encoding scheme involves injecting zeros between data bits. These extra bits are called shielding bits. As the number of shielding bits increases with DBCTM, laser power and trimming power of PNoCs also increase. Thus, there is a need to limit the number of shielding bits. We performed a sensitivity analysis using the Corona architecture with varying number of shielding bits per detecting node to quantify its effect on worst-case OSNR. We analyzed worst-case OSNR with 0%, 25%, 50%, 75% and 100% of shielding bits added to data bits for the Corona PNoC. Based on our analysis across 100 process variation maps (see Section 13.3.3), we determined $\Delta\lambda_{th}$ to be 0.45 nm, 0.88 nm, 1.25 nm and 4.25 nm, for the cases with 25%, 50%, 75% and 100% of shielding bits to data bits, respectively. Figure 13.4 shows the range of worst-case OSNR values across 100 process variation maps, for different ratios of shielding bits to data bits. From the figure it can be observed that on an average DBCTM with 25%, 50%, 75% and 100% shielding bits has 8.1%, 19.67%, 26% and 40.5% higher worst-case OSNR (note: higher OSNR is better) respectively compared to the baseline (with 0% shielding). Intuitively, higher ratios of shielding bits to data bits should result in higher worst-case OSNR, as more shielding bits can be used to shield data bits, which in turn reduces crosstalk noise in detecting MRs and improves OSNR. But, with increase in number of shielding bits, the number of MRs on the waveguides increases. High MR count on the waveguide

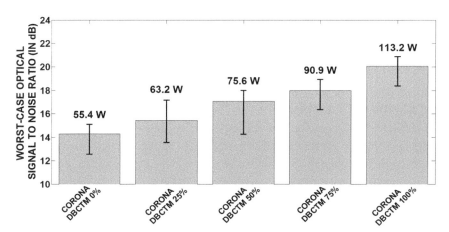

Figure 13.4 Sensitivity analysis in terms of worst-case OSNR of Corona architecture with DBCTM allowing 0%, 25%, 50% and 100% ratio of shielding bits to data bits across 100 process variation maps; average power consumption are also shown on the top of each bar.

results in higher through losses, which increases laser power consumption of PNoCs. Compensating for PV drifts of high MR counts requires high trimming/tuning power in PNoCs. Figure 13.4 shows that average power consumption of DBCTM with 25%, 50%, 75% and 100% shielding bits is 14%, 20.1%, 63.9% and 104.1% higher compared to the baseline. To balance crosstalk reliability and laser power overheads, we select the 50% shielding bits to data bits configuration, for the rest of our experiments.

To implement our DBCTM technique with 50% shielding bits on the Corona PNoC, there is a need to increase the number of MWSR waveguides in each channel from 4 to 6, to maintain the same bandwidth as in the baseline case. To distribute optical power between these waveguides using power waveguides, there is also a need to replace 1×4 splitters with 1×6 splitters with a splitting factor of R_{S16}. Additionally as explained in the previous subsection each modulating node needs to store 2,646 bits in its ROM to capture encoding requirements for all the remaining 63 detecting nodes. Power and area overheads for these modifications are presented in the experimental section (Section 13.5). Lastly, we consider up to a two cycle overhead for encoding and decoding of data in DBCTM, as per our implementations at 5 GHz. The first cycle is needed to retrieve data from ROM, whereas the second cycle is used if data is to be encoded before sending on the data waveguide. In addition, our DBCTM technique is not architecture specific and it can be easily ported to any other PNoC architectures such as Firefly [5] and Flexishare [6].

13.5 Experiments

13.5.1 Experimental setup

To evaluate our proposed DBCTM crosstalk noise mitigation technique in DWDM-based PNoCs, we implement and integrate it for the Corona [4, 16] crossbar-based PNoC architecture. We modeled and performed simulation-based analysis of the enhanced Corona PNoC using a cycle-accurate NoC simulator, for a 256 core single-chip architecture at 22 nm. As explained in the previous section, we generated 100 process variation maps to evaluate the impact of PV on our DBCTM encoding mechanism. We used real-world traffic from applications in the PARSEC benchmark suite [12] for our evaluation. GEM5 full-system simulation [17] of parallelized PARSEC applications was used to generate traces that were fed into our SystemC based cycle-accurate NoC simulator. We set a "warm-up" period of 100 million instructions and then captured traces for the subsequent 1 billion instructions. We performed geometric calculations for a 20 mm×20 mm chip size, to determine lengths of MWSR waveguides in the Corona architecture. Based on this geometrical analysis, we estimated the time needed for light to travel from the first to the last node as 8 cycles at 5 GHz clock frequency. We use a packet size of 512 bits, as advocated in the Corona PNoC.

The static and dynamic energy consumption of electrical routers and concentrators in the Corona PNoC is based on results from the Orion 2.0 [18] tool. We model and consider area, power, and performance overheads for DBCTM (50% case). The electrical area overhead was analyzed and esti-mated to be 6.24 mm^2 and power overhead is estimated to be 1.14 W, using gate-level analysis and the CACTI 6.5 [19] tool for memory and buffers. The photonic area overhead was estimated to be 9.12 mm^2 based on the physical dimensions [15] of waveguides, MRs, and splitters. For energy consumption of photonic devices, we adopt model parameters from recent work [16, 20, 21], with 0.42pJ/bit for every modulation and detection event and 0.18pJ/bit for the driver circuits of MR modulators and MR photo detectors. We used optical loss for photonic components, as shown in Table 13.1, to determine the photonic laser power budget and correspondingly the electrical laser power. Further, the MR trimming power is set to 130 μW/nm [27] for current injection (blue shift) and tuning power is set to 240 μW/nm [27] for heating (red shift).

13.5.2 Experimental Results with Corona PNoC

Utilizing the models presented in Section 13.3, we calculate the received crosstalk noise and OSNR at detectors for the node with worst-case power loss (N_{WCPL}), which corresponds to MR detectors in cluster 64 for the Corona PNoC architecture.

The first set of experiments in this work compares the baseline Corona architecture with fair token-slot arbitration [4, 16] but without any crosstalk-enhancements, with three variants of the architecture corresponding to the three crosstalk-mitigation strategies we compare: PCTM5B and PCTM6B proposed in [13] and our proposed DBCTM technique from this chapter. The worst-case OSNR for the baseline Corona PNoC occurs when all the 64-bits of a received data word in a waveguide are 1's. However, for the implementations of Corona PNoC with PCTM5B, PCTM6B and DBCTM crosstalk-mitigation techniques, this is not the case, i.e., each detector in cluster 64 has a worst-case OSNR for a different pattern of 1's and 0's in the received data word. We used our analytical models to determine these unique worst-case patterns for each of the techniques when ported to Corona, for an accurate analysis.

Figure 13.5 (a)–(d) present detector signal power loss (otherwise desired signal power received at the detector), crosstalk noise power loss, and OSNR corresponding to the detectors in the 64th cluster for the baseline and three variants of the Corona PNoC architecture for a randomly selected PV map generated using VARIUS, as explained in Section 13.3.3. Further, in this figure detectors are positioned (i.e., from left to right) in the increasing order of their designed resonance wavelengths. Figure 13.5(b) indicates that worst-case OSNR (lowest value of the bars, which represent OSNR in detectors) improves notably over the baseline in Figure 13.5(a) when using the PCTM5B technique. However, the obtained improvement is on the lower side for the remaining detectors. Figure 13.5(c) shows that the PCTM6B technique improves worst-case OSNR marginally over PCTM5B, but does a better job of improving OSNR significantly for most of the MR detectors. Figure 13.5(d) shows that the DBCTM technique results in a significant improvement in worst-case OSNR as well as OSNR for all MR detectors compared to the baseline, as well as the PCTM5B and PCTM6B techniques. The worst-case OSNR is obtained at the 42nd detector of the 64th cluster in the baseline case; whereas for the PCTM5B, PCTM6B and DBCTM configurations worst-case OSNR occurs at the 59th, 61st and 33rd detectors of the same cluster, respectively.

Figure 13.6 summarizes the worst-case OSNR results for the baseline, PCTM5B, PCTM6B, and DBCTM techniques on the Corona PNoC. From the figure, it can be surmised that Corona with DBCTM has 19.28–44.13%, 12.44–34.19% and 4.5–31.30% worst-case OSNR improvements on average, compared to baseline, PCTM5B, and PCTM6B respectively across 100 process variation maps. Both the PCTM5B and PCTM6B techniques eliminate occurrences of '111' in a data word and have limited occurrences of '11', which helps reduce crosstalk noise in the MR detectors. But, these techniques do not consider the impact of PV resonance wavelength drifts, which leads to worse OSNR degradation. The DBCTM technique reduces crosstalk noise in the detectors by using shielding bits intelligently between data bits, further it also considers the PV profile of MRs to select MRs for shielding.

Figure 13.7(a) and (b) present simulation results that quantify the average network packet latency and energy-delay product (EDP) for the four Corona configurations. Results are shown for twelve multi-threaded PARSEC benchmark applications. From Figure 13.7(a) it can be observed that on

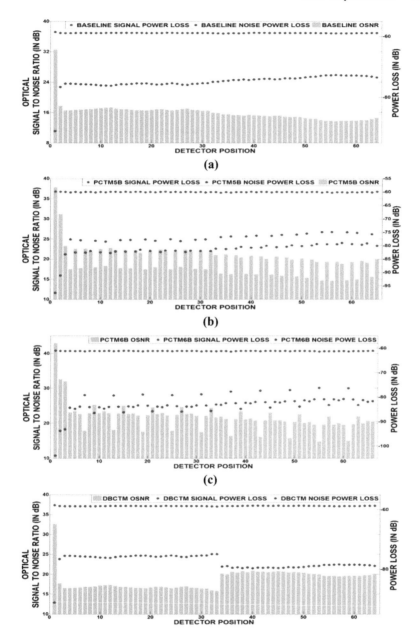

Figure 13.5 Detector-wise signal power loss, crosstalk noise power loss and minimum OSNR in worst-case power loss node for one process variation map of Corona (a) baseline with 64-detectors (b) PCTM5B with 65-detectors (c) PCTM6B with 66-detectors (d) DBCTM with 64-detectors.

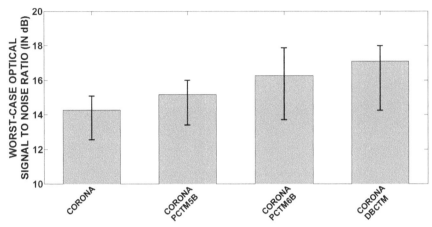

Figure 13.6 Worst-case OSNR comparison of DBCTM with PCTM5B [13] and PCTM6B [13] for Corona PNoC considering 100 process variation map.

average, the Corona configuration with DBCTM has 12.6%, 3.4% and 2.1% higher latency compared to baseline, PCTM5B and PCTM6B respectively. The additional delay due to encoding and decoding of data with DBCTM, PCTM5B and PCTM6B contributes to their increase in average packet latency. The penalty due to encoding/decoding is 1 cycle in PCTM5B and PCTM6B, whereas DBCTM has a 2 cycle penalty which increases its latency overhead.

From the results for EDP shown in Figure 13.7(b), it can be seen that on average, the Corona configuration with the DBCTM technique has 31.6% higher EDP compared to the baseline. This increase in EDP is not only due to the increase in average latency, but also due to the addition of extra shielding bits for encoding and decoding, which leads to an increase in the amount of photonic hardware in the architectures (more number of MRs, more complex splitters). This in turn increases static energy consumption of the Corona PNoC. Dynamic energy also increases in these architectures due to extra encoding bits, but by much less. However, EDP for the DBCTM technique is 16.4% lower compared to PCTM6B. Despite the higher latency overhead compared to PCTM6B, DBCTM saves considerable laser power and trimming/tuning power due to lower photonic hardware requirements than PCTM6B.

From the results presented in the previous sections, we can summarize that our proposed DBCTM helps to reduce crosstalk noise and improve OSNR in photonic data waveguides by exploiting process variations in the

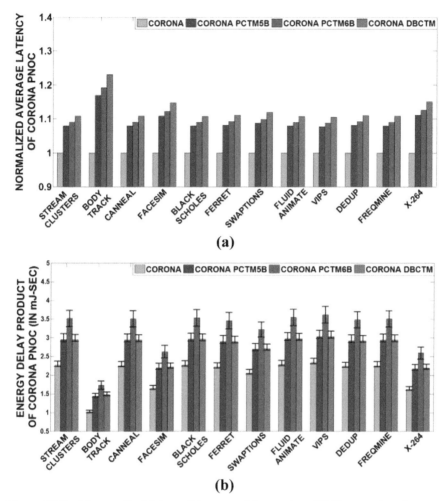

Figure 13.7 (a) normalized latency (b) energy-delay product (EDP) comparison between Corona baseline and Corona PNoC with PCTM5B, PCTM6B, and DBCTM techniques, with PARSEC benchmarks. All results are normalized to the baseline Corona architecture results.

fabrication of MRs compared to previously proposed crosstalk mitigation encoding mechanisms PCTM5B and PCTM6B. DBCTM shows 34.19% better crosstalk reliability compared to PCTM5B with similar overheads in terms of EDP. DBCTM also has 16.4% lower EDP and 31.30% better OSNR compared to the best-known crosstalk mitigation mechanism (PCTM6B).

13.6 Conclusion

We have presented a crosstalk mitigation technique DBCTM for the reduction of crosstalk noise in the detectors of dense wavelength division multiplexing (DWDM) based photonic network-on-chip (PNoC) architectures with crossbar topologies. Our DBCTM encoding mechanism show interesting trade-offs between reliability, performance, and energy overhead for the PNoC architectures. Our experimental analysis shows that the DBCTM mechanism improve worst-case OSNR by 44.13% compared to the baseline Corona architecture, and by 31.30% compared to the best known PNoC crosstalk mitigation scheme from prior work.

References

[1] L. Benini and G. De Micheli. *Networks on chip: A new paradigm for systems on chip design.* Proc. DATE, March, 2002, pp. 418–419.

[2] J. D. Owens, W. J. Dally, R. Ho, D. N. Jayasimha, S. W. Keckler and Li-Shiuan Peh. *Research challenges for on-chip interconnection networks.* IEEE Micro, September–October 2007, vol. 25, no.5, pp. 96–108.

[3] (2013). *International Technology Roadmap for Semiconductors (ITRS)* [Online]. Available: http://www.itrs.net

[4] D. Vantrease, R. Schreiber, M. Monchiero, M. McLaren, N. P. Jouppi, M. Fiorentino, A. Davis, N. Binkert, R. G. Beausoleil and J. H. Ahn. *Corona: System implications of emerging nanophotonic technology.* Proc. ISCA, June, 2008, pp.153–164.

[5] Y. Pan, P. Kumar, J. Kim, G. Memik, Y. Zhang and A. Choudhary. *Firefly: Illuminating future network-on-chip with nanophotonics.* Proc. ISCA, June, 2009, pp. 429–440.

[6] Y. Pan, J. Kim, G. Memik. *Flexishare: Channel sharing for an energy efficient nanophotonic crossbar.* Proc. HPCA, January, 2010, pp. 1–12.

[7] S. K. Selvaraja. *Wafer-Scale Fabrication Technology for Silicon Photonic Integrated Circuits.* PhD thesis, Ghent University, 2011.

[8] J. S. Orcutt, A. Khilo, C. W. Holzwarth, M. A. Popović, H. Li, J. Sun, T. Bonifield, R. Hollingsworth, F. X. Kärtner, H. I. Smith, V. Stojanović and R. J. Ram. *Nanophotonic integration in state-of-the-art cmos foundries.* Optics Express, January, 2011, vol. 19, no. 3, pp. 2335–2346.

[9] C. Batten, A. Joshi, J. Orcutt, A. Khilo, B. Moss, C. Holzwarth, M. Popovic, L. Hanqing, H. I. Smith, J. Hoyt, F. Kartner, R. Ram, V.

Stojanovic and K. Asanovic. *Building manycore processor-to-dram networks with monolithic silicon photonics*. High Performance Interconnects, August, 2008, pp. 21–30.

[10] Z. Li, M. Mohamed, C. Xi, E. Dudley, M. Ke, S. Li, A. R. Mickelson, R. Joseph, M. Vachharajani, B. Schwartz and S. Yihe. *Reliability modeling and management of nanophotonic on-chip networks*. IEEE TVLSI, December, 2010, vol. 20, no. 1, pp. 98–111.

[11] D. Vantrease, N. Binkert, R. Schreiber and M. H. Lipasti. *Light speed arbitration and flow control for nanophotonic interconnects*. Proc. IEEE/ACM MICRO, December, 2009, pp. 304–315.

[12] C. Bienia, S. Kumar, J. P. Singh and K. Li. *The PARSEC Benchmark Suit: Characterization and Architectural Implications*. PACT, October, 2008, pp. 72–81.

[13] S. V. R. Chittamuru and S. Pasricha. *Crosstalk Mitigation for High-Radix and Low-Diameter Photonic NoC Architectures*. IEEE Design and Test, March, 2015, vol. 32, no. 3, pp. 29–39.

[14] Q. Xu, B. Schmidt, S. Jagat and L. Michal. *Cascaded silicon microring modulators for WDM optical interconnection*. Optics Express, September, 2006, vol. 14, no. 20, pp. 9431–9436.

[15] S. Xiao, M. H. Khan, S. Hao and Q. Minghao. *Modeling and measurement of losses in silicon-on-insulator resonators and bends*. Optics Express, August, 2007, vol. 15, no. 17, pp. 10553–10561.

[16] L. H. K. Duong, M. Nikdast, S. Le Beux, J. Xu, X. Wu, W. Zhehui and Y. Peng. *A Case Study of Signal-to-Noise Ratio in Ring-Based Optical Networks-on-Chip*. IEEE Design and Test, October, 2014, vol. 31, no. 5, pp. 55–65.

[17] N. Binkert, B. Beckmann, B. Gabriel, S. K. Reinhardt, A. Saidi, A. Basu, J. Hestness, D. R. Hower, T. Krishna, S. Sardashti, R. Sen, K. Sewell, M. Shoaib, N. Vaish, M. D. Hill and D. A. Wood. *The gem5 Simulator*. ACM Computer Architecture News, May 2011, vol. 39, no. 2, pp. 1–7.

[18] A. B. Kahng, L. Bin, L. S. Peh and K. Samadi. *ORION 2.0: A Power-Area Simulator for Interconnection Networks*. IEEE TVLSI, March, 2011, vol. 20, no. 1, pp. 191–196.

[19] *CACTI 6.5*, http://www.hpl.hp.com/research/cacti/

[20] X. Zheng, D. Patil, J. Lexau, F. Liu, G. Li, H. Thacker, Y. Luo, I. Shubin, J. Li, J. Yao, P. Dong, D. Feng, M. Asghari, T. Pinguet, A. Mekis, P. Amberg, M. Dayringer, J. Gainsley, H. F. Moghadam, E. Alon, K. Raj, R. Ho, J. E. Cunningham and A. V. Krishnamoorthy. *Ultra-efficient*

10 Gb/s hybrid integrated silicon photonic transmitter and receiver. Optics Express, March, 2011, vol. 19, no. 6, pp. 5172–5186.

[21] P. Grani and S. Bartolini. *Design Options for Optical Ring Interconnect in Future Client Devices.* ACM JETC, May, 2014, vol. 10, no. 4, Article. 30.

[22] R. G. Beausoleil. *Large-Scale Integrated Photonics for High-Performance Interconnects.* ACM JETC, June, 2011, Vol. 7, No. 2, pp. 326–327.

[23] K. Preston, N. Sherwood-Droz, J. S. Levy, M. Lipson. *Performance guidelines for WDM interconnects based on silicon microring resonators.* IEEE CLEO, May, 2011, pp. 1–2.

[24] S. Sarangi, B. Greskamp, R. Teodorescu, J. Nakano, A. Tiwari and J. Torrellas. *Varius: A model of process variation and resulting timing errors for microarchitects.* IEEE TSM, February, 2008, vol. 21, no. 1, pp. 3–13.

[25] Y. Xum, J. Yang and R. Melhem. *Tolerating process variations in nanophotonic on-chip networks.* Proc. ISCA, June, 2012, pp. 142–152.

[26] J. Karttunen, J. Kiihamaki and S. Franssila. *Loading effects in deep silicon etching.* International Society of Optical Engineering, August, 2000, vol. 4174, pp. 90–97.

[27] C. Nitta, M. Farrens and V. Akella. *Addressing system-level trimming issues in on-chip nanophotonic networks.* Proc. HPCA, February, 2011, pp. 122–131.

[28] M. Mohamed, Z. Li, X. Chen, L. Shang, A. Mickelson, M. Vachharajani and Y. Sun. *Power-Efficient Variation-Aware Photonic On-Chip Network Management.* Proc. ISLPED, August, 2010, pp. 31–36.

[29] S. V. R. Chittamuru, I. Thakkar and S. Pasricha. *PICO: Mitigating Heterodyne Crosstalk Due to Process Variations and Intermodulation Effects in Photonic NoCs.* Proc. DAC, June 2016.

Index

3D photonics 43, 52, 65, 68

A

Add-drop filter 358, 361, 362, 373
Adiabatic tapers 41, 47, 59, 65
Antisymmetric mode 364
Application Mapping 5, 181

B

Bit error rate 116, 242
Bottom-up approach 357, 381

C

Cache-coherence traffic 192
Channel partition 21, 25, 29, 32
Channel spacing 366, 375, 395
Chip I/O bandwidth 193
Chip multi-processors 385
Computers 11, 17, 107, 108
Corner analysis 357
Coupling theory 365
Cross-layer 237, 242
Cross-over coupling
    coefficient 362, 364, 373, 374
Cross-over length 364
Crosstalk 52, 171, 177, 376

D

Design Automation 5, 137, 227
Design methodology 142,
    238, 258
Design Technology 3, 5, 355
Design tradeoffs 199

Directional coupler 41, 49,
    50, 356
Dispersion 113, 128, 359, 363

E

Effective index 49, 57, 358, 365
Eigenmode 364, 370
Eigenvalue equation 360, 364
Electrical-optical (EO)
    conversion 13, 75, 237
Electron beam (EBeam)
    lithography 377
Energy efficiency 35, 108,
    250, 306
Energy efficient 35, 175, 200, 320
Energy proportionality 123, 319

F

Fabrication non-uniformity 6,
    355, 356
Fabrication process variations 7,
    355, 381, 386
Finite Difference in
    Time Domain 56

G

Group index 231, 358, 361, 370

I

Inter-chip optical network 13,
    15, 27, 29
Interconnections 4, 41, 49, 56

About the Editors

Mahdi Nikdast received his Ph.D. degree in Electronic and Computer Engineering from The Hong Kong University of Science and Technology (HKUST) in 2014. Since 2014, he is a postdoctoral fellow in the Department of Computer and Software Engineering at Polytechnique Montréal. He authored/co-authored more than 50 papers in refereed journals and international conference publications. He is the founder and the program chair of the International Workshop on Optical/Photonic Interconnects for Computing Systems (OPTICS). Mahdi was a recipient of several awards, including the Best Paper Award from the Design, Automation and Test in Europe (DATE) Conference (DATE-2016), the Best Poster Paper Award from the Asia Communications and Photonics Conference (ACP-2015), and the Second Best Project Award from the 6th Annual AMD Technical Forum and Exhibition (AMD-TFE-2010). His research interests include silicon photonics, optoelectronics, heterogeneous computing systems, multiprocessor systems-on-chip, and interconnection networks.

Gabriela Nicolescu is a Professor at Polytechnique Montreal in the Department of Computer and Software Engineering. She obtained her B.Eng. Degree in Electrical Engineering from UPB (Polytechnic University Bucharest) in 1998 and her Ph.D. Degree in 2002 from INPG (Institut National Polytechnique de Grenoble) France. Her research interests are related to the design methodologies, programming models and security for advanced heterogeneous systems on chip integrating advanced technologies. She co-authored more than 160 papers including journal articles, conference papers, books, book chapters and patents.

Sébastien Le Beux is Associate Professor for Heterogeneous and Nanoelectronics Systems Design at Ecole Centrale de Lyon. He is currently responsible for nanoprocessors research activities at the Heterogeneous System Design group of the Lyon Institute of Nanotechnology CNRS laboratory. He obtained his Ph.D. in Computer Science from the University of Sciences and Technology of Lille in 2007. He went on to become a postdoctoral researcher at Ecole Polytechnique de Montréal, Canada 2008–2010. His research interests include design methods for emerging (nano)technologies and embedded systems, including silicon photonic interconnect and reconfigurable architectures. He has authored or co-authored over 70 scientific publications including journal articles, book chapters, patent and conference papers. He serves on the steering committees, organizing committees, and technical program committees of numerous international conferences such as DATE and CODES+ISSS.

Jiang Xu received his Ph.D. from Princeton University. He worked at Bell Labs, NEC Labs, and a startup company before joining the Hong Kong University of Science and Technology. Jiang established Big Data System Lab, Xilinx-HKUST Joint Lab, and OPTICS Lab at HKUST. He currently serves as the Associate Editor for IEEE TCAD, TVLSI, and ACM TECS. He served on the steering committees, organizing committees, and technical program committees of many international conferences, including DAC, DATE, ICCAD, CASES, ICCD, CODES+ISSS, NOCS, HiPEAC, ASP-DAC, etc. Jiang is an IEEE Distinguished Lecturer and was an ACM Distinguished Speaker. He authored and co-authored more than 100 book chapters and papers in peer-reviewed journals and international conferences. His research areas include big data system, heterogeneous computing, optical interconnection network, power delivery and management, MPSoC, low-power embedded system, hardware/software codesign.